Lecture Notes in Physics

Volume 825

For further volumes:
http://www.springer.com/series/5304

The Lecture Notes in Physics

The series Lecture Notes in Physics (LNP), founded in 1969, reports new developments in physics research and teaching—quickly and informally, but with a high quality and the explicit aim to summarize and communicate current knowledge in an accessible way. Books published in this series are conceived as bridging material between advanced graduate textbooks and the forefront of research and to serve three purposes

- to be a compact and modern up-to-date source of reference on a well-defined topic
- to serve as an accessible introduction to the field to postgraduate students and nonspecialist researchers from related areas
- to be a source of advanced teaching material for specialized seminars, courses and schools

Both monographs and multi-author volumes will be considered for publication. Edited volumes should, however, consist of a very limited number of contributions only. Proceedings will not be considered for LNP.

Volumes published in LNP are disseminated both in print and in electronic formats, the electronic archive being available at springerlink.com. The series content is indexed, abstracted and referenced by many abstracting and information services, bibliographic networks, subscription agencies, library networks, and consortia.

Proposals should be sent to a member of the Editorial Board, or directly to the managing editor at Springer:

Christian Caron
Springer Heidelberg
Physics Editorial Department I
Tiergartenstrasse 17
69121 Heidelberg/Germany
christian.caron@springer.com

Sergey Nazarenko

Wave Turbulence

Sergey Nazarenko
Mathematics Institute
University of Warwick
Gibbet Hill Road, Coventry
CV4 7AL
UK
e-mail: S.V.Nazarenko@warwick.ac.uk

ISSN 0075-8450   e-ISSN 1616-6361

ISBN 978-3-642-15941-1   e-ISBN 978-3-642-15942-8

DOI 10.1007/978-3-642-15942-8

Springer Heidelberg Dordrecht London New York

Cover design: eStudio Calamar, Berlin/Figueres

Printed on acid-free paper

Springer is part of Springer Science+Business Media (www.springer.com)

Wednesday
User Barcode: 7888
(last 4 digits)

Book Barcode:

Preface

This book has appeared as a development and extension of the material of post-graduate lecture courses given by the author at the Wolfgang Pauli Institute, Vienna, in October 2007 (4 h), The University of Warwick, Coventry, in October–December 2007 (20 h) and Institut Henri Poincaré, Paris, in April 2009 (10 h).

As a result, the book has been written and structured as a graduate text rather than a monograph, with numerous exercises offered along the way. The main aim here was to produce a compact description which would be accessible for graduate students and non-specialist mature researchers who would like to learn about Wave Turbulence and possibly apply it in their own fields.

I would like to express my warmest thanks to my teacher Vladimir Zakharov from whom I was fortunate to learn Wave Turbulence firsthand as a Ph.D. student and later as a coworker. I am grateful to all of my friends and colleagues with whom I collaborated in different areas of Wave Turbulence over 20 years: Sasha Balk, Laura Biven, Guido Boffetta, Umberto Bortolozzo, Antonio Celani, Yeontaek Choi, Colm Connaughton, Petr Denissenko, Davide Dezzani, Alex Dyachenko, Grisha Falkovich, Sebastien Galtier, Boris Gershgorin, Peter Horvai, Sang Gyu Jo, Elena Kartashova, Ho Il Kim, Sergei Lukaschuk, Victor Lvov, Yuri Lvov, Mitya Manin, Stuart McLelland, Anatolii Mikhailovskii, Balu Nadiga, Alan Newell, Miguel Onorato, Boris Pokorni, Annick Pouquet, Davide Proment, Andrei Pushkarev, Brenda Quinn, Stefania Residori, Olexii Rudenko,Alex Schekochihin, Thorwald Stein, Rob West and Norman Zabusky. I thank participants of the Wave Turbulence courses in Vienna, Warwick and Paris for their feedback which I used to improve presentation, especially Thorwald Stein who helped with processing the original lecture notes. I thank Jason Laurie, Miguel Onorato, Davide Proment and Brenda Quinn for sharing some figures obtained in their numerical simulations and for helping to correct numerous typos. Finally, I am grateful to Uriel Frisch for encouraging me to write this book.

Warwick, June 2010   Sergey Nazarenko

Contents

List of Abbreviations

1D, 2D, 3D	one-dimensional, two-dimensional, three-dimensional
BE	Bose–Einstein (distribution)
BEC	Bose–Einstein condensate
CHM	Charney–Hassegawa–Mima (equation/model)
CB	Critical Balance
DAM	Differential Approximation Model
DNS	Direct Numerical Simulation
DR	Dissipative Region
DW	Drift Wave
DWP	Drift Wave Packet
EDQNM	Eddy-Damped Quasi-Normal Markovian
GFD	Geophysical Fluid Dynamics
h.o.t.	higher-order terms
ITG	Ion Temperature Gradient (instability)
$\Im$	Imaginary Part of
IR	Infrared
JR	Jones–Roberts (soliton/vortex)
KdV	Korteweg–de Vries (equation)
K41	Kolmogorov–Obukhov 1941 (turbulence theory)
KZ	Kolmogorov–Zakharov (spectrum/state)
LHS	Left-Hand Side
LH-transition	Low-to-High confinement transition (in fusion plasmas)
MB	Maxwell–Boltzmann (distribution)
MHD	Magneto-Hydrodynamics
NLS	Nonlinear Schrödinger (equation, model)
ODE	Ordinary Differential Equation
PDE	Partial Differential Equation
PDF	Probability Density Function
QG	Quasi-Geostrophic (model)
$\Re$	Real Part of
RD	Rapid Distortion

RP	Random Phase (field)
RPA	Random Phase and Amplitude (field)
RJ	Rayleigh–Jeans (spectrum/state)
RHS	Right-Hand Side
RMHD	Reduced Magneto-Hydrodynamics
r.m.s.	root mean square
UR	Unstable Region
UV	Ultraviolet
WKB	Wentzel Kramers Brillouin (approximation)
w.r.t.	with respect to
WT	Wave Turbulence
ZF	zonal flow
ZLF-book	Book "Kolmogorov spectra of turbulence. Part 1" by Zakharov, Lvov and Falkovich

Chapter 1
Introduction

1.1 What is Wave Turbulence?

Wave turbulence (WT) can be generally defined as *out-of-equilibrium statistical mechanics of random nonlinear waves*. Often this definition is further narrowed to waves which are weakly nonlinear and dispersive, the cases when the mathematical description of WT is most systematic and unambiguous. However, we prefer to take an approach which focuses on WT as a general *physical phenomenon* without discarding a large number of systems which are commonly observed in nature but which have not been described rigorously yet. WT systems are important in a vast range of physical examples, from quantum to astrophysical scales, and this explains a huge and still growing number of situations for which WT theory approach was used. They include water surface gravity and capillary waves which are important for understanding the momentum and energy transfers from wind to ocean, as well as for navigation conditions [3–5, 27–30, 37–40, 53, 77, 92, 103, 122, 123, 127]; internal waves on density stratifications and inertial waves due to rotation are important in turbulence behavior and mixing in planetary atmospheres and oceans [20, 43, 79]; waves on quantized vortex lines which are important for understanding turbulence in superfluid helium [71, 81, 82, 94, 116, 117]; Alfvén waves which are important in turbulence of solar wind and interstellar medium [2, 12, 36, 44–49, 52, 72, 93, 111, 112]; planetary Rossby waves important for the weather and climate evolutions [6–8, 83, 86, 130]; waves in Bose–Einstein condensates and in Nonlinear Optics [35, 96, 97]; a great variety of waves in plasmas of fusion devices [41, 42, 115, 127, 132]; waves on vibrating elastic plates [31], and many other physical wave systems.

1.2 Historic Remarks

WT approach can be traced back to Rudolph Peierls who in 1929 used it to describe the kinetics of phonons in anharmonic crystals [105]. The wave kinetic

S. Nazarenko, *Wave Turbulence*, Lecture Notes in Physics, 825,
DOI: 10.1007/978-3-642-15942-8_1, © Springer-Verlag Berlin Heidelberg 2011

equation was derived for describing evolution of the wave spectrum. Remarkably, an equation for the joint probability density function (PDF) for the wave amplitudes was also obtained in this pioneering paper via averaging over random phases. This equation contains not only the kinetic equation as its first moment, but also all higher moments that allow to study deviations from gaussianity and intermittency, as well as the validity of the underlying WT assumptions. However, the main focus in the early papers was on solutions describing the thermodynamic equilibria and on small deviations from the thermodynamic states [19, 105, 108]. It became clear later that these solutions are not the most interesting and relevant to the WT systems.

The WT ideas were reinvigorated in 1960s in plasma physics [42, 115, 132] where the subject got its name, "Wave Turbulence" or sometimes "Weak Turbulence", and in the theory of water waves [9, 10, 50, 51, 122, 123]. At that time, the focus has moved from the thermodynamic equilibrium to strongly non-equilibrium turbulent states, often time and space dependent, and one of the important practical questions asked was how WT contributes to transport, e.g. the energy and particle transport across the tokamak plasma, or momentum transfer from wind to the ocean. Most important theoretical development in WT was a discovery by Vladimir Zakharov in 1965 of a new type of solutions of the kinetic equations corresponding to a constant energy flux through scales [120]. These solutions are now called Kolmogorov–Zakharov (KZ) spectra because they are analogous to the Kolmogorov spectrum of hydrodynamic turbulence describing the energy cascade state. It is these KZ solutions that firmly put WT systems into the domain of general *Turbulence, i.e. strongly non-equilibrium statistical systems with many degrees of freedom whose state is determined by a flux through phase space* rather than by temperature and thermodynamic potentials as in the case of equilibrium or weakly non-equilibrium systems. These findings inspired applications to new physical examples, including waves in anisotropic media, as well as more detailed theoretical studies, e.g. self-similar solutions of the kinetic equations, stability and locality conditions of the KZ spectra, etc. These developments were summarized first in a 1984 review by Vladimir Zakharov [121] followed by a 1994 book by Zakharov, L'vov and Falkovich (ZLF) [127], which became a classic and until now remains the main source of information for those wishing to learn about WT. However, most of the studies in that time period, and up to very recently, concentrated on the wave spectrum and the kinetic equation only, and the Peierls' PDF equation, with its great potential to study the finer statistical objects, was temporarily forgotten.

1.3 Recent Developments

Recent developments in WT, more or less sixteen years since the ZLF-book, happened in several directions which can be roughly summarized as follows.

1.3.1 Rapid Expansion of WT Applications

In the beginning of this chapter we gave a very incomplete list of physical examples of wave systems for which WT descriptions have been developed in the past. Early applications of the WT theory were introduced in 1960s in plasma turbulence and the geophysical fluid dynamics and have since become an important part of paradigm for describing the physical phenomena in these applications. Notably, the WT language, i.e. triad or quartic resonant wave interactions, is sometimes also used to qualitatively describe situations where waves are not so weak and when formally the WT theory is invalid.

Further, a large part of the above list mentions applications in new areas where WT was introduced relatively recently, and one could mention quite a few more of them. Among exotic WT applications we can mention using WT to describe a turbulent preheating stage of inflation in the early Universe [84, 85], waves in ferromagnetic fluids subject to magnetic fields [15], waves on vibrating elastic plates [31],—the later system allows us to hear the WT spectrum!

Three recent applications should be specially mentioned as the most significant ones, MHD Turbulence, Superfluid Turbulence and non-equilibrium Bose–Einstein condensation, because in all of these examples the WT theory is a part of the bigger turbulence picture, rather than a special weak-wave case with limited applicability.

In MHD Turbulence, the system of interacting Alfvén waves constitute the most essential dynamical component [12, 44–49, 93, 98]. In this case the situation is rich and interesting and it goes beyond the traditional weak-wave setup, because quasi-2D vortices play a catalytic role for the triad interactions of Alfvén waves. Thus, strongly nonlinear coherent structures, vortices, enter the game and one has a question to answer about their role in the WT cycle (see below).

In a superfluid at zero temperature, turbulence consists of very thin "spaghetti" of quantized vortex lines (see e.g. [116]). Such a spaghetti tangle behaves like a classical hydrodynamic turbulence at the scales greater than the mean inter-vortex separation. However, because there is no viscosity, the hydrodynamic Kolmogorov cascade brings the energy all the way down to the inter-vortex separation scale, which can be viewed at a classical-quantum cross-over scale. From this scale further energy cascade to even smaller scales is carried by WT of Kelvin waves, until a very small scale where all energy is lost via radiating phonons [71, 81, 82, 94, 116, 117].

In non-equilibrium Bose–Einstein condensation, initial evolution takes place as an inverse cascade driven by quartic interactions of weakly nonlinear waves [89]. When the inverse cascade reaches a certain large scale, the weak nonlinearity assumption breaks down and a coherent uniform condensate component appears [32, 35]. When the condensate grows strong, the subsequent evolution can be viewed as three-wave evolution of weak Bogoliubov sound. The most interesting stage occurs in between of the four-wave and the three-wave weak regimes: it comprises strong turbulence consisting of a gas of coherent vortices, which interact with each other and with the sound field, and the number of which is decreasing to zero in a finite time due to the pairwise vortex annihilation [96, 97]. Such transition via annihilating vortices is a realization of the Kibble-Zurek phase transition scenario first suggested in cosmology [67, 133, 134].

1.3.2 Highly Improved Quality of Experimental Data and Numerical Simulations of WT Systems

We have now a significant number of experiments and numerical simulations which were specifically focused at checking predictions of the WT theory. The picture is relatively clear in the case of the capillary waves on a fluid surface (water, ethanol, liquid hydrogen or liquid helium): both experiments [1, 16–18, 37, 40, 74, 75, 118, 119] and numerical experiments [39, 110] confirm the KZ spectrum predicted by the WT theory in this case. Especially impressive evidence of the KZ spectrum was presented in the zero-gravity experiment of [37]. For the gravity waves the picture is more complicated: numerics seem to agree with WT theory predictions for the wave spectrum [3–5, 33, 34, 70, 77, 103] but sensitivity of the system to the finite box effects was reported [77, 125]. In laboratory experiments, the finite size effects and presence of breaking coherent waves obscure the KZ state, and the latter is not observed in its pure form. However, such real life effects point at an even more interesting picture of the WT life cycle which includes coexisting random and coherent wave components (see below). An interesting extension of the WT fluid-surface experiments was set up in [15] using a ferro-fluid in an external magnetic field which modifies the linear and nonlinear properties of the capillary waves. Also, the WT theory was developed for vibrating elastic plates [31]. Once again, the real life appears to be more complex and interesting, and the experiments show deviations from the existing predictions [13, 87]. These deviations are not yet understood.

In MHD, there has been a significant recent push, using numerics [2, 90, 91] and observational data [111, 112], to resolve controversies about the turbulent energy spectrum. Note that the Iroshnikov-Kraichnan spectrum (which is widely discussed in MHD turbulence) [52, 72] qualifies to be called KZ spectrum because: (1) it assumes a (typical for WT) three-wave resonant interaction and (2) it assumes a constant flux of energy through scales. However, it also assumes turbulence isotropy, which is presently understood to be a bad assumption for MHD turbulence. Anisotropic KZ spectrum of MHD turbulence was derived in [44, 45, 98]. Thus, the three main candidates discussed and tested in numerical and observational MHD data are Kolmogorov spectrum of strong turbulence (exponent $-5/3$), Iroshnikov-Kraichnan spectrum (exponent $-3/2$ which is very close to, and hard to distinguish from, Kolmogorov $-5/3$) and the anisotropic WT spectrum (exponent -2). Also, efforts were made to check the *critical balance* hypothesis stating that the MHD anisotropy is such that the linear and nonlinear time scales are in a scale-by-scale balance [49] (we will discuss this hypothesis later in this book). Results vary for different parts of the parameter space, but it is clear that it is extremely hard to achieve the weak WT regime in the 3D simulations, namely to meet the requirements of sufficiently small non-linearity (i.e. high external magnetic field), the numerical resolution sufficient for avoiding the finite-box effects and the sufficiently long computation time. On the other hand, the WT theory was argued to be supported by the observational data [111, 112].

1.3.3   Discovery of Importance of Coherent Structures in WT Evolution

The most developed and systematic part of the WT theory deals with weakly nonlinear random waves. Some qualitative understanding has been recently achieved on the role of the coherent nonlinear structures randomly scattered through predominantly weak wave fields. The ultimate WT theory, yet to be developed in future, should include both components: incoherent weakly nonlinear waves and strongly nonlinear coherent structures. This theory should be able to describe the mechanisms of the WT cycle by which the weakly nonlinear random waves can get converted into the coherent structures and, conversely, breaking of the coherent structures with a partial return of energy to the incoherent waves. Obviously, such a theory could not be completely universal considering the fact that there is a great variety of the coherent structure types and the ways they can break depending on the particular physical system. However, some of the key qualitative features of the WT cycle appear to be surprisingly similar for different systems, and describing these universal features will be one of our goals in this book.

A particular type of coherent structure that arises in some systems with inverse cascades is the condensate [32, 35, 78, 96, 97, 128]. A very special role of this kind of structure is that it radically changes the character of the wave-wave interactions, and not only coexists with the random waves.

Note that although the largest part of WT theory deals weak random waves, we will avoid translating WT as "Weak Turbulence", because we would like to include strong coherent structures into the "Wave Turbulence", at least at a qualitative level for now and, hopefully, more rigorously later. It has become clear that these structures are typically important even if the averaged wave level is weak. For example, coherent breaking waves on water surface provide *the only* efficient mechanism of dissipation of the wave energy and, therefore, ensure existence of the statistically steady states in sea waves forced by wind. Once, again: let us treat WT as a *phenomenon* some features of which are yet to be understood better, rather than a *complete theory* or *dogma* which could only be either confirmed or disproved.

1.3.4 Theory Extension Beyond Spectra

On the theory side, an effort has been made to generalize WT description to the wave Probability Density Functions (PDF) which allows to include the non-gaussian wave fields [22–24, 100]. This provides a formalism for study of the phenomenon of turbulence intermittency, as well as the study of applicability conditions of the fundamental WT assumptions. Importantly, as we will see later such an approach allows us to understand the effect of the breaking coherent structures onto the incoherent waves and, therefore, it provides us with a natural language for describing the WT life cycle.

Also, because this extension recasts WT in such a way that all the derivation steps are clearly spelled out, we believe that it can bring WT closer to the domain of rigorous Mathematical Physics, and hopefully it will allow one to find useful connections with the other areas in the Statistical Mechanics and the Dynamical Systems. In this respect, here we would like to cite book [21] in which an effort was initiated to bring the areas of Turbulence (including WT) and the Non-equilibrium Statistical Mechanics closer together.

In the present book we will use such a generalized WT description and will explain its main points in detail.

1.3.5 Study of the Finite-Box Effects

In the weak WT description, both the large-box and the weak-nonlinearity limits are taken. It is important that the large-box limit must be taken before the weak-nonlinearity limit. Physically, this means that the nonlinear resonance broadening should be much greater than the spacing of the k-grid associated with the finite box.

It was recently realized that this condition is hard to satisfy in both the numerical and the laboratory experiments, for example for the gravity water waves and for MHD waves [28, 77, 92, 93]. If this condition is not realized, the number of active resonant triads or quartets is depleted with respect to the continuous $\mathbf{k}$-space [14, 66, 77, 92, 93, 110, 125]. This brings about some interesting new effects. When the wave amplitudes are very small, so that the nonlinear resonance broadening is much less than the inter-mode spacing, then only waves that are in exact (e.g. triad or quartet) resonance can interact and exchange energy. Finding such resonant sub-sets is an interesting Diophantine problem first formulated by Kartashova [55, 56] who found that the resonant waves are organized into non-intersecting clusters of variable sizes. The clusters were subsequently studied and classified, from individual triads or quartets, to bigger objects like e.g. "butter-flies", where two triads coupled via one common mode, to very big "monster" clusters involving a complicated (but still finite) network of interconnected triads or quartets [54–58, 62–65]. Dynamics of small clusters tends to be regular, often periodic and integrable, whereas larger clusters are believed to be chaotic [63, 66]. Interplay of the regular and chaotic behavior and transition to chaos as the cluster size increases is an interesting and still poorly studied subject.

Discreteness of the **k**-space is especially felt by the large-scale modes with wavelengths not so much smaller than the bounding box size. In particular, it was suggested in [60, 61] that dynamics of clusters of the planetary Rossby waves can explain intra-seasonal oscillations in Earth's atmosphere.

For larger amplitudes, the nonlinear resonance broadening gets bigger and it can connect different originally isolated clusters. If the wave amplitude is not strong enough, such quasi-resonant clusters still remain limited in size and they cannot pass energy from the forced waves (usually long waves) to the dissipation scales (short waves). Such non-cascading wave systems were first found by Pushkarev and Zakharov who called this phenomenon "frozen turbulence" [109, 110]. If the system of waves is forced continuously then in the frozen regime the amplitude of the modes in or/and near the forcing scales will increase. Eventually, for sufficiently large amplitudes, the resonance broadening will get large enough to connect the forcing and the dissipation scales via quasi-resonant clusters, and the energy cascade will get triggered. Note that such quasi-resonant clusters can still be sparse in the underlying **k**-space if the forcing strength is small, so that the resulting energy cascade will be very anisotropic and "bursty" reminiscent of sandpile tip-overs, as it was proposed in [92] and some resemblance of which was observed in numerical simulations [77]. Such "mesoscopic" turbulence has dynamical features of both discrete clusters and the continuous WT, and in spite of some interesting ideas expressed recently [66, 77, 92, 93, 101, 125] it is yet largely unexplored subject, theoretically, numerically and experimentally.

1.4 What is this Book About?

The main message of this book is that WT is a very *young and dynamic* rather than a completed and static subject. It has recently seen an explosion of new applications, new ideas and theoretical approaches, new field observations, experiments and numerical simulations. The point is that this is only the *beginning* of the road, and much more remains to be discovered than has already been done in all of these directions.

However, this book is not aimed at addressing *all* the recent developments in WT theory, experiment and numerics, nor at description of *all* the existing applications, nor would we like to reproduce the basic facts and methods of the WT theory given in the ZLF-book. This would probably take several volumes and it would not be too helpful for the people who would like to learn about the subject.

Instead, our aim here is to produce a compact text which would be accessible for graduate students and non-specialist mature researchers who would like to learn about WT and possible apply it in their own fields. Thus, this book is written and structured as a graduate text, with all steps carefully explained and example problems given. To be able to give detailed explanations and derivations in a compact space, we shall sacrifice generality and breadth of applications. We will try to explain the underlying assumptions and their justifications as clear as we can, with hope that this effort would move WT closer to the domains and the language of rigorous mathematics, Mathematical Physics, Dynamical Systems and Statistical Mechanics. Also, we will use an approach which was developed only recently and which is not yet so common amongst the WT users. Namely, we will include into our consideration the wave PDF's which, as we think, make the derivations more clear and elegant, and they allow to us study the problems of WT intermittency and provide a natural framework for incorporating the weak random waves and strong coherent structures into a unified picture of the WT life cycle.

After introducing the Wave Turbulence phenomenon, in part 1 we place Wave Turbulence in the general Turbulence context, and provide a Wave Turbulence "primer", i.e. a set of useful shortcut techniques for reproducing most important Wave Turbulence results (including the ones for strong Wave Turbulence) in most known applications. In part 2 we introduce the basic statistical objects, define the basic theoretical assumptions and the derivation steps in the Wave Turbulence closures with all important details explained using, in most parts, a single master example,—the Petviashvili wave model. The Petviashvili model contains many (but not all!) features typical for the WT systems. By restricting ourselves to this specific and rather compact model we aim to avoid a widespread "index phobia" problem in learning WT, which in turn has arisen from an "index mania", i.e. introduction of multiple subscripts and superscripts in a quest for generality but at expense of lesser transparency.

In part 3 we study the properties of Wave Turbulence, important solutions and predictions which arise from the Wave Turbulence closure. In part 4 we present a detailed discussion of three selected physical applications,—nonlocal Drift/Rossby wave turbulence in geophysical fluids and in plasmas, MHD turbulence and Bose–Einstein condensation. In addition to illustrating the ideas and methods explained in the preceding chapters, these descriptions will explain specific features characterizing these applications, including the nonlocality, finite-size effects, ways WT coexists with strong turbulence, zonal jets, vortices and coherent condensate. These chapters can be considered mini-reviews on the respective types of turbulence. At the same time, should this book be used for a graduate course, these three chapters can be considered model examples for projects or essays to be written by graduate students. The list of possible graduate projects, with brief descriptions and pointers to further readings, is given in the last chapter.

Finally, like most books, the present text is inevitably biased to reflecting the personal experience, work and views of the author. It is highly advisable for those who want to get a broader knowledge of the WT ideas to complement reading of this book by familiarizing themselves with other available sources. In this respect, we mention a recent compact review of WT by Newell and Rumpf [101]. A book containing reviews of several applications of WT with a strong emphasis on the experimental and observational sides is currently being written for World

Scientific. Recent book by Kartashova [59] focuses on the structure of wave resonances in bounded systems, and as such is recommended for those who are interested in better understanding of the finite-size effects in WT.

References

1. Abdurakhimov, L.V., Brazhnikov, M.Yu., Kolmakov, G.V., Levchenko, A.A., Mezhov-Deglin, L.P.: Turbulence of capillary waves on the surface of quantum liquids. J. Low Temp. Phys. **148**, 245–249 (2007). doi:10.1007/s10909-007-9377-y
2. Alexakis, A., Bigot, B., Politano, H., Galtier, S.: (P,S,G,E,B) Anisotropic fluxes and nonlocal interactions in magnetohydrodynamic turbulence. Phys. Rev. E Stat. Nonlin. Soft Matter. Phys. **76**(5–2), 056313–18233762 (2007)
3. Annenkov, S., Shrira V.: New numerical method for surface waves hydrodynamics based on the Zakharov equation. J. Fluid Mech. **449**, 341–371 (2001)
4. Annenkov, S., Shrira, V.: Role of non-resonant interactions in evolution of nonlinear random water wave fields. J. Fluid Mech. **561**, 181–207 (2006)
5. Annenkov, S., Shrira, V.: Direct numerical simulation of downshift and inverse cascade for water wave turbulence. Phys. Rev. Lett. **96**, 204501 (2006)
6. Balk, A.M., Nazarenko, S.V.: On the physical realisability of anisotropic Kolmogorov spectra of weak turbulence. Sov. Phys. JETP **70**, 1031 (1960)
7. Balk, A.M., Nazarenko, S.V., Zakharov, V.E.: On the nonlocal turbulence of drift type waves. Phys. Lett. A **146**, 217–221 (1990)
8. Balk, A.M., Nazarenko, S.V., Zakharov, V.E.: New invariant for drift turbulence, Phys. Lett. A, **152**(5–6), 276–280 (1991)
9. Benney, D.J., Saffman, P.: Nonlinear interaction of random waves in a dispersive medium. Proc. R. Soc. A **289**, 301–320 (1966)
10. Benney, B.J., Newell, A.C.: Random wave closures, studies. Appl. Math. **48**(1), 29 (1969)
11. Bigg, G.R.: Ocean-atmosphere interaction, 273 pp. Cambridge University Press, Cambridge. ISBN:0521016347 (2003)
12. Bigot, B., Galtier, S., Politano, H.: An anisotropic turbulent model for solar coronal heating. Astron. Astrophys. **490**, 325–337 (2008)
13. Boudaoud, A., Cadot, O., Odille, B. et al.: Observation of wave turbulence in vibrating plates. Phys. Rev. Lett. **100**, 234504 (2008)
14. Bourouiba, L.: Discreteness and resolution effects in rapidly rotating turbulence. Phys. Rev. E **78**, 056309 (2008)
15. Boyer, F., Falcon, E.: Wave turbulence on the surface of a ferrofluid submitted to a magnetic field. Phys. Rev. Lett. (2008)
16. Brazhnikov, M.Yu., Kolmakov, G.V., Levchenko, A.A., Mezhov-Deglin, L.P.: J. Exp. Theor. Phys. Lett. **73**, 398 (2001)
17. Brazhnikov, M.Yu., Kolmakov, G.V., Levchenko, A.A., Mezhov-Deglin, L.P.: J. Exp. Theor. Phys. Lett. **74**, 583 (2001)
18. Brazhnikov, M.Yu., Kolmakov, G., Levchenko, A., Mezhov-Deglin, L.: Observation of capillary turbulence on the water surface in a wide range of frequencies. Europhys. Lett. **58**(4), 510 (2002)
19. Brout, R., Prigogine, I.: Physica **22**, 621–636 (1956)
20. Caillol, P., Zeitlin, V.: Kinetic equations and stationary energy spectra of weakly nonlinear internal gravity waves. Dyn. Atmospheres Oceans **32**, 81 (2000)
21. Cardy, J., Falkovich, G., Gawedzki, K., Nazarenko, S., Zaboronski, O.V. (eds.): Non-equilibrium statistical mechanics and turbulence. CUP, London Mathematical Society Lecture Note Series 355.

22. Choi, Y., Lvov, Y., Nazarenko, S.V.: Probability densities and preservation of randomness in wave turbulence. Phys. Lett. A **332**(3–4), 230 (2004)
23. Choi, Y., Lvov, Y., Nazarenko, S.V., Pokorni, B.: Anomalous probability of large amplitudes in wave turbulence. Phys. Lett. A **339**(3–5), 361 (2004)
24. Choi, Y., Lvov, Y., Nazarenko, S.V.: Joint statistics of amplitudes and phases in wave turbulence. Physica D **201**, 121 (2005)
25. Choi, Y., Lvov, Y.V., Nazarenko, S.: Wave turbulence. Recent Dev. Fluid Dyn. **5** (2004) (Transworld Research Network, Kepala, India)
26. Choi, Y., Jo, G.G., Kim, H.I., Nazarenko, S.: Aspects of two-mode probability density function in weak wave turbulence. J. Phys. Soc. Jpn. **78**(8), 084403 (2009)
27. Connaughton, C., Nazarenko, S.V., Pushkarev, A.: Discreteness and quasi-resonances in capillary wave turbulence. Phys. Rev. E **63**, 046306 (2001)
28. Denissenko, P., Lukaschuk, S., Nazarenko, S.: Gravity surface wave turbulence in a laboratory flume. Phys. Rev. Lett. **99**, 014501 (2007)
29. Lukaschuk, S., Nazarenko, S., McLelland, S., Denissenko, P.: Gravity wave turbulence in wave tanks: space and time statistics. Phys. Rev. Lett. **103**, 044501 (2009)
30. Nazarenko, S., Lukaschuk, S., McLelland, S., Denissenko, P.: Statistics of surface gravity wave turbulence in the space and time domains. J. Fluid Mech. **642**, 395–420 (2010)
31. Düring, G., Josserand, C., Rica, S.: Weak turbulence for a vibrating plate: can one hear a Kolmogorov spectrum? Phys. Rev. Lett **97**, 025503 (2006)
32. Dyachenko, A., Falkovich, G.: Condensate turbulence in two dimensions. Phys. Rev. E **54**, 4431 (1996)
33. Dyachenko, A.I., Korotkevich, A.O., Zakharov, V.E.: Weak turbulent Kolmogorov spectrum for surface gravity waves. Phys. Rev. Lett. **92**, 13 (2004)
34. Dyachenko, A.I., Korotkevich, A.O., Zakharov, V.E.: Weak turbulence of gravity waves. JETP Lett. **77**(10), 546–550 (2003)
35. Dyachenko, A., Newell, A.C., Pushkarev, A., Zakharov V.E.: Optical turbulence: weak turbulence, condensates and collapsing fragments in the nonlinear Schrodinger equation. Phys. D **57**(1–2), 96 (1992)
36. Falgarone, E., Passot, T. (eds.): Turbulence and Magnetic Fields in Astrophysics. Lecture Notes in Physics. Springer, Berlin (2003)
37. Falcón, C., Falcon, E., Bortolozzo, U., Fauve, S.: Capillary wave turbulence on a spherical fluid surface in zero gravity. EPL **86**, 14002 (2009)
38. Falcon, E., Fauve S., Laroche, C.: Observation of intermittency in wave turbulence. Phys. Rev. Lett. **98**, 154501 (2007)
39. Falkovich, G.E., Shafarenko, A.B.: Sov. Phys. JETP **68**, 1393 (1988)
40. Falcon, E., Laroche, C., Fauve, S.: Observation of gravity-capillary wave turbulence. Phys. Rev. Lett. **98**, 094503 (2007)
41. Sagdeev, R.Z., Galeev, A.A.: Nonlinear plasma theory. In: O'Neil, T.M., Book, D.L. (Rev., eds.) Frontiers in Physics Series, 122 pp. W. A. Benjamin, New York (1969)
42. Galeev, A.A., Sagdeev, R.Z.: Review of plasma physics. In: Leontovich, M.A. (ed.) vol. 7, p. 307. Consultants Bureau, New York (1979)
43. Galtier, S.: Weak inertial-wave turbulence theory. Phys. Rev. E **68**, 015301 (2003)
44. Galtier, S., Nazarenko, S.V., Newell, A.C., Pouquet, A.: A weak turbulence theory for incompressible MHD. J. Plasma Phys. **63**, 447 (2000)
45. Galtier, S., Nazarenko, S.V., Newell, A.C., Pouquet, A.: Anisotropic turbulence of shear-Alfven waves. Astrophys. J. Lett. **564**, L49–L52 (2002)
46. Galtier, S., Nazarenko, S.V., Newell, A.C.: On wave turbulence in MHD. Nonlin. Process. Geophys. **8**(3), 141–150 (2001)
47. Galtier, S., Nazarenko, S.: Large-scale magnetic field re-generation by resonant MHD wave interactions. J. Turbul. **9**(40), 1–10 (2008)
48. Goldreich, P.: Incompressible MHD turbulence. Astrophys. Space Sci. **278**(1–2), 17 (2001)

49. Goldreich, P., Sridhar, S.: Toward a theory of interstellar turbulence. 2: strong Alfvénic turbulence part 1. Astrophys. J. **438**(2), 763–775 (1995) (ISSN 0004-637X)
50. Hasselmann, K.: On the nonlinear energy transfer in a gravity wave spectrum part 1. J. Fluid Mech. **12**, 481–500 (1962)
51. Hasselmann, K.: On the nonlinear energy transfer in a gravity wave spectrum. Part 2. J. Fluid Mech. **15**, 273–281 (1963)
52. Iroshnikov, R.S.: Turbulence of a conducting fluid in a strong magnetic field. Astron Zh **40**, 742 (English trans.: 1964, Sov. Astron. **7**, 566) (1963)
53. Janssen, P.: Ocean-Atmosphere Interaction. Cambridge University Press, Cambridge (2004)
54. Kartashova, E.A.: Partitioning of ensembles of weakly interacting dispersing waves in resonators into disjoint classes. Phys. D **46**(1), 43 (1990)
55. Kartashova, E.A.: On properties of weakly nonlinear wave interactions in resonators. Phys. D **54**(1–2), 125 (1991)
56. Kartashova, E.A.: Weakly nonlinear theory of finite-size effects in resonators. Phys. Rev. Lett. **72**, 2013 (1994)
57. Kartashova, E.A.: Clipping—a new investigation method for PDE-s in compact domains. Theor. Math. Phys. **99**, 675 (1994)
58. Kartashova, E.A.: Wave resonances in systems with discrete spectra. In: Zakharov, V.E. (ed.) Nonlinear Waves and Weak Turbulence, p. 95. Springer, Berlin (1998)
59. Kartashova, E.A.: Nonlinear Resonance Analysis Theory, Computation, Applications. Cambridge University Press, Cambridge (2010)
60. Kartashova, E.A., L'vov, V.S.: A model of intra-seasonal oscillations in the Earth atmosphere. Phys. Rev. Lett. **98**(19), 198501 (2007)
61. Kartashova, E.A., L'vov, V.S.: Triad dynamics of planetary waves. Europhys. Lett. **83**, 50012 (2008)
62. Kartashova, E.A., Mayrhofer, G.: Cluster formation in mesoscopic systems. Phys. A Stat. Mech. Appl. **385**, 527 (2007)
63. Kartashova, E.A., Nazarenko, S., Rudenko, O.: Resonant interactions of nonlinear water waves in a finite basin. Phys. Rev. E **78**, 016304 (2008)
64. Kartashova, E.A., Piterbarg, L.I., Reznik, G.M.: Weakly nonlinear interactions between Rossby waves on a sphere. Oceanology **29**, 405 (1990)
65. Kartashova, E.A., Reznik, G.M.: Interactions between Rossby waves in bounded regions. Oceanology **31**, 385 (1992)
66. Kartashova, E., Lvov, V., Nazarenko, S., Procacia, I.: Mesoscopic Wave Turbulence: review (2009) (in preparation)
67. Kibble, T.W.B.: Topology of cosmic domains and strings. J. Phys. A Math. Gen. **9**, 1387 (1976)
68. Kolmakov, G., Levchenko, A., Braznikov, M., Mezhov-Deglin, L., Slichenko, A., McClintock, P.: Quasiadiabatic decay of capillary turbulence on the charged surface of liquid hydrogen. Phys. Rev. Lett. **93**, 74501 (2004)
69. Kolmogorov, A.N.: On the conservation of conditionally periodic motions for a small change in Hamilton's function. Dokl. Akad. Nauk SSSR **98**, 527 (English translation in: LNP 93, 51 (Springer, 1979)) (1954)
70. Korotkevich, A.O.: Simultaneous numerical simulation of direct and inverse cascades in wave turbulence. Phys. Rev. Lett. **101**, 074504 (2008)
71. Kozik, E.V., Svistunov, B.V.: Kelvin-wave cascade and decay of superfluid turbulence. Phys. Rev. Lett. **92**, 035301 (2004)
72. Kraichnan, R.: Inertial range spectrum of hydro-magnetic turbulence. Phys. Fluids **8**, 1385–1387 (1965)
73. Krasitskii, V.P.: On reduced equations in the Hamiltonian theory of weakly nonlinear surface-waves. J. Fluid Mech. **272**, 1 (1994)
74. Lommer, M., Levinsen, M.T.: Using laser-induced fluorescence in the study of surface wave turbulence. J. Fluoresc. **12**, 45 (2002)

75. Henry, E., Alstrøm, P., Levinsen, M.T.: Prevalence of weak turbulence in strongly driven surface ripples. Europhys. Lett. **52**, 27 (2000)
76. Lvov, Y.V.: Effective five wave hamiltonian for surface water waves. Phys. Lett. A **230**, 38 (1997)
77. Lvov, Y.V., Nazarenko, S., Pokorni, B.: Discreteness and its effect on water-wave turbulence. Phys. D **218**(1), 24 (2006)
78. Lvov, Y.V., Nazarenko, S., West, R.: Wave turbulence in Bose-Einstein condensates. Phys. D **184**, 333–351 (2003)
79. Lvov, Y.V., Tabak, E.G.: Hamiltonian formalism and the Garrett and Munk spectrum of internal waves in the ocean. Phys. Rev. Lett. **87**, 169501 (2001)
80. L'vov, V.S.: Wave Turbulence Under Parametric Excitation. Series in nonlinear dynamics. Springer, Berlin (1994)
81. L'vov, V.S., Nazarenko, S.V., Rudenko, O.: Bottleneck crossover between classical and quantum superfluid turbulence. Phys. Rev. B **76**, 024520 (2007)
82. L'vov, V.S., Nazarenko, S.V., Rudenko O.: Gradual eddy-wave crossover in superfluid turbulence. J. Low Temp. Phys. **153**, 5–6 (2008). doi 10.1007/s10909-008-9844-0 arXiv:0807.1258
83. Longuet-Higgins, M.S., Gill, A.E.: Resonant interactions between planetary waves. Proc. R. Soc. Lond. Ser. A Math. Phys. Sci. **299**(1456), 120–144 (1967) (a discussion on nonlinear theory of wave propagation in dispersive systems)
84. Micha, R., Tkachev, I.I.: Relativistic turbulence: a long way from preheating to equilibrium. Phys. Rev. Lett. **90**(12), 121301–12688864 (2003) (P,S,G,E,B)
85. Micha, R., Tkachev, I.I.: Turbulent thermalization. Phys. Rev. D **70**(4), id. 043538 (2004)
86. Monin, A.S., Piterbarg, L.I.: On the kinetic-equation for Rossby-Blinova waves. Dokl. Akad. Nauk SSSR **295**, 816 (1987)
87. Mordant, N.: Are there waves in elastic wave turbulence? Phys. Rev. Lett. **100**, 234505 (2008)
88. Moser, J.: On invariant curves of area preserving mappings of an annulus. Nachr. Akad. Wiss. Goett. Math. Phys. **K1**, 1 (1962)
89. Musher, S.L., Rubenchik, A.M., Zakharov, V.E.: Hamiltonian approach to the description of nonlinear plasma phenomena. Phys. Rep. **129**(285), 285–366 (1985)
90. Muller, W.-C., Biskamp, D.: Scaling properties of three-dimensional magnetohydrodynamic turbulence. Phys. Rev. Lett. **84**, 475–478 (2000)
91. Muller, W.-C., Biskamp, D., Grappin, R.: Statistical anisotropy of magnetohydrodynamic turbulence, Phys. Rev. E **67**(1–4), 066302 (2003)
92. Nazarenko, S.: Sandpile behaviour in discrete water-wave turulence. J. Stat. Mech. LO2002 (2006)
93. Nazarenko, S.V.: 2D enslaving of MHD turbulence. New. J. Phys. **9**, 307 (2007). doi: 10.1088/1367-2630/9/8/307
94. Nazarenko, S.: Differential approximation for Kelvin-wave turbulence. JETP Lett. **83**(5), 198–200 (2005) (arXiv: cond-mat/0511136)
95. Nazarenko, S.: Kelvin wave turbulence generated by vortex reconnections. JETP Lett. **84**(11), 585–587 (2006) (Arxiv:cond-mat/0610420)
96. Nazarenko, S., Onorato, M.: Wave turbulence and vortices in Bose-Einstein condensation. Phys. D **219**, 1–12 (2006)
97. Nazarenko, S., Onorato, M.: Freely decaying turbulence and Bose-Einstein condensation in Gross-Pitaevski model. J. Low Temp. Phys. **146**(1/2), 31–46 (2007)
98. Ng, C.S., Bhattacharjee, A.: Scaling of anisotropic spectra due to the weak interaction of shear-Alfvén wave packets. Phys. Plasmas **4**, 605–610 (1997)
99. Newell, A.C.: Rossby wave packet interactions. J. Fluid Mech. **35**, 255–271 (1969)
100. Jakobsen, P., Newell, A.: Invariant measures and entropy production in wave turbulence. J. Stat. Mech. L10002 (2004)
101. Newell, A.C., Rumpf, B.: Wave turbulence. Ann. Rev. Fluid Mech. **43** (2011)

102. Ng, C.S., Bhattacharjee, A.: Interaction of shear-Alfvén wave packets: implication for weak magnetohydrodynamic turbulence in astrophysical plasmas. Astrophys. J. **465**, 845 (1996)
103. Onorato, M., Osborne, A., Serio, M., Resio, D., Pushkarev, A., Zakharov, V.E., Brandini, C.: Freely decaying weak turbulence for sea surface gravity waves. Phys. Rev. L **89**(14) (2002)
104. Pedlosky, J.: Geophysical Fluid Dynamics. Springer, Berlin (1987)
105. Peierls, R.: Annalen Physik **3**, 1055 (1929)
106. Phillips, O.M.: Theoretical and experimental studies of gravity waves interactions. Proc. R. Soc. London Ser. A **299**(1456), 105 (1967)
107. Piterbarg, L.I.: Hamiltonian formalism for Rossby waves. In: Zakharov, V.E. (ed.) Nonlinear Waves and Weak Turbulence, p. 131. Springer, Berlin (1998)
108. Prigogine, I.: Nonequilibrium Statistical Mechanics. Wiley, New York (1962)
109. Pushkarev, A.N.: On the Kolmogorov and frozen turbulence in numerical simulation of capillary waves. Eur. J. Mech. B/Fluids **18**(3), 345 (1999)
110. Pushkarev, A.N., Zakharov, V.E.: Turbulence of capillary waves—theory and numerical simulation. Phys. D **135**(1–2), 98 (2000)
111. Saur, J., Politano, H., Pouquet, A., Matthaeus, W.H.: Evidence for weak turbulence in Jupiter's middle magnetosphere. Astron. Astrophys. **386**(2), 699 (2002)
112. Saur, J.: Turbulent heating of Jupiter's middle magnetosphere. Astrophys. J. Lett. **602**, L137–L140 (2004). doi:10.1086/382588
113. Silberman, I.: Planetary waves in atmosphere. Meteorology **11**, 27 (1954)
114. Tanaka, M., Yokoyama, N.: Fluid Dyn. Res. **34**, 216 (2004)
115. Vedenov, A.A.: Theory of weakly turbulent plasma. In: Leontovich, M.A. (ed.) Reviews of Plasma Physics, vol. 3, p. 229. Consultants Bureau, New York (1967)
116. Vinen, W.F., Niemela, J.J.: J. Low Temp. Phys. **128**, 167 (2002)
117. Vinen, W.F., Tsubota, M., Mitani, A.: Phys. Rev. Lett. **91**, 135301 (2003)
118. Wright, W.B., Budakian, R., Putterman, S.J.: Diffusing light photography of fully developed isotropic ripple turbulence. Phys. Rev. Lett. **76**, 4528 (1996)
119. Wright, W.B., Budakian, R., Pine, D.J., Putterman, S.J.: Imaging of intermittency in ripple-wave turbulence. Science **278**, 1609 (1997)
120. Zakharov, V.E.: Weak turbulence in media with decay spectrum. Zh. Priklad. Tech. Fiz. **4**, 35–39 (1965) [J. Appl. Mech. Tech. Phys.**4**, 22–24 (1965)]
121. Zakharov, V.E.: Kolmogorov spectra in weak turbulence problems. In: Galeev, A.A., Sudan, R.N. (eds.) Handbook of Plasma Physics, vol. 2, Basic Plasma Physics, pp. 3–36. North-Holland, Elsevier (1984)
122. Zakharov, V.E., Filonenko, N.N.: Energy spectrum for stochastic oscillations of a fluid surface. Dokl. Acad. Nauk SSSR **170**, 1292–1295 (1966) (translation: Sov. Phys. Dokl. **11**, 881–884 (1967))
123. Zakharov, V.E., Filonenko, N.N.: Weak turbulence of capillary waves. Zh. Prikl. Mekh. Tekh. Phys. **4**(5), 62 (1967) (in Russian: J. Appl. Mech. Tech. Phys. **4**, 506)
124. Zakharov, V.E., Guyenne, P., Pushkarev, A.N., Dias, F.: Wave turbulence in one-dimensional models. Phys. D **152–153**, 573 (2001)
125. Zakharov, V.E., Korotkevich, A.O., Pushkarev, A.N., Dyachenko, A.I.: Mesoscopic wave turbulence. JETP Lett. **82**(8), 491 (2005)
126. Zakharov, V.E., L'vov, V.S.: The statistical description of the nonlinear wave fields. Izv. Vuzov, Radiofizika **18**(10), 1470–1487 (1975)
127. Zakharov, V.E., L'vov, V.S., Falkovich, G.: Kolmogorov Spectra of Turbulence 1: Wave Turbulence. Springer, Berlin (1992)
128. Zakharov, V.E., Nazarenko, S.V.: Dynamics of the Bose-Einstein condensation. Phys. D **201**, 203–211 (2005)
129. Zakharov, V.E., Piterbarg, L.I.: Sov. Phys. Dokl. **32**, 560 (1987)
130. Zakharov, V.E., Piterbarg, L.I.: Canonical variables for Rossby waves and plasma drift waves. Phys. Lett. A **126**(8–9), 497 (1988)

131. Zakharov, V.E., Sagdeev, R.Z.: Spectrum of acoustic turbulence. Sov. Phys. Dokl. **15**, 439 (1970)
132. Zaslavskii, G.M., Sagdeev, R.Z.: Sov. Phys. JETP **25**, 718 (1967)
133. Zurek, W.H.: Cosmological experiments in superfluid helium? Nature **317**, 505 (1985)
134. Zurek, W.H.: Acta. Phys. Polonica B**24**, 1301 (1993)

Part I
Primer on Wave Turbulence

In this part, we will start with a basic description of hydrodynamic turbulence, find common and distinct features in the hydrodynamic and the wave turbulence, and thereby put WT in the context of the general subject of Turbulence.

We will continue by presenting shortcuts and tricks using which one can recover most of the key results in the WT theory. Namely, we will explain how to determine the order of the resonant interaction based on the shape of the dispersion relation and the nonlinearity degree, how to find spectra of weak WT using the dimensional analysis, and we will apply the critical balance ideas for strong WT.

This part is intended for a quick introduction or/and a quick recovery of the known results without having to use more advanced and rigorous approach of the following chapters. This part covers the great majority of the known WT applications, and as such it can be used for quick reference.

Chapter 2
Wave Turbulence as a Part of General Turbulence Theory

2.1 Basic Facts about Hydrodynamic Turbulence

The name "Wave Turbulence" may appear paradoxical to people unfamiliar with the subject, because traditionally turbulence is associated mostly with vortices, and the waves are only secondary. We will try to explain that, on the contrary, a system of weakly nonlinear waves behaves very similar to the classical turbulence system, even in absence of vortices. For this, let us first briefly review the classical results in turbulence. For a more detailed discussion of the hydrodynamic turbulence see book [1].

Even though the fluid equations are deterministic, turbulence is a chaotic motion, and it is difficult to obtain a complete detailed information about the fluid flow. Indeed, individual paths of the fluid particles in a turbulence are sensitive to slight changes in initial conditions and imperfections in repeated experiments. However, averaged quantities in turbulence are well defined and can be studied, which is the subject of the *statistical theory* of Turbulence.

As with any statistical system, one might be tempted to approach turbulence using thermodynamics. Unfortunately, a simple-minded straightforward application of the classic thermodynamic theory would predict that turbulent systems have extremely high temperatures (energy per degree of freedom), which is absurd. Instead, rather than through temperature or chemical potential, turbulence is better described by the *energy flux through scales*. If such a flux is *local*, i.e. takes place as a sequence of transfers between eddies of similar sizes, it is called the *energy cascade*. However, as we will see later, there is some space for the thermodynamical solutions too.

The energy spectrum of turbulence is a mean quantity and is one of the main objects studied in turbulence theory. We write this spectrum as follows

$$E^{(3D)}(\mathbf{k}) = \frac{1}{2} \int_{\mathbb{R}^3} \langle \mathbf{u}(\mathbf{x}) \cdot \mathbf{u}(\mathbf{x} + \mathbf{r}) \rangle e^{-i\mathbf{k}\cdot\mathbf{r}} \frac{d\mathbf{r}}{(2\pi)^3}, \qquad (2.1)$$

S. Nazarenko, *Wave Turbulence*, Lecture Notes in Physics, 825,
DOI: 10.1007/978-3-642-15942-8_2, © Springer-Verlag Berlin Heidelberg 2011

and we define *homogeneous turbulence* as turbulence for which the energy spectrum is independent of $\mathbf{x}$. We also define *isotropic turbulence* as a system where the energy spectrum is independent of the direction of the wave number $\mathbf{k}$, i.e. $E^{(3D)}(\mathbf{k}) = E^{(3D)}(k)$ with $k = |\mathbf{k}|$. The angular bracket denotes a suitable average, ensemble, volume or space (here we will omit discussion of relevance of different averaging procedures).

The super-script $(3D)$ refers to the fact that $E^{(3D)}$ represents the kinetic energy density in the 3D $\mathbf{k}$-space, i.e.

$$\frac{1}{2}\langle u^2 \rangle = \int_{\mathbb{R}^3} E^{(3D)}(\mathbf{k})d\mathbf{k}. \tag{2.2}$$

On the other hand, for isotropic spectra the same information is contained in a 1D spectrum which is obtained from $E^{(3D)}$ by integration over the unit sphere in the 3D $\mathbf{k}$-space. This gives

$$E^{(1D)}(k) = 4\pi k^2 E^{(3D)}(k)$$

so that $E^{(1D)}(k)$ represents the energy density over $k = |\mathbf{k}|$,

$$\frac{1}{2}\langle u^2 \rangle = \int_0^{+\infty} E^{(1D)}(k)dk. \tag{2.3}$$

2.1.1 Richardson Cascade

So, nat'ralists observe, a flea
Hath smaller fleas that on him prey;
And these have smaller yet to bite 'em,
And so proceed ad infinitum.

This is a verse from the poem "On Poetry: a Rhapsody" written by Jonathan Swift in 1733. It was not about insects but, rather, it was meant to be a metaphor with a rather pessimistic conclusion in the lines that immediately follow:

Thus every poet, in his kind,
Is bit by him that comes behind:
Who, though too little to be seen,
Can teaze, and gall, and give the spleen.

Swift liked "self-similar" examples, suffices to recall the "Gulliver Travels" with its little people and big people, and it is precisely the self-similarity of the "flea" verse that fascinated Lewis Richardson who in 1922 came up with his own version for turbulence:

Fig. 2.1 Richardson's
energy cascade

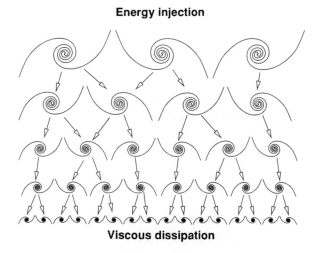

Energy injection

Viscous dissipation

Big whorls have little whorls,
Which feed on their velocity,
And little whorls have lesser whorls,
And so on to viscosity.

Richardson pictured that small vortices obtain their energy from break-ups of larger ones, only to find themselves breaking to even smaller ones, and so on in a self-similar way, see Fig. 2.1. The largest vortices in the Richardson cascade picture obtain their energy from an external forcing (e.g. a mechanical forcing or an instability mechanism to release the internal energy into the kinetic one) and the smallest vortices are dissipated by viscosity [2]. The total rate of the energy injection at large scales is equal on average to the energy dissipation rate at small scales, so that a statistically steady turbulent state forms.

The Richardson cascade is best represented in Fourier **k**-space, see Fig. 2.2. Here, the lengthscale is $1/k$ (where $k = |\mathbf{k}|$), so that we have our turbulence source at small k_f, and the energy cascade is in the positive k-direction towards the dissipation scale at large k_ν.

2.1.2 Kolmogorov–Obukhov Theory

In 1941, Kolmogorov [3, 4] and Obukhov [5] introduced the universality hypothesis for the inertial range, i.e. for $k_f \ll k \ll k_\nu$. The idea is that far away from the source and the sink turbulence properties only depend on the energy cascade rate (equal to the energy dissipation rate in the steady state), and not on details of the forcing or the dissipation of energy. This is because the Richardson

Fig. 2.2 Energy cascade in the k-space

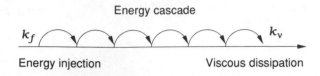

cascade is assumed to be *local*, i.e. transferring energy in many steps each involving transfers among eddies with similar sizes only.

The Kolmogorov–Obukhov idea leads to a dimensional argument, where the energy dissipation rate ε and wavenumber k are assumed to be the only relevant dimensional quantities in the inertial range $k_f \ll k \ll k_v$. In particular, the viscosity v is irrelevant.

We find for the dimensions

$$[\varepsilon] = \left[\frac{\mathbf{u}^2}{t}\right] = \frac{l^2}{t^3},$$

$$\left[E^{(1D)}\right] = \left[\frac{\mathbf{u}^2}{k}\right] = \frac{l^3}{t^2}. \tag{2.4}$$

With these in mind, we see that the only combination of ε and k that results in the correct dimension of $E^{(1D)}$ is

$$E^{(1D)} = C\varepsilon^{2/3}k^{-5/3}, \tag{2.5}$$

which is known as the Kolmogorov–Obukhov spectrum, often also referred to as Kolmogorov spectrum, or K41 [3–5]. Here constant $C \sim 1.6$ is called the Kolmogorov constant. In spite of such a simple dimensional derivation, this spectrum represents the strongest result in turbulence which is supported by numerous experimental confirmations in laboratory and in field observations. However, there have also been observed some deviations from the Kolmogorov scaling, which are small for the spectrum (and other second-order moments) but become significant for the high-order moments of the velocity field. These deviations are associated with the turbulence intermittency phenomenon, which has been attracting a considerable attention in the modern turbulence research [1].

The Richardson cascade picture and the Kolmogorov–Obukhov spectrum are of fundamental importance for the WT theory, because similar cascade states are typical for WT, as we will see later. Moreover, deviations from the Kolmogorov-type scalings associated with WT intermittency are also interesting and they will be discussed in the present book.

2.1.3 2D Turbulence

It is often possible to find more than one positive conserved quantity in WT systems. This will modify the cascade picture presented above. We shall show that the behavior of these systems is similar to 2D turbulence, where two invariants are transferred in the opposite directions in the scale space, as discovered by Fjørtoft [6]. Kraichnan found Kolmogorov-like spectra for each of the two cascades [9]. Thus, let us review the results concerning 2D turbulence.

Let us consider a 2D flow. For instance, horizontal motions in the planetary atmosphere stretch over far larger scales than the atmosphere height, and this system can therefore be approximated by a 2D flow. Rapidly rotating fluids provide another example (via Taylor–Proudmann two-dimensionalization). Charged particles in a very strong external magnetic field give yet another example [7].

2.1.3.1 Energy and Enstrophy

From a basic Fluid Dynamics course we learn that the 2D incompressible ideal flow conserves two quadratic quantities, energy and enstrophy. Naturally, for infinite homogeneous turbulence these two quantities are infinite, and it only makes sense to talk about their spatial density. We can write both in terms of the energy spectrum $E^{(1D)}(k)$:

$$E = \frac{\text{Energy}}{\text{Area}} = \frac{1}{2}\langle \mathbf{u}^2 \rangle = \int_0^\infty E^{(1D)}(k)dk, \tag{2.6}$$

$$\Omega = \frac{\text{Enstrophy}}{\text{Area}} = \frac{1}{2}\langle \omega^2 \rangle = \int_0^\infty k^2 E^{(1D)}(k)dk \tag{2.7}$$

where we have used the relation between the velocity and the vorticity in Fourier space, $\hat{\omega}_k = i\mathbf{k} \times \hat{\mathbf{u}}_k$.

2.1.3.2 Dual Cascade Behavior

Let us introduce the enstrophy production rate η. In statistically stationary turbulence, the dissipation rate is equal to the production rate. Let us consider turbulence excited near wavenumber k_f and dissipated at very small wavenumbers $k_- \ll k_f$ and at very large wavenumbers $k_+ \gg k_f$, and let there be neither forcing nor dissipation at wavenumbers such that $k_- < k < k_f$ or $k_f < k < k_+$, see Fig. 2.3. These intervals are called the *inverse and the direct cascade inertial ranges*

Fig. 2.3 2D turbulence: dual cascade behavior in the k-space

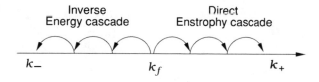

Inverse Energy cascade   Direct Enstrophy cascade

k_- k_f k_+

respectively. The relation between η and ε follows from (2.6) and (2.7) which dictate $\eta \sim k_f^2 \varepsilon$.

Let us use an *ad absurdum* argument to find the directions of the energy and the enstrophy cascades. This will constitute so-called Fjørtoft argument [6].

If energy was dissipated at k_+ at the rate comparable to the injection rate ε then enstrophy would be dissipated at a rate $\sim k_+^2 \varepsilon \gg k_f^2 \varepsilon \sim \eta$, which is a contradiction, because in a steady state it is impossible to dissipate enstrophy at a rate which is higher than the injection rate. Therefore, energy must be dissipated at k_-, and we call the energy cascade from k_f to k_- the *inverse energy cascade*. In a picture like Richardson's, we would see vortices merging rather than breaking up, i.e. energy is transferred from small to large vortices.

We can use a similar argument for the enstrophy, assuming *ad absurdum* that it is dissipated at $k_- \gg k_f$ at the rate comparable to the production rate η. But this would imply that the energy is dissipated at k_- at the rate $\sim \eta/k_-^2$ which is much greater than the energy production rate $\varepsilon \sim \eta/k_f^2$ which is impossible in stationary turbulence. Therefore, the enstrophy must dissipate at k_+. We call this transfer of enstrophy from k_f to k_+ the *direct enstrophy cascade*. In a picture like Richardson's, the enstrophy is transferred into long and thin vorticity filaments during the vortex merging process. Such thin filaments also occur when weaker vortices are stretched by stronger ones.

2.1.3.3 Fjørtoft Argument in Terms of Centroids

Note that there exist several versions of the Fjørtoft argument, a couple of which were presented by Fjørtoft himself in his original paper [6], some other variations can be found e.g. in [8]. These different versions vary in degree of rigor from being a mathematical theorem to a less rigorous physical speculation. Here we will present yet another (probably new) way to formulate Fjørtoft in terms of the energy and the enstrophy centroids in the k and the l (the length scale) spaces. This formulation will be rigorous and quite useful for visualizing the directions of transfer of the energy and the enstrophy. In contrast with the above version, this formulation is for a non-dissipative evolving turbulence rather than for a forced/dissipated system. First let us define the centroids.

Definition 2.1 The energy and the enstrophy k-centroids are defined respectively as

$$k_E = \int_0^\infty k E^{(1D)}(k)dk/E, \qquad (2.8)$$

$$k_\Omega = \int_0^\infty k^3 E^{(1D)}(k)dk/\Omega, \qquad (2.9)$$

and the energy and the enstrophy l-centroids are defined respectively as

$$l_E = \int_0^\infty k^{-1} E^{(1D)}(k) dk / E, \qquad (2.10)$$

$$l_\Omega = \int_0^\infty k E^{(1D)}(k) dk / \Omega \equiv k_E E / \Omega, \qquad (2.11)$$

Theorem 2.1 *Assuming that the integrals defining $E, \Omega, k_E, k_\Omega, l_E$ and l_Ω converge, the following inequalities hold,*

$$k_E \leq \sqrt{\Omega/E}, \qquad (2.12)$$

$$k_\Omega \geq \sqrt{\Omega/E}, \qquad (2.13)$$

$$k_E k_\Omega \geq \Omega/E, \qquad (2.14)$$

$$l_E \geq \sqrt{E/\Omega}, \qquad (2.15)$$

$$l_\Omega \leq \sqrt{E/\Omega}, \qquad (2.16)$$

$$l_E l_\Omega \geq E/\Omega. \qquad (2.17)$$

Exercise 2.1 *Prove this theorem using Cauchy–Schwartz inequality.*

We see that according to inequalities (2.12) and (2.13), during the system's evolution the energy centroid $k_E(t)$ is bounded from above and the enstrophy centroid $k_\Omega(t)$ is bounded from below (both by the same wavenumber $k = \sqrt{\Omega/E}$), as one would expect from Fjørtoft argument. Further, inequality (2.14) means that if $k_E(t)$ happened to move to small k's then $k_\Omega(t)$ *must* move to large k's, that is roughly, *there cannot be inverse cascade of energy without a forward cascade of enstrophy*. Note that there is no complimentary restriction which would oblige $k_E(t)$ to become small when $k_\Omega(t)$ goes large, so the k-centriod part of the Fjørtoft argument is asymmetric, and one has to consider the l-centroids to make it symmetric. Indeed, in addition to conditions (2.15) and (2.16) which are similar to (2.12) and (2.13), we have inequality (2.17) meaning that if $l_\Omega(t)$ happened to move to small l's then $l_E(t)$ *must* move to large l's, i.e. *any forward cascade of enstrophy must be accompanied by an inverse cascade of energy*.

Importantly, we do not always have $k_E \sim 1/l_E$ and $k_\Omega \sim 1/l_\Omega$, as illustrated in the following exercise.

Exercise 2.2 *Consider a state with spectrum $E^{(1D)}(k) \sim k^{-5/3}$ for $k_a < k < k_b$ (with $k_b \gg k_a$) and $E^{(1D)}(k) \equiv 0$ outside of this range. Show that for this state $k_E \sim k_b^{1/3} k_a^{2/3}$ and $l_E \sim 1/k_a$ (i.e. $k_E \nsim 1/l_E$) and $k_\Omega \sim 1/l_\Omega \sim k_b$.*

2.1.3.4 Spectra of 2D Turbulence

Robert Kraichnan [9] realized that, provided the energy and the enstrophy cascades are local, the energy spectrum in the inverse cascade range will only be determined by the energy flux, whereas in the direct cascade range—by the enstrophy flux. He used a Kolmogorov-style dimensional argument and assumed that ε and k are the only relevant dimensional quantities that can enter the expression for the energy spectrum in the inverse cascade, and for the direct cascade these relevant quantities are respectively η and k.

The dimensional argument for the energy spectrum is basically identical to the one presented above for 3D turbulence, so we have [9]

$$E^{(1D)}(k) = C_\varepsilon \varepsilon^{2/3} k^{-5/3}.$$

Note, however, that the value of the dimensionless constant $C_\varepsilon \sim 6$ is not determined by the dimensional argument and, unsurprisingly, is different from the Kolmogorov constant C for 3D turbulence.

To find the enstrophy cascade spectrum, we note that the dimension for η is

$$[\eta] = [k^2][\varepsilon] = \frac{1}{l^2}\frac{l^2}{t^3} = \frac{1}{t^3}. \tag{2.18}$$

Thus, we find for the spectrum

$$E^{(1D)} = C_\eta \eta^{2/3} k^{-3}, \tag{2.19}$$

which is called the Kraichnan spectrum [9].[1] Here $C_\eta \sim 1.9$ is yet another dimensionless constant. Note a rather large difference ~ 3 between the values for C_ε and C_η. For an explanation of this large difference see paper [11].

The Kraichnan dual-cascade picture presented above was recently confirmed numerically in high-resolution (16384^2) DNS in [12].

2.1.3.5 Yet Another Argument about the Cascade Directions

In Sects. 2.1.3.2 and 2.1.3.3 we already presented two versions of the arguments predicting the energy and the enstrophy cascade directions, and here we will give yet another simple argument.

Let us consider a power-law spectrum $n_k = Ak^{-x}$ (not necessarily a stationary one) and let us construct plausible plots for the energy flux ε and the enstrophy flux η as functions of x, see Fig. 2.4.

We know that $\varepsilon(x)$ will cross zero for both of the exponents corresponding to the thermodynamic states with an energy equipartition, $x = -1$ (point TE on

---

[1] Later [10], Kraichnan found this spectrum to be marginally nonlocal, and remedied it with introducing a log-correction, $E^{(1D)} = C_\eta \, \eta^{2/3} \, k^{-3} \ln^{-1/3}(k/k_f)$.

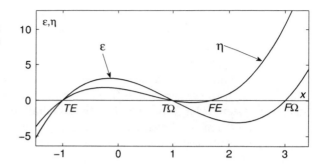

Fig. 2.4 The energy and the enstrophy fluxes as a function of the spectral index x for the 2D turbulence

Fig. 2.4), and an enstrophy equipartition, $x = 1$ (point $T\Omega$ on Fig. 2.4). Also, $\varepsilon(x)$ will cross zero at the enstrophy cascade solution, $x = 3$ (point $F\Omega$ on Fig. 2.4).

In turn, $\eta(x)$ will cross zero at both of the thermodynamics solutions (points TE and $T\Omega$ on Fig. 2.4) and on the energy cascade spectrum, $x = 5/3$ (point FE on Fig. 2.4).

It is clear that the fluxes will be both positive if x is very large, see Fig. 2.4. Thus, starting with large x and tracing the plots $\varepsilon(x)$ and $\eta(x)$ toward smaller x and making the necessary zero crossings as described above, we uniquely determine the x-ranges for which these fluxes are positive and negative. In particular, we find that the energy flux ε is negative at point FE (inverse energy cascade), and the enstrophy flux η is positive at point $F\Omega$ (direct enstrophy cascade).

2.2 Placing Wave Turbulence in the Context of General Turbulence

Now we are in the position to discuss similarities and differences between WT and hydrodynamic turbulence and to place WT in the context of the general Turbulence theory. Let us recall the definition of WT which we gave in the beginning of this book:

Wave Turbulence is out-of-equilibrium statistical mechanics of random nonlinear waves.

This definition refers to two features common for all turbulent systems, the *randomness* and the *non-equilibrium character*, as well as to properties specific for WT only, particularly the fact that it is a system of *waves rather than vortices*. Let us discuss these common and the distinct properties in greater detail.

2.2.1 Common Turbulence Properties

Randomness and large deviation from the thermodynamic equilibrium are important ingredients of the above definition which make WT similar to the other turbulent systems. Randomness is related to (but not completely predetermined by) the fact that there are many wave modes excited in the system, covering a wide range of dynamical length- and time-scales. Thus, the dynamics of these modes is chaotic and should be described by a statistical rather than a deterministic theory.

It is important to understand that the set of waves form an open system, with sources and sinks of energy at different (and often well separated) scales. This statistical system is far from thermodynamical equilibrium. The most typical states for such a system are determined by a flux of energy (or another conserved quantity) through scales rather than temperature or any other thermodynamic potentials. This property is common for all turbulent systems, and the most known example here is Richardson cascade and the Kolmogorov–Obukhov spectrum which corresponds to this statistical state. In WT, the analogs of the Kolmogorov–Obukhov spectrum are Kolmogorov-Zakharov (KZ) spectra. As we will see later, KZ spectra can also be obtained from a dimensional analysis which strengthens their similarity to the Kolmogorov–Obukhov spectrum of hydrodynamic turbulence. It is reasonable to consider the prevalence of the flux/cascade dominated statistical states as a definition of Turbulent systems. In this case, one could mention another non-equilibrium system which qualifies to be called turbulence under this definition: a set of sticky particles coagulating upon collision which also exhibit cascades [13–15]. In this case the cascade takes form of a mass flux in the space of particle sizes.

Finally, some WT systems have several positive conserved quantities and, like 2D turbulence, exhibit a dual cascade behavior. Also like in 2D turbulence, inverse cascades may lead to accumulation of turbulence at the largest available scales, followed by breakdown of the local cascade picture and onset of the nonlocal interaction. We will see later that this happens, in particular, in the nonlinear Schrödinger (NLS) model, where this process corresponds to a non-equilibrium Bose–Einstein condensation (BEC). We will see that during this process WT changes its nature from being a four-wave to a three wave process, with a non-trivial phase transition between these states involving strong turbulence and vortices [16]. Further examples of inverse WT cascades can be found in the water gravity waves (swell effect) and the Rossby/Drift waves (leading to formation of zonal jets). Later in this book, we will discuss these wave systems too.

2.2.2 Distinct Properties of WT

One obvious difference between the hydrodynamic turbulence and WT is that in the former case the basic type of motion is a hydrodynamic vortex whereas in the latter case it is a propagating wave. Note that in many physical examples there are both vortices and waves which co-exist and interact with each other, for example compressible fluids, MHD turbulence, fluids in rotating and stratified media, quantum turbulence. In these cases WT is only a part of the whole turbulent system.

In most of the WT theory, it is further assumed that the waves are weakly nonlinear and dispersive. These assumptions allow a systematic mathematical treatment of WT. Indeed, weakly nonlinear waves can, for short time intervals, be approximated by independent linear waves. Their amplitudes are sufficiently small and time-independent over such scales. At long times however, the wave amplitudes change, but the large difference in scale between this nonlinear evolution and the linear wave period allows one to average over the fast linear times. It should be noted that the WT description is not a perturbation theory, for even slow nonlinear evolution may lead to order-one changes of wave amplitudes.

In WT, the wave dispersion is important, since non-dispersive waves have constant (k-independent) group velocities and the wave packets can strongly affect each other because they "stick together" for an infinite time. However, in exceptional cases it is possible for WT to be applied to non-dispersive systems, e.g. in Alfvén waves where co-propagating waves do not interact, or in 3D sound where divergence of waves propagating in various space directions plays a similar role as the wave dispersion (this is not the case for the sound in lower-dimensional spaces!).

A remarkable distinct feature of weak WT is that the KZ spectra corresponding to the cascade states can be obtained as exact analytical solutions of the WT kinetic equation and not only dimensionally. Further, stability of the solution and locality of interaction can also be examined analytically. This is a luxury which is not available in the theory of hydrodynamic turbulence, and it brings about yet another important role of WT as a testing ground for universal concepts proposed in the context of more complex turbulent systems, and for developing new ideas that could later be applied more broadly in turbulence.

References

1. Frisch, U.: Turbulence: The Legacy of A.N. Kolmogorov. Cambridge University Press, Cambridge (1995)
2. Richardson, L.F.: Proc. R. Soc. Lond. Ser. A **110**, 709 (1926)
3. Kolmogorov, A.N.: The local structure of turbulence in incompressible viscous fluid for very large Reynolds numbers, Dokl. Akad. Nauk SSSR **30**, 9 (1941)
4. Kolmogorov, A.N.: Dissipation of energy in a locally isotropic turbulence, Dokl. Akad. Nauk SSSR **32**, 141 (1941) (English translation in: American Mathematical Society Translations 1958, Series 2, **8**, 87, Providence, RI)
5. Obukhov, A.M.: On the distribution of energy in the spectrum of turbulent flow. Dokl. Akad. Nauk SSSR **32**(1), 22–24 (1941)
6. Fjørtoft: On changes in the spectral distribution of kinetic energy for two-dimensional non-divergent flow. Tellus **5**, 225 (1953)
7. Driscoll, C.F., Fine, K.S.: Experiments on vortex dynamics in pure electron plasmas. Phys. Fluids B **2**, 1359 (1990)
8. Rhines, P.B.: Geostrophic turbulence. Ann. Revs. Fluid Mech. **11**, 404–441 (1979)
9. Kraichnan, R.H.: Inertial ranges in two-dimensional turbulence. Phys. Fluids **10**(7), 1417–1423 (1967)
10. Kraichnan, R.: J. Fluid Mech. **47**, 525 (1971)

11. Lvov, V.S., Nazarenko, S.V.: Differential model for 2D turbulence. JETP Lett. **83**(12), 635–639 (2006)
12. Boffetta, G.: Energy and enstrophy fluxes in the double cascade of two-dimensional turbulence. J. Fluid Mech. **589**, 253 (2007)
13. Connaughton, C., Rajesh, R., Zaboronski, O.: Cluster-cluster aggregation as an analogue of a turbulent cascade: Kolmogorov phenomenology, scaling laws and the breakdown of self-similarity. Phys. D **222**, 97 (2006)
14. Horvai, P., Nazarenko, S.V., Stein, T.: Coalescence of particles by differential sedimentation. J. Stat. Phys. **130**(6), 1177–1195 (2007). doi:10.1007/s10955-007-9466-y
15. Cardy, J., Falkovich, G., Gawedzki, K., Nazarenko, S., Zaboronski, O.V. (eds.): Non-equilibrium Statistical Mechanics and Turbulence, pp. 132–142. CUP, London Mathematical Society Lecture Note Series 355 (2008)
16. Nazarenko, S., Onorato, M.: Wave turbulence and vortices in Bose–Einstein condensation. Phys. D **219**, 1–12 (2006)

Chapter 3
For the Impatient: A WT Cheatsheet

In this chapter, we will present a shortcut derivation of WT spectra based on a version of dimensional analysis which was originally used for MHD turbulence by Kraichnan [1] and was also outlined in book [2]. It was further developed and applied to a wide variety of WT systems in [3]. We will follow the approach developed in paper [3] and we will add some new examples.[1] Moreover, following [4], we will extend this approach to strongly nonlinear systems which are in the state of so-called critical balance (when the linear and the nonlinear time scales are of the same order on the scale-by-scale basis).

The shortcut approach of this chapter is limited in what can be achieved, and it has to be used with caution because it implicitly uses some unchecked assumptions (like the turbulence locality or an assumption about isotropy or a particular type of anisotropy). However, this drawback is compensated by the method's great simplicity, and the author of this book has been frequently using it for quick estimates in new applications or/and as a quick reference for recovery of already known results.

We will deal with wave systems whose wave behavior is completely determined by a *single dimensional physical quantity*. This condition might seem restrictive, but it appears to be satisfied for majority of important wave systems. For example:

1. Surface tension coefficient σ determines both linear and nonlinear properties of the capillary waves;
2. Speed of sound plays the same role for the acoustic waves;
3. Alfvén velocity determines behavior of MHD waves;
4. Gravitational constant g is the key constant for the water gravity waves;
5. Angular velocity of rotation Ω—for the inertial waves in rotating fluids;

---

[1] We will deal here only with KZ spectra, whereas [3] also considered scalings of the interaction coefficients and the WT breakdown scales. Please read this paper if you are interested in these problems.

S. Nazarenko, *Wave Turbulence*, Lecture Notes in Physics, 825,
DOI: 10.1007/978-3-642-15942-8_3, © Springer-Verlag Berlin Heidelberg 2011

6. For Langmuir waves in plasmas the relevant constant turns out to be $v_{th}\,\lambda_D$ where v_{th} is the thermal velocity and λ_D is Debye length;
7. Quantum of circulation $\kappa = h/m$ is a determining constant for the Kelvin waves on quantized vortex lines in superfluids (here h is the Plank's constant and m is the atom's mass, e.g. of ^{4}He).

3.1 Weak Wave Turbulence

First of all, we note that the determining constant (as introduced above) is an extra dimensional entity with respect to the usual dimensional quantities present in the classical K41 turbulence (the latter being the energy spectrum, the energy dissipation rate and the wavenumber). Thus, the standard K41 dimensional argument presented in the previous chapter is not applicable because there is now an infinite number of ways to obtain the energy spectrum from the other dimensional quantities. On the other hand, the character of the nonlinear wave interactions can be classified by the number of waves N involved in the resonant interaction, and this fact poses an extra condition which restores uniqueness of the dimensional argument. But before we show this, let us try to clarify what N is.

3.1.1 Three-Wave? Four-Wave? N-Wave?

N is the minimal number of waves for which:

• The N-wave resonant conditions for the wavevectors and the frequencies are satisfied simultaneously for a non-trivial set of wavenumbers,

$$\omega(k_1) \pm \omega(k_2) \pm \cdots \pm \omega(k_N) = 0, \tag{3.1}$$

$$k_1 \pm k_2 \pm \cdots \pm k_N = 0, \tag{3.2}$$

where $\omega(k)$ is the frequency of the wave with wavenumber k. Signs plus or minus depend on the type of the N-wave process, e.g. for the three-wave processes $2 \rightarrow 1$ two of the signs must be "+" and one sign "−", and for the four-wave processes $2 \rightarrow 2$ there must be two "+"es and two "−"es.

In some cases there are simple rules to determine whether or not a particular set of resonant conditions can be satisfied or not. For example, in 2D for cases $\omega \sim k^\alpha$ three-wave resonances can exist only for $\alpha \geq 1$, which has a simple graphical proof suggested by Vedenov [5] and shown in Fig. 3.1 and 3.2. For, $\alpha \geq 1$, in Fig. 3.1 the resonant solutions correspond to the intersection of the frequency surfaces marked by a dashed line. For, $\alpha < 1$, in Fig. 3.2 there are no nontrivial resonant solutions because the frequency surfaces do not intersect except for one point (the latter corresponds to a trivial solution $k_1 = k_3, \; k_2 = 0$).

Fig. 3.1 Graphical solution
for the three-wave resonant
condition: solutions exist for
$\alpha \geq 1$

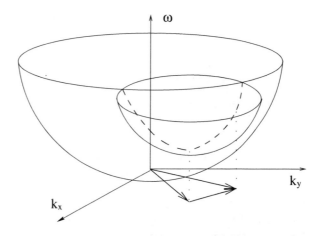

Fig. 3.2 Graphical proof that
for $\alpha < 1$ there is no solution
to the three-wave resonant
condition

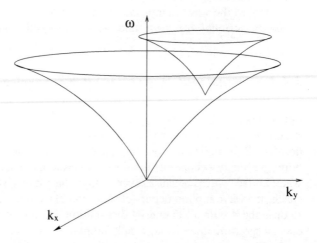

Exercise 3.1 *Show that in 1D there are no nontrivial four-wave resonances for* $\omega \sim k^{\alpha}$ *with* $\alpha \geq 1$ *(this is why e.g. there are no four-wave resonances for Kelvin waves with* $\omega \sim k^{2}$*).*

Exercise 3.2 *Show that in 2D there are no nontrivial four-wave resonances of type* $3 \to 1$ *for* $\omega \sim k^{\alpha}$ *with* $0 < \alpha < 1$*. Can such resonances exist for* $\alpha \geq 1$*? (this could be relevant to the systems with cubic nonlinearities).*

- The N-wave interaction coefficient is not zero for this set of resonant wave-numbers. (Important: the N-wave interaction coefficient must include the lower-order non-resonant contributions. For example, in a system where the lowest order resonances is four, but the nonlinearity is quadratic in the original wave variables, the order of the process is likely to be $N = 4$ even if the cubic

nonlinearity is absent in the original wave variables. This is because the cubic
nonlinearity will arise in the new canonical wave variables after a nonlinear
change of variables designed to remove the quadratic nonlinearity, see Appendix
A3 in book [2]).

In an N-wave process, we have the following rate relation,

$$\varepsilon \sim \dot{E} \sim E^{N-1} \tag{3.3}$$

For example, the three-wave process could be compared to a binary chemical
reaction where two wave-packets/particles collide to produce a third wave-packet/
particle ($2 \rightarrow 1$ process). Admittedly, such a "chemical reaction" interpretation is
not so useful for higher-order processes, e.g. a four-wave process ($2 \rightarrow 2$) because
the rate of the process is proportional not only to the concentrations of "reactants"
but also to the ones of the reaction "products". However, for now we will take the
relation (3.3) for granted (its validity will become obvious to us later when we will
have derived the wave kinetic equation). *It is this relation that restores uniqueness
of the dimensional argument*, and we will proceed with this argument below.

3.1.2 Dimensional Derivation of KZ Spectra

Let us first of all find the physical dimension of the energy spectrum for general
d-dimensional wave systems. Let us suppose that that the d-dimensional energy
density, E, is finite in physical space. For example, $d = 3$ for acoustic waves
homogeneously occupying the three-dimensional space. However, $d = 2$ for the
water surface waves because, even though the entire system is embedded in the 3D
space, the wave motion occurs close to the 2D water-air interface and it is natural
to describe it with a 2D energy distribution. Here is a shortcut way of finding the
energy spectrum dimension, which might look "dodgy" at first, but actually makes
sense. In an incompressible fluid, we have constant density $\rho = \frac{m}{V}$ with mass m and
volume V. In this case one can simply set $\rho = 1$ which means that we have chosen
the system of units where m and V have the same physical dimension. Thus, let the
dimension of mass be l^3, i.e. we will "measure mass in liters". For the physical
dimension of energy in the system we write

$$[E] = \left[\frac{1}{2}mv^2\right] = \frac{[l]^5}{[t]^2}. \tag{3.4}$$

For the energy density (energy per unit volume or area or length) we then write

$$\frac{[E]}{[l^d]} = \left[\int E_k^{(1D)} dk\right] = \left[E_k^{(1D)}\right] [l]^{-1}. \tag{3.5}$$

This relation and (3.4) give

$$\left[E_k^{(1D)}\right] = \frac{[l]^{6-d}}{[t]^2}. \tag{3.6}$$

This and the energy balance equation, $\dot{E}_k^{(1D)} + \partial_k \varepsilon = 0$, give us the dimension for the energy flux,

$$[\varepsilon] = \frac{[l]^{5-d}}{[t]^3}. \tag{3.7}$$

Let us consider a physical wave system characterized by a single dimensional parameter, e.g. a system of capillary waves, acoustic waves, Alfvén waves, water gravity waves, Langmuir waves in plasmas, spin waves in solids, inertial waves in rotating fluids or Kelvin waves on quantized vortex lines (see the list at the beginning of this chapter). For each of these cases, there exists a dispersion relation

$$\omega = \lambda k^\alpha,$$

with both α and λ determined by the respective relevant dimensional parameter. Conversely, we could take λ to be the defining physical parameter whose dimension will be related to α as

$$[\lambda] = [\omega][k]^{-\alpha} = [t]^{-1}[l]^\alpha. \tag{3.8}$$

The original physical constant (surface tension, gravity etc.) will then be uniquely determined in terms of λ.

Relation (3.3) and expressions (3.6)–(3.8) make the scaling for the energy spectrum unique, and we have

$$E_k^{(1D)} \sim \lambda^x \varepsilon^{1/(N-1)} k^y, \tag{3.9}$$

where

$$x = 2 - \frac{3}{N-1} \quad \text{and} \quad y = d - 6 + 2\alpha + \frac{5 - d - 3\alpha}{N-1}. \tag{3.10}$$

Remark 3.1.1 Often in WT, the dimensional argument also uses the homogeneity degree β of the nonlinear interaction coefficient (which we have not even introduced yet!). Our dimensional argument here has not used this coefficient which might seem odd at first sight. The trick is that the coefficient β is not independent in the systems with the single relevant dimensional parameter: β itself is fixed by the same dimensional parameter. The reader is referred to paper [3] for details on how to find β in such cases, as well as for finding the WT breakdown scale from the same kind of dimensional analysis.

3.1.2.1 Waveaction Cascade Spectra

For N-wave systems with an even N, the total waveaction, or "particle number",

$$\mathcal{N} = \int \frac{E_k^{(1D)}}{\omega_k} d\mathbf{k},$$

is also conserved for $N/2 \to N/2$ processes. Roughly, this means conservation of the total number of wavepackets during the process $N/2 \to N/2$, i.e. when $N/2$ wavepackets collide and disappear while producing $N/2$ new wavepackets. As a result, there can also exist a constant-flux spectrum carrying a flux of particles in addition to the energy cascade. Thus, such even-N systems are characterized by a dual cascade behavior, and applying the standard Fjørtoft argument of the previous chapter immediately leads to the conclusion that for $\alpha > 0$ the energy cascade is direct (unlike 2D turbulence) and the waveaction cascade is inverse. Further discussion of the dual cascades will be given in Sect. 8.2.1. We will also see later in Chap. 15 that such inverse cascades can lead to Bose-Einstein condensation, which in turn can completely reorganize the WT structure (e.g. lead to growth of a coherent phase and changes from a 4-wave to 3-wave interaction).

Exercise 3.3 *Prove that for* $\alpha > 0$ *the energy cascade is direct and the waveaction cascade is inverse. For this, use the Fjørtoft argument of the previous chapter (introduced there for 2D turbulence)—either in its standard* ad absurdum *form for the steady-state or in the centroid form for the evolving turbulence.*

Let us use the same dimensional analysis as above in this chapter to derive the energy spectrum corresponding to the state with the inverse waveaction cascade. Keeping in mind that for the dimension of the waveaction flux (dissipation rate) ζ is $[\zeta] = [\varepsilon/\omega] = l^{5-d} t^{-2}$ and using the relation $\zeta \sim E^{N-1}$ (which is similar to (3.3)) we get

$$E_k^{(1D)} \sim \lambda^x \zeta^{1/(N-1)} k^y$$

$$\Rightarrow x = 2 - \frac{2}{N-1} \Rightarrow y = d - 6 + \frac{5 - d + 2\alpha(N-2)}{N-1}. \qquad (3.11)$$

Let us now consider examples.

3.1.3 Examples

3.1.3.1 Capillary Waves

For capillary waves, we have $N = 3$, $d = 2$ and $\omega = \sigma^{1/2} k^{3/2}$, where σ the surface tension coefficient, and thus $\alpha = 3/2$. Using (3.9) we find the Zakharov-Filonenko spectrum [6, 7]:

$$E_k^{(1D)} \sim \sqrt{\varepsilon}\sigma^{1/4}k^{-7/4}. \tag{3.12}$$

3.1.3.2 Acoustic Turbulence. Waves in Isotropic Elastic Media

For acoustic turbulence, we have $N = 3$, $d = 3$ and $\omega = c_s k$, where $c_s = \sqrt{\gamma p/\rho}$ is the speed of sound (p is pressure, ρ is density), and thus $\alpha = 1$ and $\lambda = c_s$. Using (3.9) we may write the Zakharov-Sagdeev spectrum [8]:

$$E_k^{(1D)} \sim \sqrt{\varepsilon c_s}k^{-3/2}. \tag{3.13}$$

Similar situation arises in wave turbulence in isotropic elastic media. In this case the waves have the same dispersion relation and the energy spectrum with pressure γp replaced by the relevant elastic constant in the expression for the wave speed c_s, i.e. $\Lambda_1 + 2\Lambda_2$ for the (longitudinal) P-waves and Λ_2 for the the (transverse) S-waves, where Λ_1 and Λ_2 are the first and the second Lamé moduli.

3.1.3.3 Alfvén Waves

For Alfvén waves, we have $N = 3$, $d = 3$ and $\omega = c_A k_\parallel$, where c_A is Alfvén velocity and $k_\parallel$ is the wavevector projection onto the external magnetic field. With $\alpha = 1$ and $\lambda = c_A$ and using (3.9) we get the Iroshnikov-Kraichnan spectrum [1, 9]:

$$E_k^{(1D)} \sim \sqrt{\varepsilon c_A}k^{-3/2}. \tag{3.14}$$

However, this spectrum ignores anisotropy. In reality, the MHD turbulence of this kind tends to very anisotropic states with $k_\perp \gg k_\parallel$, i.e. it becomes nearly 2D in the plane transverse to the external magnetic field. Thus, it makes sense to take $d = 2$, which leads to spectrum [10, 11]:

$$E_k^{(1D)} \sim \sqrt{\varepsilon c_A}\,k_\perp^{-2}. \tag{3.15}$$

3.1.3.4 Surface Gravity Waves

Surface gravity wave turbulence is 2D, $d = 2$, and 4-wave, $N = 4$ ($2 \rightarrow 2$). Thus this is a dual cascade system. The dispersion relation for the surface gravity waves is $\omega = \sqrt{gk}$, i.e. $\lambda = \sqrt{g}$ and $\alpha = 1/2$, so that we find from (3.9) for the direct energy cascade spectrum

$$E_k^{(1D)} \sim g^{1/2}\varepsilon^{1/3}k^{-5/2}. \tag{3.16}$$

For the inverse waveaction cascade we have from (3.11),

$$E_k^{(1D)} \sim g^{2/3}\zeta^{1/3}k^{-7/3}. \tag{3.17}$$

Spectrum (3.16) was obtained by Zakharov and Filonenko [12]. Spectrum (3.17) was obtained by Zakharov and Zaslavskii [13].

A curious reduction of the gravity wave turbulence arises if all the waves have wavevectors which are parallel to each other. This WT system was considered in [14] and it was shown to be governed by a five-wave process, $N = 5$. In spite of the collinear geometry, we should still take $d = 2$ because the wave energy is distributed in the 2D space. Using (3.9) with $N = 5$, $d = 2$ and $\alpha = 1/2$, we get

$$E_k^{(1D)} \sim g^{5/8}\varepsilon^{1/4}k^{-21/8}, \tag{3.18}$$

which is a spectrum obtained in [14].

3.1.3.5 Langmuir Waves in Isotropic Plasmas, Spin Waves

Langmuir waves are described by the dispersion relation $\omega_k = \omega_p + \frac{3}{2}v_{th}\lambda_D k^2$ where ω_p, v_{th} and λ_D are the plasma frequency, thermal velocity and Debye length, respectively. Langmuir WT was considered in [15]. Magnetic spin waves in solids also obey a dispersion relation of this type, but the physical meaning of the dimensional parameters is different [16]. This system is 4-wave, $N = 4$ ($2 \rightarrow 2$), so the constant factor ω_p cancels out of both sides of the 4-wave resonance condition so that the effective dispersion is $\omega_k = \frac{3}{2}v_{th}\lambda_D k^2$. Thus taking $\lambda = \frac{3}{2}v_{th}\lambda_D$, $\alpha = 2$ and $d = 3$ and using (3.9) we get for the direct energy cascade spectrum

$$E_k^{(1D)} \sim v_{th}\lambda_D \varepsilon^{1/3}k^{-1/3}, \tag{3.19}$$

and using (3.11) we get for the inverse waveaction cascade spectrum

$$E_k^{(1D)} \sim (v_{th}\lambda_D)^{4/3}\zeta^{1/3}k^{1/3}. \tag{3.20}$$

These spectra were originally derived by Zakharov [15].

3.1.3.6 Kelvin Waves on Vortex Filaments

Weak WT for spiral Kelvin waves propagating on quantized vortex lines was introduced by Kozik and Svistunov [17]. In this case the dispersion relation is (omitting log factors) $\omega_k = \kappa k^2$, where κ is is the circulation quantum (see the beginning of this chapter). Thus $\alpha = 2$, and $d = 1$. In the Kozik-Svistunov theory, the leading process is six-wave, $N = 6$ ($3 \rightarrow 3$), so this is a dual cascade system. Using (3.9) we get for the direct energy cascade spectrum

$$E_k^{(1D)} \sim \kappa^{7/5}\varepsilon^{1/5}k^{-7/5}, \tag{3.21}$$

which is Kozik-Svistunov spectrum [17].

Using (3.11) we get for the inverse waveaction cascade spectrum

$$E_k^{(1D)} \sim \kappa^{8/5}\zeta^{1/5}k^{-1}. \tag{3.22}$$

This spectrum was discussed in [18].

However, both spectra (3.21) and (3.22) (as all KZ spectra) imply *locality of interactions* when mode k effectively interacts with modes of similar scales $k' \sim k$ and not with $k' \ll k$ or $k' \gg k$. (Later we will discuss locality of WT spectra in a great detail). The locality assumption for spectra (3.21) and (3.22) was recently proven wrong [19], and therefore these spectra are irrelevant and cannot be realized in Nature.

It was shown in [19] that spectrum (3.22) is nonlocal only marginally and it can be "fixed" by a log correction:

$$E_k^{(1D)} \propto k^{-1}\ln^{-1/5}(k\ell), \tag{3.23}$$

where ℓ is the mean inter-vortex separation.

On the other hand, spectrum (3.21) cannot be "fixed", because of the divergence of interaction at $k' \ll k$, but one can develop a theory where the interaction is nonlocal and dominated by interactions with the largest scales $k' \ll k$. This was recently done by Lvov and Nazarenko [20] who showed that the resulting turbulence can be viewed as Kelvin waves propagating on vortex lines with random large-scale curvature and interaction via a four-wave ($1 \leftrightarrow 3$) process (i.e. $N = 4$). They derived a four-wave kinetic equation and obtained a new stationary spectrum

$$E_k^{(1D)} \sim \varepsilon^{1/3}\ln(\ell/a)\,\kappa\Phi^{-2/3}k^{-5/3}, \tag{3.24}$$

where a is the vortex line radius and Φ is the mean square large-scale angle of the vortex lines. (The fact that the exponent is the same as Kolmogorov 5/3 is, of course, purely coincidental).

It is interesting that the Lvov-Nazarenko spectrum can also be obtained directly from the dimensional approach we have followed in this section by simply replacing $N = 6$ to $N = 4$ in the beginning to this subsection and again using (3.9). Note that the dimensionless factor $\ln(\ell/a)\,\Phi^{-2/3}$ is not recovered in such an approach.

3.1.3.7 Inertial Waves in Rotating Fluids

For waves in uniformly rotating fluids, the crucial dimensional quantity is the angular velocity of rotation Ω. The dispersion law is anisotropic,

$$\omega_k = 2\Omega \frac{k_\parallel}{|k|},$$

(here $\parallel$ refers to the rotation axis) but we will first of all make a (wrong) assumption that turbulence is isotropic and $k_\parallel \sim k_\perp$. This means $\lambda \sim \Omega$ and $\alpha = 0$. Taking $d = 3$, $N = 3$ (it is 3-wave process) and using (3.9) we get for the energy cascade spectrum

$$E_k^{(1D)} \sim \Omega^{1/2} \varepsilon^{1/2} k^{-2}, \tag{3.25}$$

which was originally obtained by Zeman [21] and Zhou [22].

However, like in MHD turbulence, the true spectrum tends to become strongly anisotropic, $k_\perp \gg k_\parallel$, which looks like a set of vertically elongated "columns". Therefore it is reasonable to take $d = 2$ (as we did for the Alfvén turbulence before) and $\alpha = -1$ (because $k_\parallel$ stays approximately constant during the cascade). Then (3.9) gives

$$E_k^{(1D)} \sim \Omega^{1/2} \varepsilon^{1/2} k_\perp^{-5/2}, \tag{3.26}$$

which is the Galtier spectrum [23].

3.1.3.8   Internal Waves in Stratified Fluids

Internal waves have the dispersion relation $\omega_k = \Omega_N \frac{k_H}{|k|}$, where Ω_N is the buoyancy frequency and k_H is the horizontal projection of the wavevector. It is a 3-wave system, $N = 3$. Similarly to the Alfvén and the inertial waves considered above, the simplest approximation is to assume isotropy, $k_H \sim k_V \sim |k|$ (k_V is the vertical wavevector component), which immediately makes the consideration identical to the isotropic inertial WT considered above. For the energy cascade spectrum we have

$$E_k^{(1D)} \sim \Omega_N^{1/2} \varepsilon^{1/2} k^{-2}, \tag{3.27}$$

which is a scaling which was previously discussed in [24–26].

However, again, the isotropy assumption is unrealistic: in this case turbulence tends to a state with $k_V \gg k_H$, which looks like a set of flat horizontal "pancakes". One would think that the scaling for such anisotropic state could be obtained by assuming $d = 1$ and $\alpha = -1$. However, this would not lead to the correct anisotropic spectra obtained in [24–26]. We see that our simple dimensional method works most of the time but not on this particular occasion! The reason is probably that it would not be consistent in this case to take $\alpha = -1$ because k_H might change during the cascade along with k_V in certain proportion.

3.1.3.9 Waves on Elastic Plates

Wave turbulence on an elastic plate, like a metal sheet or a gong, were considered theoretically in [27] and experimentally in [28, 29]. For waves which are much longer than the plate thickness h (i.e. $kh \ll 1$) the system is effectively two-dimensional, $d = 2$, and the dispersion relation is

$$\omega_k = \sqrt{\frac{Eh^2}{12(1 - \sigma^2)\rho}} \; k^2,$$

where E is the Young modulus, ρ is density and σ is the Poisson ratio. We see that $\alpha = 2$. It might seem natural to assume that in the long-wave limit $kh \ll 1$ the thickness h would not enter the answer independently, and the only defining physical parameter should be $\lambda = \sqrt{\frac{Eh^2}{12(1-\sigma^2)\rho}}$ which is an effective coefficient measuring response of the plate to an external force. This way, we could proceed with our dimensional approach as usual, assume following [27] that the leading process is 4-wave, $N = 4$, and use (3.9) to obtain the direct energy cascade spectrum

$$E_k^{(1D)} \sim \lambda \varepsilon^{1/3} k^{-1}. \tag{3.28}$$

Further, using (3.11) we get for the inverse waveaction cascade spectrum

$$E_k^{(1D)} \sim \lambda^{4/3} \zeta^{1/3} k^{-1/3}. \tag{3.29}$$

These spectra are different from the one obtained by a proper theoretical derivation in [27], $E_k^{(1D)} \sim \varepsilon^{1/3} k^1$ (we ignore the log correction arising from the fact that the same power law corresponds to the thermodynamic energy equipartition state) and $E_k^{(1D)} \sim \zeta^{1/3} k^0$ respectively. A closer look at this problem reveals that our simple dimensional argument is inapplicable in this case, because h does indeed appear in the nonlinear term as an extra independent dimensional parameter. Experimentally observed direct-cascade spectrum [28, 29] (if one translates from the measured frequency spectrum to $E_k^{(1D)}$ using the linear-wave dispersion relation) is approximately $E_k^{(1D)} \sim \varepsilon^{0.7} k^{-1/5}$, which very different from the WT predictions of [27] (for reasons yet to be understood).

3.1.3.10 WT in Nonlinear Optics and in Bose-Einstein Condensates

Wave turbulence in Nonlinear Optics and in Bose-Einstein condensates is often described by the Nonlinear Schrödinger (NLS) model, see (5.1). This is another example where (like for the elastic plate) our simple dimensional analysis is inapplicable, because there are more than one relevant dimensional parameters in the problem which independently determine the scalings of the linear and the nonlinear terms respectively. WT arising in the NLS model will be considered in detail in Chap. 15.

3.2 Strong Wave Turbulence and Critical Balance

Strong wave turbulence is of course complicated and its description is far from being completed. Some of its features are non universal and may include inter-actions with turbulent structures which are not waves, e.g. hydrodynamic eddies, collapsing bursts, etc. However, in many applications remarkably similar approaches were suggested to describe strong wave turbulence, mostly indepen-dently and without reference to each other, and called by different names in different research communities.

Namely, it is often assumed that the energy spectrum will saturate when the nonlinear interaction time becomes of the same order as the linear wave period *over a wide range of turbulent scales*. Then the prediction about the turbulent states is obtained simply from equating the nonlinear term to the linear one in Fourier space for each k. We will call these states generally as a *critical balance (CB)*, the name that was introduced in the context of MHD turbulence by Goldreich and Shridhar [30]. Let us go over several examples. Below, we will follow the approach of paper [4], with some modifications and extensions.

3.2.1 MHD Turbulence

In MHD turbulence CB, as introduced in [30], means that the Alfvén wave period $\tau_L = 2\pi/(c_A k_{\parallel})$ is of the same size (scale by scale) as the hydrodynamic eddy turnover time, $\tau_{NL} = \sqrt{E_k^{(1D)} k^3}$. It is usually further assumed that turbulence is strongly anisotropic $k_{\perp} \gg k_{\parallel}$ and that the energy spectrum has Kolmogorov scaling $E_k^{(1D)} \sim \varepsilon^{2/3} k^{-5/3}$, in which case the balance condition $\tau_L \sim \tau_{NL}$ gives [30]:

$$c_A k_{\parallel} \sim \varepsilon^{1/3} k_{\perp}^{2/3}. \tag{3.30}$$

Relation (3.30) is a statement about the position of the spectrum's maximum in the $(k_{\perp}, k_{\parallel})$ space.

The CB condition in MHD can form in two different ways, see Fig. 3.3. In the first scenario, one could start with a weak Alfvén wave turbulence which develops a KZ cascade, which proceeds toward larger $k_{\perp}$ with almost no cascade in $k_{\parallel}$. As we will see in Chap. 14, the nonlinearity increases as the energy flux proceeds to higher $k_{\perp}$ and at some large $k_{\perp}$ the CB condition is inevitably reached. In the second scenario, one could start from the opposite end, i.e. with a strongly tur-bulent pure 2D state. For simplicity, let us think of a state where the initial

Fig. 3.3 Two ways to set the CB in turbulence: starting from weak WT and from 2D turbulence

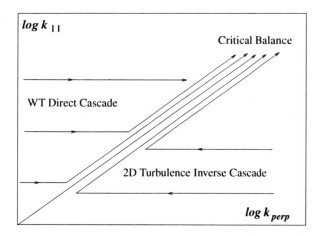

magnetic field is absent. Such a state would at first exhibit properties of 2D turbulence, e.g. an inverse energy cascade, but the pure 2D motions will be unstable with respect to Cherenkov-type emission of Alfvén waves (provided that the inverse eddy turnover time exceeds the minimal Alfvén wave frequency $c_A\, k_{\|min}$). The Alfvén wave emission will continue until the CB condition will be reached,—now from the opposite end. This will stop the inverse energy cascade to smaller $k_{\perp}$s and from now on the energy of the coupled eddy and wave components will cascade to higher $k_{\perp}$ along the CB curve (3.30), see Fig. 3.3. Note that further CB cascade will become more and more anisotropic, $k_{\perp}/k_{\|} \to \infty$.

Thus, the energy cascade undergoes a reversal, which is especially interesting in the case of a continuous forcing of 2D turbulence: the net (integrated over $k_{\|}$) flux in the resulting steady state will be zero in between the forcing and the cascade reversal scales, even though locally the flux is finite in the two-dimensional ($k_{\|}$, $k_{\perp}$) space.

3.2.2 Gravity Water Waves

CB approach for the gravity water waves, or equating the linear and the nonlinear timescales, leads to the famous Phillips spectrum [31],

$$E_k^{(1D)} \sim g k^{-3}. \tag{3.31}$$

This is easy to understand if we remember that both the linear and the nonlinear scalings in this system are completely determined by the same single dimensional parameter g. Thus, balancing the linear and the nonlinear terms in this case is equivalent to saying that g is the only dimensional constant on which the energy spectrum can depend (not ε in this case!), and this is precisely how this spectrum was originally found by Phillips [31].

Independence of Phillips spectrum of the energy input rate ε suggests presence of wave breaking which occurs when the wave steepness reaches a limiting slope (whose value is independent of the forcing). Indeed, the well-known Stokes wave solution has a limiting slope $\tan(\pi/6) \approx 0.58$ for which the wave crests become sharp (with a discontinuous slope). At the limiting amplitude, the downward acceleration of some fluid particles reach g (which, again, is equivalent to the CB condition), and exceeding this amplitude would mean that these fluid particles would not be able to stay attached to the surface. Therefore, the wave slope is limited, at each scale, by the critical value ≈ 0.6, which implies the CB state.

3.2.3 Stratified Turbulence

For turbulence in a stratified fluid, like in the surface gravity wave example, one can assume that there is only one relevant dimensional parameter,—in this case the buoyancy frequency Ω_N. CB-equating of the linear and nonlinear dynamical terms is equivalent to saying that Ω_N will be the only dimensional parameter defining the energy spectrum. Taking into account that strong stratification leads to strong anisotropy with pancake-like structures, $k_V \gg k_H$, we get

$$E_k^{(1D)} \sim \Omega_N^2 k_V^{-3}. \tag{3.32}$$

This spectrum was obtained by Dewan [32] and Billand and Chomaz [33]. Here, like for the surface gravity waves, formation of the CB state can be associated with a wave breaking process.

On the other hand, like in the MHD example, one also can establish a CB relation between the vertical and the horizontal wavevectors. For this, we need to equate the linear wave frequency, $\Omega_N k_H / k_V$, to the horizontal eddy inverse turn-over time, $u_H k_H$ (u_H is the characteristic horizontal velocity at scale $1/k_H$), which gives relation

$$\Omega_N \sim u_H k_V,$$

which is a scaling that was put forward in [32]. Assuming further, following Lindborg [34], that u_H satisfies a Kolmogorov-type scaling, $u_H \sim \varepsilon^{1/3} k_H^{1/3}$, this relation becomes

$$k_H \sim l_o^2 k_V^3, \tag{3.33}$$

where $l_o = \sqrt{\varepsilon/\Omega_N^3}$ is the Ozmidov scale. This condition was discussed by Lindborg [34]. Note that it is a direct analog of the MHD CB condition (3.30).

Like in MHD, the energy cascade in this case is direct, with the coupled wave and eddy components cascading together, and this property was discussed in [34]. Also similarly to MHD, the CB state may get established in the system if one starts with either weak WT or a purely 2D initial condition, like in Fig. 3.3, and the assumed anisotropy condition (in this case $k_V \gg k_H$) is satisfied better and better as the energy cascades to higher ks along the CB curve. The difference with MHD is that the WT cascade is likely to proceed over both the perpendicular and the parallel scales, so the arrows on the WT side of Fig. 3.3 are not necessarily horizontal in this case.

3.2.4 Rotating Turbulence

In rotating turbulence, the determining dimensional quantity is the rotation frequency Ω, and the anisotropy is of type $k_\perp \gg k_\parallel$. Then, the (critical) balancing the linear and nonlinear timescales is equivalent to writing the energy spectrum in terms of Ω only. This gives

$$E_k^{(1D)} \sim \Omega^2 k_\perp^{-3}, \tag{3.34}$$

which is Smith-Waleffe spectrum [35]. In the context of rotating turbulence, the $k_\perp^{-3}$ spectrum was also discussed in [36] where it was interpreted as the direct enstrophy cascade in 2D turbulence (thus it would have a different pre-factor, $\zeta^{2/3}$), because, they argued, under strong rotation turbulence should become purely 2D via the Taylor-Proudman theorem. Note however that CB state is very different from 2D turbulence because (like in MHD and in stratified turbulence): the energy cascade in CB is direct and not inverse as in pure 2D.

Finally, like in MHD or in stratified turbulence, let us use CB to predict the relation between $k_\perp$ and $k_\parallel$. Note that this will be a different CB, somewhat contradictory to the one of the previous paragraph, because it will assume K41 scaling in the perpendicular scales. Let us equate the linear wave frequency, $\Omega k_\parallel / k_\perp$, to the horizontal eddy inverse turnover time, $u_\perp \sim \varepsilon^{1/3} k_\perp^{2/3}$ (like in K41). This gives

$$\Omega k_\parallel \sim \varepsilon^{1/3} k_\perp^{5/3}, \tag{3.35}$$

which is another CB-type prediction [4].

Again like in MHD, the CB state may get established in the system if one starts with either weak WT or a purely 2D initial condition, cf. Fig. 3.3. However, there is an essential difference with MHD turbulence. In MHD, cascading along the CB scales means that turbulence becomes more and more anisotropic. In rotating turbulence, cascading along the CB scales leads to less anisotropy. At some point the cascade will cross the scale $k_V \sim k_H$, and the local Rossby number, which is a measure of turbulent velocity strength compared to the background rotation, reaches one, $Ro = u_H k_H / \Omega \sim 1$. After this the effect of the external rotation will be unimportant, and further cascade to higher k's will proceed isotropically with $k_V \sim k_H$.

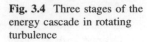

Fig. 3.4 Three stages of the energy cascade in rotating turbulence

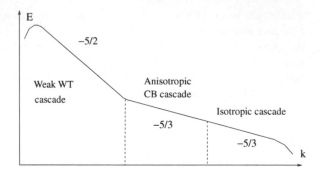

Thus, if we start with weak WT corresponding to low initial Rossby numbers, $Ro \ll 1$, then the spectrum will be as schematically shown as in Fig. 3.4. Namely, there will be three different ranges: the weak WT spectrum $k_H^{-5/2}$ will be followed by anisotropic CB spectrum $k_H^{-5/3}$ at higher k's which in turn will be followed by an isotropic spectrum $k^{-5/3}$ at even higher k's. It is interesting that a similar change of slope, from ~ -2.2 to $\sim -5/3$, was indeed observed in the rotating turbulence experiments, when the initial Ro was small, see Figure 3.5b in [37]. However it would be premature to say if these results are definitely related to the CB transition or merely to an experimental noise at high k's.

3.2.5 Quasi-Geostrophic Turbulence

The defining parameter for the large-scale quasi-geostrophic (QG) turbulence is the gradient of the projection of the planetary rotation frequency Ω onto the (2D) motion plane, which is $\sim \Omega/R$, where R is the planet's radius. In this case, CB-type balancing of the linear and the nonlinear timescales is equivalent to writing the energy spectrum in terms of Ω/R only,

$$E_k^{(1D)} \sim (\Omega/R)^2 k^{-5}. \tag{3.36}$$

This spectrum was put forward by Rhines [38] who then immediately discarded it arguing that such a steep spectrum would mean strongly nonlocal interaction.

Rhines spectrum is isotropic, whereas QG turbulence is known to be strongly anisotropic with strongly dominating zonal component, $k_{\text{zonal}} \gg k_{\text{meridional}}$, which is another reason why it is unrealistic. However, it was later re-introduced by Galperin, Sukoriansky and Huang [39], but now for the zonal-flow component,

$$E_k^{(1D)} \sim (\Omega/R)^2 k_{\text{zonal}}^{-5}, \tag{3.37}$$

and they argued that this spectrum is observed on giant planets and in numerical simulations.

Fig. 3.5 QG Turbulence: a
CB-type curve (half of the
"lazy 8") separating the
quasi-isotropic inverse cascade of strong hydrodynamic
turbulence and weak WT

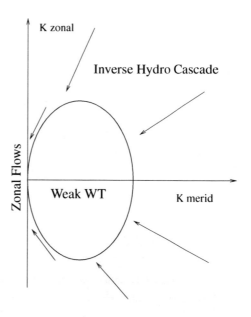

As for relating the different wavevector components, CB-type equating the
linear and nonlinear terms leads to a "lazy 8" shaped curve in the 2D **k**-space [40],
half of which is shown in Fig. 3.5 (the other half-plane, $k_{\mathrm{meridional}} < 0$, just
duplicates the shown region because the original wave variable is real in this case,
so that all the spectra and the fluxes are completely symmetric with respect to
k → − **k**). This curve separates the eddy- and the wave-dominated regions in the
inverse-cascade setting, and it marks the scales at which such a cascade becomes
anisotropic. It would be quite in the spirit of the CB philosophy to think that upon
reaching the lazy-8 curve the inverse cascade of energy turns and proceeds to the
zonal scales along this curve, see Fig. 3.5 It is probably true at least partially
because it is natural to assume a slowdown of the inverse cascade rate at the
weakly-nonlinear scales inside the lazy-8 with respect to the strongly nonlinear
ones outside of this curve. However, recent numerical simulations with initial
turbulence concentrated in CB scales (i.e. where the nonlinear time is of order of
the linear one) showed that the nonlinearity degree is decreasing with time, which
can be interpreted as a turbulent cascade entering inside the lazy-8 [41]. Thus, CB
is not well-sustained by QG turbulence.

3.2.6 Kelvin Waves

For 1D Kelvin waves on the quantum filaments, the relevant dimensional quantity
is the quantum circulation κ, and the CB spectrum that depends only on κ is

$$E_k^{(1D)} \sim \kappa^2 k^{-1}, \tag{3.38}$$

which is the Vinen spectrum [42]. Note coincidence of the spectrum's exponent with the one of the inverse cascade spectrum (3.22) (but the pre-factor is different!).

The mechanism limiting the Kelvin wave amplitudes at the CB level is provided by a vortex reconnection process. Indeed, reconnections occur when the vortex line bending angle becomes of order one, which corresponds to a state with the nonlinear timescale of the same order as the linear wave period. Thus we see that the reconnections play the same role for the Kelvin waves as the wave breaking for the gravity water waves.

3.2.7 Burgers? KdV?

Finally, would CB work for the most basic of the nonlinear models, Burgers equation? This equation is

$$u_t + uu_x = \nu u_{xx},$$

where ν is viscosity. As first sight this approach is promising because, as we know, this model predicts formation of shock waves, coherent structures in which the nonlinear and the linear terms are in balance with each other, in the spirit of CB.

Thus, let us assume that ν is the defining parameter for the energy spectrum, and also take into account that, even though Burgers equation is 1D, the corresponding energy density is defined in a 3D physical space (i.e. the velocity is defined in a 3D space, but it depends on one coordinate only). This gives[2] $E = \nu^2 k$. But this is different from the well-known answer, Burgers spectrum, k^{-2}, see [43]. We know that Burgers spectrum arises from the Fourier asymptotics of shocks, and inside the shock structures the nonlinear terms are balanced by the linear (viscosity) terms, so why does not the CB estimate work in this case? This is because the interaction is nonlocal in the Fourier space, as easily verified from the Burgers equation in the Fourier space by substituting the shock solution, whereas CB assumes locality.

Let us now consider Korteweg-de Vries (KdV) equation,

$$u_t + uu_x + \mu u_{xxx} = 0,$$

where μ is a dispersion coefficient. In this case, balancing scale-by-scale the nonlinear and the linear terms (the second and the third terms respectively) we get $E \sim k^3$. On the other hand, we know that the KdV solution in the long-time regime

---

[2] Note that circulation κ and viscosity ν have the same physical dimensions, but the resulting spectrum for Burgers is different from the Vinen spectrum of Kelvin waves because of the difference in the energy density: in Burgers it is a 3D density whereas for Kelvin it is a 1D density.

consists of a number of isolated coherent structures—solitons. Again, the linear and the nonlinear terms are balanced in solitons, and naively one could think that they are consistent with CB. However, this is not the case. If the separations between the solitons are much greater than their widths then for the scales intermediate between the soliton width and the inter-soliton separation the spectrum is basically the one of a "gas" of delta-function peaks. This corresponds to the spectrum $E \sim$ const, which is very different from the CB prediction. Like in the Burgers example, the reason for this is nonlocality of the scale interactions.

Thus the Burgers and the KdV examples teach us that one has to use CB ideas with caution because of various implicit assumptions which may not hold in particular situations (like locality in these cases).

References

1. Kraichnan, R.: Inertial range spectrum of hydromagnetic turbulence. Phys. Fluids **8**, 1385–1387 (1965)
2. Zakharov, V.E., L'vov, V.S., Falkovich, G.: Kolmogorov Spectra of Turbulence 1: Wave Turbulence. Series in Nonlinear Dynamics. Springer (1992)
3. Connaughton, C., Nazarenko, S., Newell, A.C.: Dimensional analysis and weak turbulence. Physica D **184**, 86–97 (2003)
4. Nazarenko, S.V., Schekochihin, A.A.: Critical Balance in Magnetohydrodynamic, Rotating and Stratified Turbulence: Towards a Universal Scaling Conjecture (submitted to JFM, arXiv:0904.3488) (2009)
5. Vedenov, A.A.: Theory of weakly turbulent plasma. In: Leontovich, M.A. (ed.) Reviews of Plasma Physics, vol. 3, pp. 229–276. Consultants Bureau, New York (1967)
6. Zakharov, V.E., Filonenko, N.N.: Weak turbulence of capillary waves. Zh. Prikl. Mekh. Tekh. Phys. 4(5), 62 (1967) [in Russian]
7. Zakharov, V.E., Filonenko, N.N.: J. Appl. Mech. Tech. Phys. **4**, 506 (1967)
8. Zakharov, V.E., Sagdeev, R.Z.: Spectrum of acoustic turbulence. Sov. Phys. Dokl. **15**, 439 (1970)
9. Iroshnikov, R.S.: Turbulence of a conducting fluid in a strong magnetic field. Astron. Zh. **40**, 742 (1963) (English trans.: 1964, Sov. Astron. **7**, 566)
10. Ng, C.S., Bhattacharjee, A.: Scaling of anisotropic spectra due to the weak interaction of shear-AlfvTn wave packets. Phys. Plasmas **4**, 605–610 (1997)
11. Galtier, S., Nazarenko, S.V., Newell, A.C., Pouquet, A.: A weak turbulence theory for incompressible MHD. J. Plasma Phys. **63**, 447 (2000)
12. Zakharov, V.E., Filonenko, N.N.: Energy spectrum for stochastic oscillations of a fluid surface. Dokl. Acad. Nauk SSSR **170**, 1292–1295 (1966) [in Russian]
13. Zakharov, V.E., Zaslavskii, M.M.: Izv. Acad. Nauk SSSR Atmos. Ocean Phys. **18**, 970 (1982)
14. Dyachenko, A.I., Lvov, Y.V., Zakharov, V.E.: Five-wave interaction on the surface of deep fluid. Physica D **87**, 233 (1995)
15. Zakharov, V.E.: Sov. Phys. JETP **35**, 908 (1972)
16. L'vov, V.S.: Nonlinear Spin Waves. Nauka, Moscow (1987) [in Russian]
17. Kozik, E.V., Svistunov, B.V.: Phys. Rev. Lett. **92**, 035301 (2004)
18. Nazarenko, S.: Differential approximation for Kelvin-wave turbulence. JETP Lett. 83(5), 198–200 (2005)
19. Laurie, J., L'vov, V.S., Nazarenko, S., Rudenko, O.: Interaction of Kelvin waves and non-locality of the energy transfer in superfluids. Phys. Rev. B. **81**, 104526 (2010)

20. L'vov, V.S., Nazarenko, S.: Spectrum of Kelvin-wave turbulence in superfluids. JETP Lett. (Pis'ma v ZhETF) **91**, 464–470 (2010)
21. Zeman, O.: Phys. Fluids **6**, 3221 (1994)
22. Zhou, Y.: Phys. Fluids **7**, 2092 (1995)
23. Galtier, S.: Weak inertial-wave turbulence theory. Phys. Rev. E **68**, 015301 (2003)
24. Pelinovsky, E.N., Raevsky, M.A.: Weak turbulence of the internal waves in the ocean. Atmos. Ocean Phys. Izvestija **13**, 187–193 (1977)
25. Caillol, P., Zeitlin, Z.: Kinetic equations and stationary energy spectra of weakly nonlinear internal gravity waves. Dyn. Atmos. Ocean **32**, 81–112 (2000)
26. Lvov, Y.V., Tabak, E.G.: Hamiltonian formalism and the Garrett-Munk spectrum of internal waves in the ocean. Phys. Rev. Lett.**87**, 168501 (2001)
27. Düring, G., Josserand, C., Rica, S.: Weak turbulence for a vibrating plate: can one hear a Kolmogorov spectrum? Phys. Rev. Lett. **97**, 025503 (2006)
28. Mordant, N.: Are there waves in elastic wave turbulence?. Phys. Rev. Lett. **100**, 234505 (2008)
29. Boudaoud, A., Cadot, O., Odille, B., Touzé, C.: Observation of wave turbulence in vibrating plates. Phys. Rev. Lett. **100**, 234504 (2008)
30. Goldreich, P., Sridhar, S.: Toward a theory of interstellar turbulence. 2: strong Alfvénic turbulence. Astrophys. J. Part 1 **438**(2), 763–775 (1995) (ISSN 0004-637X)
31. Phillips, O.M.: The equilibrium range in the spectrum of wind generated waves. J. Fluid Mech. **4**, 426–434 (1958)
32. Dewan, E.M.: Saturated-cascade similitude theory of gravity wave spectra. J. Geophys. Res. **102**(D25;29), 799–817 (1997)
33. Billant, P., Chomaz, J.M.: Self-similarity of strongly stratified inviscid flows. Phys. Fluids **13**, 1645–1651 (2001)
34. Lindborg, E.: The energy cascade in a strongly stratified fluid. J. Fluid Mech. **550**, 207–242 (2006)
35. Smith, L.M., Waleffe, F.: Transfer of energy to two-dimensional large scales in forced, rotating three-dimensional turbulence. Phys. Fluids **11**(6), 1608–1622 (1999)
36. Cambon, C., Rubinstein, R., Godeferd, F.S.: Advances in wave turbulence: rapidly rotating flows. New J. Phys. **6**, 73 (2004). doi:10.1088/1367-2630/6/1/073
37. Morize, C., Moisy, F., Rabaud, M.: Decaying grid-generated turbulence in a rotating tank. Phys. Fluids **17**, 095105 (2005)
38. Rhines, P.B.: Waves and turbulence on a beta-plane. J. Fluid Mech. **69**, 417 (1975)
39. Galperin, B., Sukoriansky, S., Huang, H.-P.: Universal n^{-5} spectrum of zonal flows on giant planets. Phys. Fluids **13**, 1545–1548 (2001)
40. Holloway, G.: Contrary roles of planetary wave propagation in atmospheric predictability. In: Holloway, G., West, B.J. (eds.) Predictability of Fluid Motions, pp. 593–599. American Institute of Physics, New York (1984)
41. Nazarenko S., Quinn B.: Tripple cascades and formation of zonal jets in Rossby/drift turbulence. Phys. Rev. Lett. **103**, 118501 (2009)
42. Vinen, W.F.: Phys. Rev. B **61**, 1410 (2000)
43. Burgers, J.M.: Koninkl. Ned. Akad. Wetenschap., **53**, 247, 393 (1950); **B57**, 45, 159, 403 (1954)

Chapter 4
Solutions to Exercises

4.1 Fjørtoft Argument in Terms of Centroids: Exercise 2.1

We are going to use Cauchy–Schwartz inequality, which says that

$$\left| \int_0^\infty f(k)g(k)\,dk \right| \le \left| \int_0^\infty f^2(k)\,dk \right|^{1/2} \left| \int_0^\infty g^2(k)\,dk \right|^{1/2}$$

for any functions $f(k)$, $g(k) \in L^2$. We will only deal with positive functions, so the absolute value brackets may be omitted. Being in L^2 in our case means that all the relevant integrals converge, as suggested in the statement of the problem.

First, let us consider integral $\int k\, E\, dk$ and apply the Cauchy–Schwartz inequality as follows:

$$\int_0^\infty kE\,dk = \int_0^\infty (kE^{1/2})(E^{1/2})\,dk \le \left(\int_0^\infty k^2 E\,dk \right)^{1/2} \left(\int_0^\infty E\,dk \right)^{1/2}.$$

This immediately yields (2.12) and (2.16).

Second, let us consider $\Omega = \int k^2 E\,dk$ and split it as

$$\int_0^\infty k^2 E\,dk = \int_0^\infty (k^{3/2}E^{1/2})(k^{1/2}E^{1/2})\,dk \le \left(\int_0^\infty k^3 E\,dk \right)^{1/2} \left(\int_0^\infty kE\,dk \right)^{1/2}.$$

S. Nazarenko, *Wave Turbulence*, Lecture Notes in Physics, 825,
DOI: 10.1007/978-3-642-15942-8_4, © Springer-Verlag Berlin Heidelberg 2011

This immediately yields (2.14). Combining (2.14) with (2.12) gives (2.13).

Now, let us split $\int k\,E\,dk$ in a different way,

$$\int_0^\infty kE\,dk = \int_0^\infty (k^{3/2}E^{1/2})(k^{-1/2}E^{1/2})\,dk \le \left(\int_0^\infty k^3 E\,dk\right)^{1/2}\left(\int_0^\infty k^{-1}E\,dk\right)^{1/2}.$$

This immediately yields (2.17). Combining (2.17) with (2.16) gives (2.15).

4.2 k-Centroids versus l-Centroids: Exercise 2.2

Substituting spectrum $E = A\ k^{-5/3}$ (with $A = $ const) into the definitions of the k- and l-centroids, we have (taking into account that $k_b \gg k_a$):

$$k_E = \frac{A\int_{k_a}^{k_b} k^{-2/3}\,dk}{A\int_{k_a}^{k_b} k^{-5/3}\,dk} = -\frac{2\big[k^{1/3}\big]_{k_a}^{k_b}}{\big[k^{-2/3}\big]_{k_a}^{k_b}} \approx 2k_b^{1/3}k_a^{2/3};$$

$$l_E = \frac{A\int_{k_a}^{k_b} k^{-8/3}\,dk}{A\int_{k_a}^{k_b} k^{-5/3}\,dk} = \frac{2\big[k^{-5/3}\big]_{k_a}^{k_b}}{5\big[k^{-2/3}\big]_{k_a}^{k_b}} \approx \frac{2}{5}k_a^{-1};$$

$$k_\Omega = \frac{A\int_{k_a}^{k_b} k^{4/3}\,dk}{A\int_{k_a}^{k_b} k^{1/3}\,dk} = \frac{4\big[k^{7/3}\big]_{k_a}^{k_b}}{7\big[k^{4/3}\big]_{k_a}^{k_b}} \approx \frac{4}{7}k_b;$$

$$l_\Omega = \frac{A\int_{k_a}^{k_b} k^{-2/3}\,dk}{A\int_{k_a}^{k_b} k^{1/3}\,dk} = \frac{4\big[k^{1/3}\big]_{k_a}^{k_b}}{\big[k^{4/3}\big]_{k_a}^{k_b}} \approx 4k_b^{-1}.$$

4.3 Four-Wave Resonances in 1D Systems: Exercise 3.1

Fact that in 1D there is no nontrivial four-wave resonances for $\omega \sim k^\alpha$ with $\alpha \ge 1$ was mentioned in [1]. It is easy to prove using a graphical illustration (see Fig. 4.1). To find a solution graphically, one has to draw curve $\omega = k^\alpha$, and then two more curves with the same shape and with the lowest points shifted along the first curve to points $(k_1, \omega(k_1))$ and $(k_3, \omega(k_3))$ respectively (with $k_1 \neq k_3$). The latter two curves will intersect once and the intersection point k^* will give us the solution: $k_2 = k^* - k_1$ and $k_4 = k^* - k_3$. Easy to see that this is *the only* intersection of these curves, which means that there is only one solution. On the other hand, since we already know that there is a *trivial* solution $k_2 = k_3$ and $k_4 = k_1$, and therefore this is exactly the solution we have just found graphically, and there is no other (nontrivial) solutions.

Fig. 4.1 Graphical solution of the four-wave resonance conditions in 1D for $\omega \sim k^{\alpha}$ with $\alpha \geq 1$

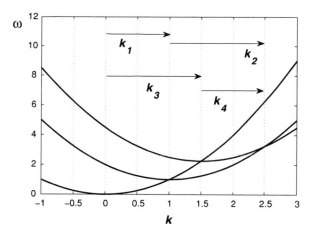

Fig. 4.2 Graphical proof that for $\alpha < 1$ there is no solution to the resonant condition of type 3 → 1

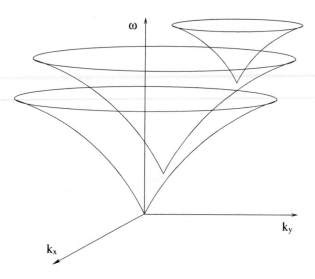

4.4 Four-Wave 3 → 1 Resonances in 2D Systems: Exercise 3.2

For the first part: see graphical solution in Fig. 4.2. The three "funnels" in this Figure are nested so that the second funnel is standing with its vertex on the first funnel and the third funnel has its vertex on the second funnel. Easy to see that the third funnel does not have points of intersection with the first one, which means that in 2D there is no solutions for the frequency resonance condition $\omega_1 + \omega_2 + \omega_3 = \omega_4$ for the wavevectors satisfying $\mathbf{k}_1 + \mathbf{k}_2 + \mathbf{k}_3 = \mathbf{k}_4$ for $\omega \propto k^{\alpha}$ with $\alpha < 1$.

For the second part: the solution exists because when one of the wavenumbers is much smaller than the remaining ones, the solution reduces to the respective three-wave problem. Think of three paraboloids, two of which are close together. By continuity, the solution will persist when we move these paraboloids apart.

Reference

1. Zakharov, V.E., Guyenne, P., Pushkarev, A.N., Dias, F.: Wave turbulence in one-dimensional models. Phys. D **152–153**, 573 (2001)

Part II
Wave Turbulence Closures

In this part we will introduce the main statistical objects in Wave Turbulence such as wave PDF's, moments and spectra. We will present definition of fields with random phases and amplitudes and explain their use in Wave Turbulence closures. We will derive the evolution equations for the one-mode and the multi-mode PDF's and the kinetic equations for the spectra and the higher moments of the wave amplitudes.

Chapter 5
Statistical Objects in Wave Turbulence

Before developing the WT description, we need to do some preliminary work by defining the statistical objects used in WT, and to introduce some definitions about the properties of these objects. There are several complimentary ways to describe WT statistics. Here, we will follow an approach which was recently developed in [1–6] and which has an advantage of being able to go beyond describing the wave spectrum and extending the theory to the higher wave moments and to PDF's.

First, we need to start with a particular wave model based on a nonlinear Hamiltonian PDE in the physical **x**-space. For example, most of our theory will be presented below using the example of Petviashvilli equation describing a particular limit of the geophysical Rossby waves or the plasma drift waves, see (6.1) below. Another example (which we will also consider later) is the nonlinear Schrödinger equation (NLS), which describes waves in nonlinear optics, and Bose–Einstein condensates:

$$i\,\partial_t \psi + \nabla^2 \psi \pm |\psi|^2 \psi = 0, \tag{5.1}$$

with wave function $\psi(\mathbf{x}, t) : \mathbb{R}^d \times \mathbb{R}^+ \to \mathbb{C}$, with dimension $d = 2$ or 3. Here $\mathbf{x}$ is the physical coordinate and t is time.

5.1 Statistical Variables

We will generally find that the wave function $\psi(\mathbf{x}, t)$ is complex, like in the NLS equation (5.1), although in some applications (e.g. Petviashvilli equation (6.1) to be considered later) it is purely real. Let the system be in a periodic box, $\mathbf{x} \in \mathbb{T}^d$, with period L in all directions. We will use Fourier transform of ψ

$$\psi(\mathbf{x}, t) = \sum_{\mathbf{k}} \hat{\psi}(\mathbf{k}, t)\, e^{i\mathbf{k}\cdot\mathbf{x}}, \tag{5.2}$$

S. Nazarenko, *Wave Turbulence*, Lecture Notes in Physics, 825,
DOI: 10.1007/978-3-642-15942-8_5, © Springer-Verlag Berlin Heidelberg 2011

with Fourier coefficients

$$\hat{\psi}(\mathbf{k}, t) = \frac{1}{L^d} \int\limits_{\text{Box}} \psi(\mathbf{x}, t) e^{-i\mathbf{k}\cdot\mathbf{x}} d\mathbf{x}. \tag{5.3}$$

The pre-factor is chosen so that the Parseval's theorem for the volume averaged intensity does not involve L,

$$\frac{1}{L^d} \int\limits_{\text{Box}} |\psi(\mathbf{x})|^2 d\mathbf{x} = \sum_{\mathbf{k}} \left|\hat{\psi}_k\right|^2.$$

As an example, for periodic box $\mathbb{T}^d$, we can rewrite the NLS equation (5.1) in the Fourier space,

$$i\dot{\hat{\psi}}(\mathbf{k}) - k^2 \hat{\psi}(\mathbf{k}) \pm \sum_{\mathbf{k}_1 + \mathbf{k} = \mathbf{k}_2 + \mathbf{k}_3} \hat{\psi}^*(\mathbf{k}_1)\hat{\psi}(\mathbf{k}_2)\hat{\psi}(\mathbf{k}_3) = 0, \tag{5.4}$$

where "dot" means the time derivative.

The wavenumbers in (5.3) are discrete, $\mathbf{k}_l = \frac{2\pi \mathbf{l}}{L}$, where $\mathbf{l} \in \mathbb{Z}^d$. For simplicity we will assume that the total number of modes is finite and the wavenumber is bounded by the maximum value $k_{\max}$ in each of its components. The value of $k_{\max}$ may represent, for instance, a dissipation cutoff at high wavenumbers. Respectively, the total number of modes is

$$N = \left(\frac{k_{\max}}{k_{\min}}\right)^d = \left(\frac{k_{\max}L}{2\pi}\right)^d.$$

We denote by B_N the set of all wavenumbers $\mathbf{k}_l$ inside the $\mathbf{k}$-space box of volume $(2k_{\max})^d$, see Fig. 5.1.

We write the wave function in terms of its amplitude and phase

$$\hat{\psi}_l = \hat{\psi}(\mathbf{k}_l, t) = \sqrt{J_l}\phi_l, \tag{5.5}$$

Fig. 5.1 Set of active wave modes in discrete Fourier space, $B_N \subset \mathbb{Z}^d$

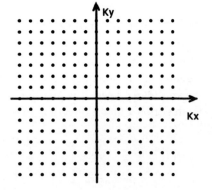

Fig. 5.2 Space of phase
factors: $\mathbb{S}^1$

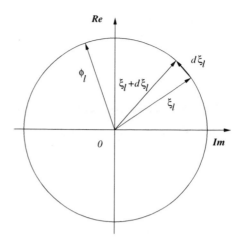

where $J_l \in \mathbb{R}^+$ is the intensity of the mode l, and $\phi_l \in \mathbb{S}^1$ is the phase factor of the
mode l. By $\mathbb{S}^1$ we mean the unit circle in the complex plane, i.e. $\phi_l = e^{i\varphi_l}$, see
Fig. 5.2. We write the set of all J_l and ϕ_l with l such that $\mathbf{k}_l \in B_N$ as

$$\{J_l, \phi_l;\ \mathbf{k}_l \in B_N\} \text{ or for short: } \{J, \phi\}.$$

5.2 Probability Density Functions

We may now define the Probability Density Function (PDF).

Definition 5.1 The probability of finding J_l inside $(s_l, s_l + ds_l) \subset \mathbb{R}^+$ and finding
ϕ_l in the sector $(\xi_l, \xi_l + d\xi_l) \subset \mathbb{S}^1$ (see Fig. 5.2) is given in terms of the joint PDF
$\mathscr{P}^{(N)}\{s, \xi\}$ as

$$\mathscr{P}^{(N)}\{s, \xi\} \prod_{\mathbf{k}_l \in B_N} ds_l |d\xi_l|. \tag{5.6}$$

For a function f we write $f\{\hat{\psi}\} = f\{J, \phi\}$, and we can find its mean value as

$$\langle f\{J, \phi\}\rangle = \left(\prod_{\mathbf{k}_l \in B_N} \int_0^\infty ds_l \oint_{\mathbb{S}^1} |d\xi_l| \right) \mathscr{P}^{(N)}\{s, \xi\} f\{s, \xi\} = \oint \mathscr{P}^{(N)}\{s, \xi\} f\{s, \xi\},$$

$$\tag{5.7}$$

where we introduced notation for the multiple ($2N$-fold) integral over all the phase and the amplitude variables of the modes in B_N.

We see that in the infinite box limit $L \to \infty$

$$\mathscr{P}^{(N)}\{s, \phi\} \to \mathscr{P}[s(\mathbf{k}), \xi(\mathbf{k})],$$

and the averages can be written via Feynman path integration,

$$\left\langle f(\hat{\psi}) \right\rangle = \int \mathscr{D}s_k \oint |\mathscr{D}\xi_k| \mathscr{P}(s_k, \phi_k) f(\psi_k). \tag{5.8}$$

We can consider reduction of the full N-mode PDF to an M-mode joint PDF ($M < N$):

$$\mathscr{P}^{(M)}_{j_1 j_2, \dots j_M} = \left(\prod_{l \neq j_1, j_2, \dots j_M} \int_0^\infty ds_l \oint_{\mathbb{S}^1} |d\xi_l| \right) \mathscr{P}^{(N)}\{s, \xi\}, \tag{5.9}$$

where we have integrated out all phases and amplitudes not contained in the set $\{J, \phi\}_M = \{J_l, \phi_l; \, \mathbf{l} = \mathbf{j}_1, \mathbf{j}_2, \dots, \mathbf{j}_M\}$.

Obviously, we could define objects where we integrate out an arbitrary ($<N$) number of the phases and an arbitrary ($<N$) number of the amplitudes, not necessarily equal in number or belonging to the same modes. A particularly important example is the N-mode amplitude-only PDF obtained by integrating out all the phases ϕ,

$$\mathscr{P}^{(N,a)}\{s\} = \left(\prod_{\mathbf{k}_l \in B_N} \oint_{\mathbb{S}^1} |d\xi_l| \right) \mathscr{P}^{(N)}\{s, \xi\}. \tag{5.10}$$

Finally, we can write the single-mode amplitude PDF

$$\mathscr{P}^{(1,a)} \equiv \mathscr{P}^{(a)}_j(s_j), \tag{5.11}$$

where we have integrated out all phases ϕ, and all amplitudes J but one.

5.3 Random Phase and Amplitude Fields

Definition 5.2 A random phase (RP) field is such a field in which all ϕ are independent random variables (i.r.v.) each of whom is uniformly distributed on the unit circle $\mathbb{S}^1$.

Using this definition, we can write the PDF for a RP field

$$\mathscr{P}^{(N)}\{s, \xi\} = \frac{1}{(2\pi)^N} \mathscr{P}^{(N,a)}\{s\}. \tag{5.12}$$

Remark 5.3.1 We will see below that RP (in addition to the weak nonlinearity) is enough for the lowest level WT closure which results in an equation for the N-point amplitude-only PDF. However, it is not sufficient for the one-point WT closure, in particular for the wave kinetic equation, and we need to assume something about the amplitudes too.

Remark 5.3.2 For RP fields, it would be misleading to think of $L \to \infty$ as a "continuous" limit: the spacing between adjacent modes is decreasing but their phase does not tend to a continuous function in the **k**-space remaining completely random.

Definition 5.3 Random Phase and Amplitude (RPA) field is a field with the following properties:

1. All the amplitudes and all the phases are i.r.v.,
2. All the phases are uniformly distributed on $\mathbb{S}^1$.

For RPA fields we can write the PDF in a product-factorized form as follows,

$$\mathscr{P}^{(N)}\{s, \xi\} = \frac{1}{(2\pi)^N} \prod_{\mathbf{k}_l \in B_N} \mathscr{P}_j^{(a)}(s_j). \qquad (5.13)$$

Remark 5.3.3 We have changed the standard meaning of RPA which usually stands for "Random phase approximation". In our definition of RPA:

1. The amplitudes are random, not only the phases.
2. RPA is defined as a property of the field, not an approximation.

Remark 5.3.4 RPA does not mean Gaussianity because we have not assumed anything about the form or shape of the PDF $\mathscr{P}_j^{(a)}(s_j)$. On the other hand, for Gaussian fields this PDF is fixed and known as the Rayleigh distribution

$$\mathscr{P}^{(a)}(s_j) = \frac{1}{\langle J_j \rangle} \exp\left[-\frac{s_j}{\langle J_j \rangle}\right]. \qquad (5.14)$$

Remark 5.3.5 Wave turbulence does not need Gaussianity, only RPA. This is useful, for it means we can study non-Gaussian wave fields. As a result, we can study turbulence intermittency, i.e. anomalously high (w.r.t. Gaussian) probability of strong events. For example, for water waves this would be the probability of freak waves.

5.4 Generating Functions

Let us define the N-mode generating function which is, as we will see later, a very useful auxiliary object for statistical derivations.

Definition 5.4 The N-mode generating function is defined as

$$\mathscr{L}^{(N)}\{\lambda,\mu\} = \left\langle \prod_{k_l \in B_N} \exp\left[\left(\frac{L}{2\pi}\right)^d J_l \lambda_l\right] \phi_l^{\mu_l} \right\rangle, \tag{5.15}$$

where $\{\lambda, \mu\}$ is a set of $2N$ real numbers λ_j and μ_j; $(j = 1, \dots, N)$.

In terms of the PDF, we can write

$$\mathscr{L}^{(N)}\{\lambda,\mu\} = \oint \mathscr{P}^{(N)}\{s,\xi\} \prod_{k_l \in B_N} \exp\left[\left(\frac{L}{2\pi}\right)^d J_l \lambda_l\right] \phi_l^{\mu_l}$$

$$= \int_0^\infty \int_0^\infty \dots \int_0^\infty ds_1 ds_2 \dots ds_N$$

$$\times \int_0^{2\pi} \int_0^{2\pi} \dots \int_0^{2\pi} d\varphi_1 d\varphi_2 \dots d\varphi_N \mathscr{P}^{(N)}\{s,\xi\} e^{\sum_l \left[\left(\frac{L}{2\pi}\right)^d J_l \lambda_l + i\mu_l \varphi_l\right]}. \tag{5.16}$$

We can see now that $\mathscr{L}$ is the Laplace transform of $\mathscr{P}$ over all amplitude variables J, and the Fourier transform over all phases φ. Thus, we can find the PDF $\mathscr{P}$ from the generating function $\mathscr{L}$ using the inverse Fourier and Laplace transforms. We note that the study of this multi-mode generating function is important when we strive to prove the underlying assumptions of wave turbulence. Namely, we will see later that RPA, if present initially, survives through the long-time nonlinear evolution, i.e. the PDF remains product factorized in some (coarse-grain) sense.

We can obtain the one-mode generating function from (5.15) by setting all $\lambda_{l \neq j} = 0$ and all $\mu_l = 0$,

$$\mathscr{L}_j^{(a)}\{\lambda_j\} = \left\langle \exp\left[\left(\frac{L}{2\pi}\right)^d J_j \lambda_j\right] \right\rangle, \tag{5.17}$$

so that we can find the 1-mode PDF $\mathscr{P}_j^{(a)}(s_j)$ by taking the inverse Laplace of $\mathscr{L}_j^{(a)}$.

Exercise 5.1 *Derive an expression for $\mathscr{L}_j^{(a)}$ for Gaussian fields by applying the Laplace transform to the Rayleigh distribution.*

5.5 Wave Spectrum, Higher Moments and Structure Functions

Definition 5.5 The *wave spectrum* is defined as follows

$$n_k = \left(\frac{L}{2\pi}\right)^d \langle J_k \rangle. \tag{5.18}$$

Here we do not specify what kind of spectrum this is (e.g. energy, enstrophy or waveaction). However, in most that follow we will deal with variables $\hat{\psi}$ and $\hat{\psi}^*$ which represent canonically conjugated Hamiltonian variables, and for these cases n_k stands for the waveaction spectrum.

For the infinite-box limit, $L \to \infty$, we can use standard rules of thumb $\delta(0) = \left(\frac{L}{2\pi}\right)^d$ and $\psi_k \to \left(\frac{L}{2\pi}\right)^d \psi_k$ when replacing Fourier series by Fourier transform (thus replacing $\frac{1}{L^d}$ with $\frac{1}{(2\pi)^d}$ in (5.3)). This leads to a familiar definition for the spectrum in the infinite space:

$$\langle \psi_k, \psi_{k'}^* \rangle = n_k \delta(\mathbf{k} - \mathbf{k}'),$$

where $\delta(\mathbf{x})$ is the Dirac's delta function.

In terms of the generating function and the PDF, the wave spectrum can be expressed as follows,

$$n_k = \left[\partial_{\lambda_k} \left\langle e^{\lambda_k \left(\frac{L}{2\pi}\right)^d J_k} \right\rangle \right]_{\lambda_k = 0} = \left[\partial_\lambda \mathscr{Z}^{(a)}(\lambda_k) \right]_{\lambda_k = 0}$$

$$= \left(\frac{L}{2\pi}\right)^d \int_0^\infty s_k \mathscr{P}^{(a)}(s_k) ds_k. \tag{5.19}$$

Along the same lines, we can define higher-order one-mode moments of the wave amplitudes.

Definition 5.6 The *higher one-mode moments* are defined as:

$$M_k^{(p)} = \left(\frac{L}{2\pi}\right)^{pd} \langle J_k^p \rangle. \tag{5.20}$$

In terms of the generating function and the PDF:

$$M_k^{(p)} = \left[\partial_\lambda^p \mathscr{Z}^{(a)}(\lambda_k) \right]_{\lambda_k = 0}$$

$$= \left(\frac{L}{2\pi}\right)^{pd} \int_0^\infty s_k^p \mathscr{P}^{(a)}(s_k) ds_k. \tag{5.21}$$

The higher moments are of interest since we can use them to study fluctuations of the wave spectra. For instance, for $p = 2$, we write

$$M_k^{(2)} = \left(\frac{L}{2\pi}\right)^{2d} \langle J_k^2 \rangle, \tag{5.22}$$

In terms of this quantity, we can obtain the standard deviation for the wave intensity,

$$\sigma_k = \left(\frac{L}{2\pi}\right)^{2d} \left(\langle J_k^2 \rangle - \langle J_k \rangle^2 \right) = M_k^{(2)} - n_k^2. \tag{5.23}$$

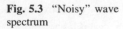

Fig. 5.3 "Noisy" wave
spectrum

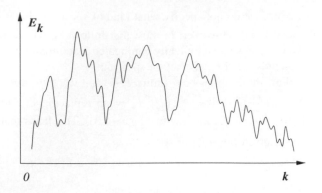

The quantity $\sqrt{\sigma_k}$ measures the strength of fluctuations of the wave intensity in the **k**-space about its mean value, the spectrum, see Fig. 5.3. Note that most experimentally and numerically obtained spectra look like in Fig. 5.3, i.e. very noisy, unless averaged over many realizations, time, space or/and angle.

If the one-mode PDF for the wave mode **k** is Rayleigh, $\mathscr{P}^{(a)}(s_j) = \frac{1}{\langle J_j \rangle} \exp\left[-\frac{s_j}{\langle J_j \rangle}\right]$, then one can easily express the higher moments $M_k^{(p)}$ in terms of the spectrum n_k, as formulated in the following exercise.

Exercise 5.2 *Derive for Gaussian fields:*

$$M_k^{(p)} = p!n_k^p.$$

Note that the one-mode PDF's are Rayleigh for Gaussian fields, but not only— since the phase statistics should also be prescribed for the Gaussian fields.

We see immediately that for Gaussian fields $\sigma_k = n_k^2$ so that the fluctuations are equal to the mean, see Fig. 5.3. On the other hand, if the one-mode PDF's decay slower than the Rayleigh (exponential) distribution, for example as a power-law function, then there will be an enhanced probability of strong waves. This phenomenon is called *intermittency* and is of a general interest in the turbulence theory (see e.g. book [7]).

Another example of use of the higher moments is in analysis of structure functions in the **x**-space.

Definition 5.7 We define the *p*-th order structure function as

$$\mathscr{S}^{(p)} = \langle |\psi(\mathbf{x}, t) - \psi(\mathbf{x} - \mathbf{r}, t)|^p \rangle. \tag{5.24}$$

This definition is similar to the one used in the strong hydrodynamic turbulence (see e.g. book [7]) in which case one distinguishes between the parallel and transverse structure functions by taking respectively

$$\psi = u_\| = \frac{\mathbf{u} \cdot \mathbf{r}}{|r|}$$

or

$$\psi = u_\perp = \frac{|\mathbf{u} \times \mathbf{r}|}{|r|},$$

where $\mathbf{u}$ is the fluid velocity field. We can write these structure functions in terms of the moments (5.21) if the fields are of RPA type, as we will see below.

5.6 RPA Averaging

Since RPA fields have random phases and random amplitudes, we need to average over both of these quantities separately. Let us recall (5.5)

$$\hat{\psi}_l = \hat{\psi}(\mathbf{k}_l, t) = \sqrt{J_l}\phi_l,$$

and we will also use the Kronecker delta in the following discussion

$$\delta_2^1 = 1 \quad \text{if } \mathbf{k}_1 = \mathbf{k}_2, \quad \delta_2^1 = 0 \quad \text{otherwise.} \tag{5.25}$$

Now we can write for the average

$$\left\langle \hat{\psi}_1 \hat{\psi}_2 \hat{\psi}_3^* \hat{\psi}_4^* \right\rangle = \left\langle \left\langle \hat{\psi}_1 \hat{\psi}_2 \hat{\psi}_3^* \hat{\psi}_4^* \right\rangle_\phi \right\rangle_J = \left\langle \sqrt{J_1 J_2 J_3 J_4} \right\rangle_J \left\langle \phi_1 \phi_2 \phi_3^* \phi_4^* \right\rangle_\phi, \tag{5.26}$$

where the average over ϕ is defined as

$$\langle f \rangle_\phi = \frac{1}{2\pi^N} \left(\prod_{\mathbf{k}_l \in B_N} \oint_{S^1} |d\xi_l| \right) f. \tag{5.27}$$

In order to find the average over the phases, let us first assume that $\mathbf{k}_1$, $\mathbf{k}_2$, $\mathbf{k}_3$, and $\mathbf{k}_4$ are all different wavenumbers. We then use (5.27) in which the integrands are equal to 1 for all the integrals, except when we hit $\mathbf{k}_1$, $\mathbf{k}_2$, $\mathbf{k}_3$, or $\mathbf{k}_4$. So we get something like

$$\frac{1}{2\pi^N} \oint 1 |d\xi_1| \dots \oint \xi_{k_1} |d\xi_{k_1}| \dots \oint \xi_{k_3}^* |d\xi_{k_3}| \dots$$

Note that $\oint 1 |d\xi| = 2\pi$ and that since ξ is an odd function, we have $\oint \xi |d\xi| = 0$. Similarly, the complex conjugate $\oint \xi^* |d\xi| = 0$. Therefore, if at least one integral of type $\oint \xi_j |d\xi_j| = 0$ (or its conjugate) is present then the result for the respective correlator is zero. Thus, we get zero when $\mathbf{k}_1$, $\mathbf{k}_2$, $\mathbf{k}_3$ and $\mathbf{k}_4$ are all different wavenumbers.

Let us now assume that $\mathbf{k}_1 = \mathbf{k}_2$. The integrals corresponding to $\mathbf{k}_1$, $\mathbf{k}_2$ become $\oint \xi_{k_1}^2 |d\xi_{k_1}|$. However, ξ^2 is still an odd function, and the integral is zero. Obviously, we can draw a similar conclusion for $\mathbf{k}_3 = \mathbf{k}_4$.

Finally, let us assume $\mathbf{k}_1 = \mathbf{k}_3$ and $\mathbf{k}_2 = \mathbf{k}_4$ (or $\mathbf{k}_1 = \mathbf{k}_4$ and $\mathbf{k}_2 = \mathbf{k}_3$). We now obtain nonzero factors, e.g.

$$\oint \xi_{k_1} \xi_{k_3}^* |d\xi_{k_1}| = \delta_1^3 \oint |\xi_{k_1}|^2 |d\xi_{k_1}| = \delta_1^3 \oint 1 |d\xi_{k_1}| = 2\pi \delta_1^3.$$

Thus, we need to pair wavenumbers belonging to ξ's and to ξ^*'s in order to obtain a finite average. We find

$$\left\langle \phi_1 \phi_2 \phi_3^* \phi_4^* \right\rangle_\phi = \delta_3^1 \delta_4^2 + \delta_4^1 \delta_3^2 - \delta_2^1 \delta_3^1 \delta_4^1 . \tag{5.28}$$

The last term on the RHS accounts for the fact that the case $\mathbf{k}_1 = \mathbf{k}_2 = \mathbf{k}_3 = \mathbf{k}_4$ corresponds to one term only, i.e. the RHS must give 1 as the answer for this case. We can now rewrite (5.26) as follows,

$$\left\langle \hat{\psi}_1 \hat{\psi}_2 \hat{\psi}_3^* \hat{\psi}_4^* \right\rangle = \langle J_1 J_2 \rangle_J \left(\delta_3^1 \delta_4^2 + \delta_4^1 \delta_3^2 - \delta_2^1 \delta_3^1 \delta_4^1 \right). \tag{5.29}$$

Again, we differentiate between the case when $\mathbf{k}_1 = \mathbf{k}_2$ and when $\mathbf{k}_1 \neq \mathbf{k}_2$. For RPA (but not for RP!) we have

$$\langle J_1 J_2 \rangle_J = \begin{cases} \langle J_1 \rangle \langle J_2 \rangle & \text{if } \mathbf{k}_1 \neq \mathbf{k}_2, \\ \langle J_1^2 \rangle & \text{if } \mathbf{k}_1 = \mathbf{k}_2, \end{cases} \tag{5.30}$$

and we can write the the final expression for our fourth-order correlator of an RPA field as

$$\left\langle \hat{\psi}_1 \hat{\psi}_2 \hat{\psi}_3^* \hat{\psi}_4^* \right\rangle = \langle J_1 \rangle \langle J_2 \rangle \left(\delta_3^1 \delta_4^2 + \delta_4^1 \delta_3^2 \right)$$
$$+ \left(\langle J_1^2 \rangle - 2\langle J_1 \rangle^2 \right) \delta_2^1 \delta_3^1 \delta_4^1. \tag{5.31}$$

The second term on RHS (with the product of the three delta functions) is called the fourth-order cumulant: it is zero for Gaussian fields as it is clear from Exercise 5.2.

Finally, we generalize our result and obtain a rule of thumb—known as **Wick's contraction rule** which can be formulated as a theorem.

Theorem 5.1 *(For RP and RPA fields) Correlator* $\left\langle \hat{\psi}_{l_1} \hat{\psi}_{l_2} \ldots \hat{\psi}_{m_1}^* \hat{\psi}_{m_2}^* \right\rangle$ *is zero unless the number of ψ's in it equals the number of ψ^*'s. If the number of ψ's* *equals the number of ψ^*'s, the answer will be equal to the sum of terms in each of whom the wavenumbers of ψ's matched to the wavenumbers of ψ^*'s in all possible ways.*[1]

(For RPA fields, not for RP) If for a particular contraction combination all the wavenumbers of contracted pairs are different from each other, then the respective terms will involve only the spectrum n_k. If some of the contracted pairs are themselves matched by wavenumbers then these combinations lead to higher moments $M_k^{(p)}$ (where p will be the number of pairs with coinciding wavenumbers).

Like for the fourth order, the higher order moments will in general contain the Gaussian part and a cumulant. To achieve full understanding of the Wick's contraction rule, we suggest to do the following exercise.

Exercise 5.3 *Consider a (possibly non-Gaussian) RPA field ψ and derive expression for the correlator $\left\langle \hat{\psi}_1 \hat{\psi}_2 \hat{\psi}_3 \hat{\psi}_4^* \hat{\psi}_5^* \hat{\psi}_6^* \right\rangle$.*

We can also express the structure functions in terms of the moments (5.21) using the RPA averaging in a similar way. For instance for $p = 4$, we can formulate it as an exercise:

Exercise 5.4 *Derive the following expression for the fourth-order structure function,*

$$\mathscr{S}^4(\mathbf{r}) = 2\left[\mathscr{S}^{(2)}(\mathbf{r})\right]^2 + 16\left(\frac{2\pi}{L}\right)^{2d}\sum_k\left(M_k^{(2)} - 2n_k^2\right)\left|\sin\frac{\mathbf{k}\cdot\mathbf{r}}{2}\right|^4. \tag{5.32}$$

The first part of this expression is the Gaussian factorization in terms of the second-order structure function, whereas the second part is the cumulant.

Remark 5.6.1
- The above expression is an example of useful relationships between the multi-point x-space correlators and the one-mode $\mathbf{k}$-space moments. Indeed, the study of $M_k^{(p)}$'s is easier because of the smaller number of independent variables involved (just one wavevector!)
- However, the cumulant part is vanishing in the limit $L \to \infty$. Therefore, the finite box is essential for being able to "probe" the multi-point correlators with the one-mode $\mathbf{k}$-space moments.
- According to (5.32), the cumulant part is not decaying as the function of $\mathbf{r}$ in the directions of the excited $\mathbf{k}$'s for which $(M_k^{(2)} - 2n_k^2) \neq 0$. Thus, in RPA fields the correlations decay much slower across the wave crests than along them (which is somewhat counter-intuitive).

---

[1] For Gaussian fields, this is where the standard Wick's rule ends, and the final expression will involve only the spectrum n_k. That is, for every pairwise contraction: $\hat{\psi}_k \hat{\psi}_k^* \to \langle J_k \rangle$. But for non-Gaussian fields, Wick's rule has to be slightly extended.

Remark 5.6.2 Gaussianity arises naturally for the one-mode statistics for short-correlated in the x-space fields in the $L \rightarrow \infty$ limit due to the central limit theorem. To see that, just split the L-box Fourier transform into a sum of Fourier transforms over smaller boxes whose size remains greater than the correlation length. The terms in the sum will be i.r.v.'s and the sum will tend to a Gaussian distribution in the large sum limit (which is generally not RPA like, e.g., in strong turbulence). Thus, for such fields too one has to keep L finite in order to detect intermittency with the one-mode moments.

References

1. Choi, Y., Lvov, Y., Nazarenko, S.V.: Probability densities and preservation of randomness in wave turbulence. Phys. Lett. A **332**(3–4), 230 (2004)
2. Choi, Y., Lvov, Y., Nazarenko, S.V., Pokorni, B.: Anomalous probability of large amplitudes in wave turbulence. Phys. Lett. A **339**(3–5), 361 (2004)
3. Choi, Y., Lvov,Y., Nazarenko, S.V.: Joint statistics of amplitudes and phases in wave turbulence. Phys. D **201**, 121 (2005)
4. Choi, Y., Lvov, Y.V., Nazarenko, S.: Wave turbulence. In: Recent Developments in Fluid Dynamics, vol. 5. Transworld Research Network, Kepala, India (2004)
5. Choi, Y., Jo, G.G., Kim, H.I., Nazarenko, S.: Aspects of two-mode probability density function in weak wave turbulence. J. Phys. Soc. Japan **78**(8), 084403 (2009)
6. Jakobsen, P.P., Newell, A.: Invariant measures and entropy production in wave turbulence. J. Stat. Mech. L10002 (2004)
7. Frisch, U.: Turbulence: The Legacy of A. N. Kolmogorov. Cambridge University Press, Cambridge (1995)

Chapter 6
Wave Turbulence Formalism

The most systematic part of the WT description deals with weakly nonlinear waves. In this chapter, we will present the main concepts and techniques of weak WT using Petviashvili equation as a master example. This will be followed by considering the four-wave, five-wave and six-wave systems, with respective examples discussed along the way: NLS model, 2D water gravity waves, 1D water gravity waves and 1D optical turbulence. More examples will be discussed in Parts 3 and 4 of this book. As in the previous chapter, we will continue following the approach developed in [1–6].

6.1 Dharmachakra of Wave Turbulence: Main Steps, Ideas and Building Blocks

First of all, we would like to outline the "Wheel of Law" of WT, like Dharma-chakra in Buddhism, i.e. the grand scheme of WT derivations, their main steps, building blocks, rules and techniques. Further like in Buddhism, we can distinguish a comprehensive and a reduced schemes, Mahayana and Hinayana (the big and the small vehicles respectively). The comprehensive approach is the one of "an ideal world", including description of the multi-mode statistical objects, which could be used to provide a posteriori justification for the statistical assumptions, namely for RPA. Such a comprehensive programme has not been completed yet, and developing this approach is important for better understanding of the fundamental statistical processes in WT, and for building firmer mathematical basis for a rigorous WT theory. The reduced description deals with the one-mode objects only, such as the one-mode amplitude PDF $\mathscr{P}_k^{(a)}$ or/and one-mode moments $M_k^{(p)}$, i.e. a more modest reduced scheme which is sufficient for most practical applications. In the most reduced version (and yet the only one developed for most applications) the WT theory deals with the wave spectrum only.

S. Nazarenko, *Wave Turbulence*, Lecture Notes in Physics, 825,
DOI: 10.1007/978-3-642-15942-8_6, © Springer-Verlag Berlin Heidelberg 2011

Let us now consider both the comprehensive and the reduced schemes, see Figs. 6.1 and 6.2 respectively.

6.1.1 Mahayana (Comprehensive Scheme)

Here we will outline the comprehensive WT scheme which is presented diagrammatically in Fig. 6.1. Do not worry if you do not understand some of terminology here, it will be explained in great detail later.

- **Step 1**. Start with a *nonlinear wave equation*. In a great number of cases, this equation will be in a diagonal Hamiltonian form, but not always. For example, the system of Alfvén waves in incompressible fluids is best considered in a non-Hamiltonian form in terms of so-called Elsässer variables.
- **Step 2**. Further, one has to rewrite the nonlinear equation at hand in the **k**-space using the *interaction representation*. In some cases (e.g. for four-wave systems

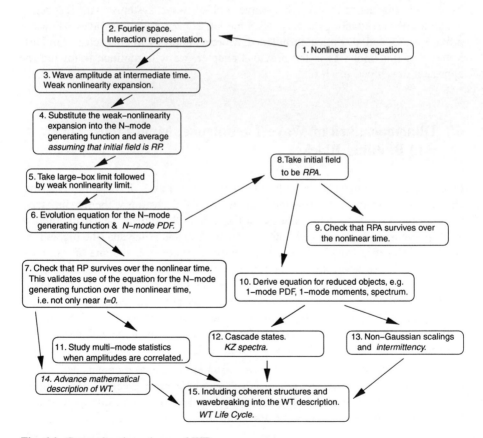

Fig. 6.1 Comprehensive scheme of WT

Fig. 6.2 Reduced scheme of
WT

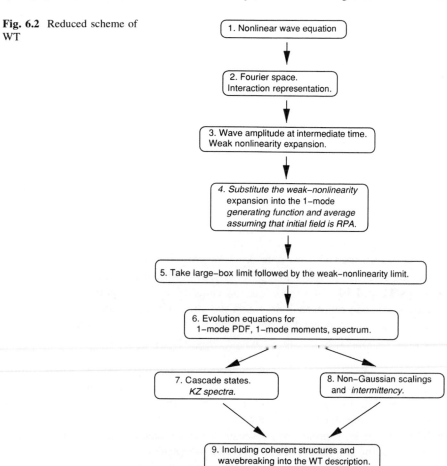

and not three-wave systems, as we will see later) a *frequency re-normalization*
must be done at this point.

- **Step 3**. Use weakness of nonlinearity and separation of the linear and the
  nonlinear time scales. Namely, chose an auxiliary intermediate time T which is
  much longer than the wave period but much shorter than the nonlinear time and
  develop *perturbation series in the small nonlinearity parameter* ϵ for the
  complex wave amplitude at $t = T$ in terms of its value at $t = 0$.
- **Step 4**. Substitute this solution into the joint N-mode generating function
  $\mathscr{L}^{(N)}\{\lambda, \mu\}$. Perform *statistical averaging* assuming that the initial wave field is
  of RP type, i.e. by assuming the phase factors to be random and independent, but
  without assuming anything about statistics of the amplitudes or about their
  statistical independence.

- **Step 5**. Take the infinite box and the weak nonlinearity limits, $L \to \infty$ and $\epsilon \to 0$ respectively. As we will see later, the order of the limits is important: the $L \to \infty$ has to be taken first and $\epsilon \to 0$ – second.
- **Step 6**. The previous step will result in a closed equation for $\mathscr{Z}^{(N)}\{\lambda, \mu\}$ and, via the inverse Laplace transform, an equation for the n-mode joint PDF. However this equation is based on evaluation of the time derivative of $\mathscr{Z}^{(N)}\{\lambda, \mu\}$ at $t = 0$, assuming initial RP field, and validity of the resulting equation for $t > 0$ depends on whether the RP property survives over this time. This is to be checked in step 7.
- **Step 7**. Check that RP survives over the nonlinear time, i.e. over the characteristic time of the evolution equations, e.g. for $\mathscr{Z}^{(N)}\{\lambda, \mu\}$ (we will see that this time is ϵ^{-2} times greater than the linear wave period for the three-wave systems, ϵ^{-4} for the four-wave systems etc). At this step we simply have to check that the RHS of the evolution equation for the N-mode PDF is independent of the phase variables (in the right order in ϵ), so that the PDF remains phase independent (which corresponds to the independent phases with uniform distributions on $\mathbb{S}^1$).
- **Steps 11 and 14**. After the evolution equation for the N-point objects are validated, one can study the multi-mode statistics, without assuming that the modes are statistically independent and search for mathematical justifications for the transition from the N-mode to the 1-mode description, find rigorous ways for transition from the discrete to the continuous Fourier space, etc. This is where WT is expected to converge with the other statistical systems where similar questions are studied, e.g. transition from the N-particle Liouville equation to the 1-particle Vlasov equation in the gas of charged particles. A lot of problems still remain unsolved in these directions.
- **Steps 8 and 9**. Alternatively, one can further assume that the initial statistics is RPA (i.e. the amplitudes are statistically independent too) and to check that the RPA property can survive over the nonlinear time. This is done by checking that the multi-mode PDF's remain product factorized by the 1-mode PDF's (in the limit of large box and small ϵ) if such product factorization is valid for the initial waves. We will see that the product factorization survives only in a "coarse-grained" sense, i.e. only for the M-mode PDF's with $M \ll N$.
- **Step 10**. If RPA is proven to be valid over the nonlinear time, one can integrate out $N - 1$ modes and derive a closed equation for the 1-mode PDF. In turn, this yields a kinetic equation for the wave spectrum and for the higher 1-mode moments of the wave amplitudes.
- **Step 12**. Given the kinetic equation for the energy spectrum, one can study solutions of this equation. Most important of these solutions is the KZ spectrum corresponding to the energy (or some other invariant's) cascade through scales. Finding KZ spectra is at focus of most works in WT.
- **Step 13**. Given the evolution equation for the 1-mode PDF, one can study its solutions, in particular the ones that correspond to deviations from the Gaussian statistics and turbulence intermittency.

- **Step 15**. The "intermittent" solutions for the 1-mode PDF appear to provide a useful language for describing interaction of weak incoherent waves with strong coherent waves and thereby describe the *WT life cycle*. Probably, knowledge of the correlated multi-mode dynamics (steps 11 and 12) can provide an additional insight about interaction with the coherent waves (since the k-modes are correlated in these waves), but this remains a poorly studied subject.

6.1.2 Hinayana (Reduced Scheme)

The reduced scheme of WT is shown in Fig. 6.2. Basically, it differs from the comprehensive scheme in Fig. 6.1 by taking away all studies of the multi-mode statistical objects. The 1-mode objects are derived directly after step 4 by assuming RPA. Obviously, one cannot check anymore whether RPA survives over the nonlinear time because the N-point PDF is not available. Thus, in this reduced approach RPA remains an assumption (sometimes referred to as an approximation).

However, the reduced scheme does allow one to study many important properties of WT, particularly the KZ spectra, non-Gaussian wave fields corresponding to turbulence intermittency, as well as the WT cycle of interaction of incoherent and coherent waves.

In fact, the reduced scheme provides "good value for money" because it allows to present most of WT ideas and to obtain many important results avoiding lengthy technical details associated with derivations of equations for the N-point objects. Thus, in what follows we will start by presenting derivations for the 1-mode objects, namely the one-mode amplitude generating function $\mathscr{Z}_k^{(a)}$, the 1-mode PDF $\mathscr{P}_k^{(a)}$, the 1-mode moments and the wave spectrum. This will be followed by deriving WT for the N-mode statistics, a part which could be skipped by those wanting to avoid extra technicalities at this point. Derivations for the N-mode objects may be found in papers [1, 3, 4, 6] and they could be a good theme for an advanced graduate project.

6.2 Master Example: Petviashvilli Equation

Below, we will illustrate the WT formalism and present all the essential derivation steps using a relatively simple particular example. Namely, we will study the Petviashvilli equation [7], which is relevant for drift waves in fusion plasmas, and Rossby waves in geophysical fluids:

$$\partial_t \psi = \nabla^2 \partial_x \psi - \psi \partial_x \psi, \tag{6.1}$$

where $\psi = \psi(x, y, t)$ is a real function of two space coordinates (x, y) and time t. Note that this equation is a 2D generalization of the famous KdV equation. In plasmas, ψ would be the electrostatic potential, whereas in geophysics, it is the velocity stream function. The nonlinear term $\psi \partial_x \psi$ in (6.1) is called "scalar nonlinearity" to distinguish it from the "vector nonlinearity" of Charney-Hassegawa-Mima equation (6.13) which is another model for the drift and Rossby waves.

6.2.1 Conservation Laws

Petviashvilli equation (6.1) conserves "energy":

$$E = \frac{1}{2\rho^2} \int \psi^2 \, d\mathbf{x} = \text{const}, \tag{6.2}$$

where ρ is a characteristic radius, i.e. Rossby deformation radius in the geophysical fluids and the ion Larmor radius in plasmas (pre-factor $1/\rho^2$ is added to make the correct physical dimension of E, assuming that ψ is a stream function).

Petviashvilli equation (6.1) also conserves "potential enstrophy":

$$\Omega = \frac{1}{2} \int \left[(\nabla \psi)^2 + \frac{1}{3} \psi^3 \right] d\mathbf{x} = \text{const}. \tag{6.3}$$

The "energy" and the "potential enstrophy" are the conventional physical interpretations of these invariants (agreeing with the respective physical invariants of the Charney-Hassegawa-Mima equation (6.13), but they should be treated with caution, because the Hamiltonian is actually given by Ω, not E. Indeed, we can write (6.1) in a Hamiltonian form as[1]

$$\partial_t \psi = -\partial_x \frac{\delta \Omega}{\delta \psi}. \tag{6.4}$$

The Fourier-space Hamiltonian formulation will be discussed later. Moreover, as we will see later, in this example E will correspond to one of the standard WT invariants called the x-component of momentum.

---

[1] Note the Gardner type of the Poisson bracket, just like for the Hamiltonian form of the KdV equation.

Exercise 6.1 *Prove conservation of E and Ω given in (6.2) and (6.3).*

6.2.2 Fourier Space

We can represent the wave field in the 2D **k**-space using Fourier coefficients

$$\hat{\psi}_k = \frac{1}{L^2} \int_{\text{Box}} \psi(x,y)e^{-i\mathbf{k}\cdot\mathbf{x}}d\mathbf{x}, \tag{6.5}$$

and rewrite (6.1) as

$$\dot{\hat{\psi}}_k + ik_xk^2\hat{\psi}_k + i\sum_{1,2}\hat{\psi}_1\hat{\psi}_2 k_{2x}\delta^k_{12} = 0, \tag{6.6}$$

where dot denotes the time derivative.

First of all, let us take the linear limit, assuming that the wave amplitude is very small $\hat{\psi} \to 0$. Ignoring the nonlinear term, we obtain a basic linear ODE

$$\dot{\hat{\psi}}_k + ik_xk^2\hat{\psi}_k = 0, \tag{6.7}$$

which has the fundamental solution

$$\hat{\psi}_k = A_k e^{-i\omega_k t}, \tag{6.8}$$

where $A_k \in \mathbb{C}$ is the amplitude of the wave with wavevector **k**, and the frequency of this wave, ω_k, is given by the following dispersion relation,

$$\omega_k = k_xk^2. \tag{6.9}$$

By the superposition principle, we know that any linear combination of waves with different wavevectors **k** is a solution, and there is no energy exchange between modes with different **k**'s. In the other words, the amplitudes A_k are time independent in the linear approximation.

Let us now re-introduce nonlinearity, which describes the energy exchange among the **k**-modes. For the moment, the nonlinearity is arbitrary and not necessarily small. By symmetrizing the nonlinear term in (6.6), we rewrite it as

$$i\dot{\hat{\psi}}_k = \omega_k\hat{\psi}_k + \sum_{1,2}\hat{\psi}_1\hat{\psi}_2 k_{2x}\delta^k_{12} + \sum_{1,2}\hat{\psi}_1\hat{\psi}_2 k_{1x}\delta^k_{12}$$

$$= \omega_k\hat{\psi}_k + \frac{k_x}{2}\sum_{1,2}\hat{\psi}_1\hat{\psi}_2\delta^k_{12}. \tag{6.10}$$

Now, let us introduce the *waveaction variable* $a_k = \frac{\hat{\psi}_k}{|k_x|^{1/2}}$ and rewrite (6.10) as

$$i\dot{a}_k = \omega_k a_k + \text{sign}(k_x)\sum_{1,2}V_{12k}a_1a_2\delta^k_{12}, \tag{6.11}$$

where we have introduced the interaction coefficient

$$V_{12k} = \frac{1}{2} \sqrt{|k_x k_{1x} k_{2x}|}. \tag{6.12}$$

Equation (6.11) is now in a form which is generic for the wave systems with quadratic nonlinearity, and whose wave functions ψ have real-valued x-space amplitudes. We will show later that this kind of equations can be written in a diagonal Hamiltonian form, see (6.64). Another equation of this kind is given in the following exercise.

Exercise 6.2 *Consider Charney-Hassegawa-Mima Equation*

$$\left(\nabla^2 \psi - \rho^{-2}\psi\right)_t + \beta \partial_x \psi + (\partial_x \psi)(\partial_y \nabla^2 \psi) - (\partial_y \psi)(\partial_x \nabla^2 \psi) = 0, \tag{6.13}$$

where ρ and β are constants. This equation is a model for planetary Rossby waves and drift waves in inhomogeneous plasmas. It is a more common model than the simpler Petviashvilli equation which is a master model in our derivations (the nonlinearity here is sometimes called the "vector nonlinearity" because it can be written as a vector product, $[\nabla \psi \times \nabla \nabla^2 \psi]_z$). Show that (6.13) can be written in the generic form of (6.11).

6.2.3 Interaction Representation

The idea of the interaction representation is to separate the time scales by introducing variables which do not change in the linear approximation (i.e. in absence of interaction). These variables correspond to the amplitudes A_k in the linear solution (6.8). Namely, let us define the interaction representation variables as

$$b_k = \frac{a_k}{\epsilon} e^{i\omega_k t}, \tag{6.14}$$

where we introduced parameter $\epsilon \in \mathbb{R}^+$, which, for now, is arbitrary.[2] This is an auxiliary parameter the sole purpose of which is an easier power count in the weakly nonlinear expansion later when we will take $\epsilon \ll 1$ and $b_k \sim 1$ which corresponds to $|a_k| \ll 1$. Note that b_k is time-independent in the linear limit. We now rewrite (6.11) as

$$i\dot{b}_k = \epsilon \, \mathrm{sign}(k_x) \sum_{1,2} V_{12k} b_1 b_2 \delta_{12}^k e^{i\omega_{12}^k t}, \tag{6.15}$$

where we have introduced a shorthand notation $\omega_{12}^k = \omega_k - \omega_1 - \omega_2$. Note that the linear term $\omega_k a_k$ is gone, and that there is an explicit time-dependence in the

---

[2] Please do not confuse ϵ with ε which we have reserved for the energy flux notation.

nonlinear term. We emphasize that we have not made any approximations yet, and the interaction representation (6.15) is identical to the original Petviashvili equation (6.1).

6.3 Weak Nonlinearity Expansion

6.3.1 Solution for the Wave Amplitudes at Intermediate Times

Now assume that the nonlinearity is small, i.e. the energy exchanges between modes are much slower than the periods of the linear waves,

$$T_k = \frac{2\pi}{\omega_k}. \tag{6.16}$$

The long nonlinear time for the Petviashvili model (as well as for any other three-wave system) is $\tau_{NL} = \frac{2\pi}{\epsilon^2 \omega_k}$. This value for τ_{NL} will be verified *a posteriori*. Thus the linear and nonlinear time scales are separated,

$$\tau_L \sim T_k \ll \tau_{NL}.$$

We are going to exploit this time-scale separation and filter out the fast oscillatory motions. For this, let us introduce an auxiliary intermediate time T such that

$$\tau_L \ll T \ll \tau_{NL}.$$

In particular, one can simply take $T \sim \frac{2\pi}{\epsilon \omega_k}$.

Let us find a solution for the wave amplitude b_k at time $t = T$ using a formal expansion in the small nonlinearity parameter $\epsilon \ll 1$:

$$b_k(T) = b_k^{(0)} + \epsilon b_k^{(1)} + \epsilon^2 b_k^{(2)} + \cdots. \tag{6.17}$$

We emphasize that T is an auxillary parameter needed to filter out the fast linear oscillations, and it will not enter into the final WT equations.

The first term on the RHS in (6.17) corresponds to the linear approximation ($\epsilon = 0$) in which the interaction representation amplitude is time-independent and equal to its initial value at $t = 0$, i.e.

$$b_k^{(0)} = b_k(0).$$

We can find the second term by looking for the factors of order ϵ^1 in $\dot{b}_k(T)$, which can simply be obtained by an iterative substitution of the previous order into the RHS of (6.15), namely

$$ib_k^{(1)} = \text{sign}(k_x) \sum_{1,2} V_{12k} b_1^{(0)} b_2^{(0)} \delta_{12}^k e^{i\omega_{12}^k t}.$$ (6.18)

We integrate this equation and we obtain

$$b_k^{(1)}(T) = -i\,\text{sign}(k_x) \sum_{1,2} V_{12k} b_1^{(0)} b_2^{(0)} \delta_{12}^k \Delta_T(\omega_{12}^k),$$ (6.19)

with

$$\Delta_T(\omega_{12}^k) = \int_0^T e^{i\omega_{12}^k t} dt = \frac{e^{i\omega_{12}^k T} - 1}{i\omega_{12}^k}.$$ (6.20)

Let us now find the terms of order ϵ^2 by performing the next iteration. This gives for $b_k^{(2)}$:

$$ib_k^{(2)} = 2\,\text{sign}(k_x) \sum_{1,2} V_{12k} b_1^{(1)} b_2^{(0)} \delta_{12}^k e^{i\omega_{12}^k t},$$ (6.21)

where factor 2 arises due to the symmetry with respect to changing indices $1 \leftrightarrow 2$. We now use our formula for $b_k^{(1)}$ from (6.19) and write

$$ib_k^{(2)} = -2i \sum_{1,2,3,4} \text{sign}(k_x k_{1x}) V_{12k} V_{341} b_2^{(0)} b_3^{(0)} b_4^{(0)} \delta_{12}^k \delta_{34}^1 \Delta_T(\omega_{34}^1) e^{i\omega_{12}^k t}.$$ (6.22)

Integrating this equation gives

$$b_k^{(2)}(T) = -2 \sum_{1,2,3,4} \text{sign}(k_x k_{1x}) V_{12k} V_{341} b_2^{(0)} b_3^{(0)} b_4^{(0)} \delta_{34}^1 \delta_{12}^k E(\omega_{34}^1, \omega_{12}^k),$$ (6.23)

with

$$E(\omega_{34}^1, \omega_{12}^k) = \int_0^T \Delta_T(\omega_{34}^1) e^{i\omega_{12}^k t} dt.$$ (6.24)

As we will see, a nontrivial closure arises in the order ϵ^2 in this (three-wave) case. Thus we do not need to find the higher-order terms in our ϵ-expansion.

6.3.2 Weak Nonlinearity Expansion for the Generating Function

We will now substitute the ϵ-expansion for $b_k(T)$ found above into (5.17) and thereby find the weak nonlinearity expansion for the 1-mode generating function for this system at the intermediate time T, $\mathscr{Z}_k(\lambda_k, T)$. First, we recall that $d = 2$ and we will use $J_k = |b_k|^2$ to rewrite (5.17) as

$$\mathscr{Z}_k(\lambda_k, T) = \left\langle \exp\left(\tilde{\lambda}_k |b_k(T)|^2\right)\right\rangle, \tag{6.25}$$

where for convenience we have introduced $\tilde{\lambda}_k = \lambda_k \left(\frac{L}{2\pi}\right)^2$. We will find an expression for $\mathscr{Z}_k(T)$, and use this to find the evolution equation for the generating function. In turn, this will help us obtain equations for the 1-mode PDF, for the moments of the wave amplitudes and the kinetic equation for the energy spectrum.

Substituting (6.17) into (6.25) we have

$$\mathscr{Z}_k(T) \approx \left\langle \exp\left(\tilde{\lambda}_k |b_k^{(0)} + \epsilon b_k^{(1)} + \epsilon^2 b_k^{(2)}|^2\right)\right\rangle$$
$$= \left\langle \exp\left(\tilde{\lambda}_k \left[|b_k^{(0)}|^2 + \epsilon(b_k^{(0)} b_k^{(1)*} + \text{c.c.}) + \epsilon^2 |b_k^{(1)}|^2 + \epsilon^2(b_k^{(0)} b_k^{(2)*} + \text{c.c.})\right]\right)\right\rangle, \tag{6.26}$$

where we kept terms up to order ϵ^2 only, because, as we will see later, these terms provide the leading contribution into the evolution of the averaged objects. We now do further expansion of (6.26) in ϵ and find

$$\mathscr{Z}_k(T) \approx \left\langle e^{\tilde{\lambda}_k J_k^{(0)}} \left\{ 1 + \tilde{\lambda}_k(b_k^{(0)} b_k^{(1)*} + \text{c.c.})\epsilon \right.\right.$$
$$\left.\left. + [\tilde{\lambda}_k(|b_k^{(1)}|^2 + b_k^{(2)} b_k^{(0)*} + \text{c.c.}) + \frac{\tilde{\lambda}_k^2}{2}(b_k^{(0)} b_k^{(1)*} + \text{c.c.})^2]\epsilon^2 \right\} \right\rangle, \tag{6.27}$$

where we have used $|b_k^{(0)}|^2 = J_k^{(0)}$.

We can immediately see that the $\epsilon^{(0)}$ order gives $\mathscr{Z}_k(\lambda_k, 0)$. To find closed expressions in the next orders in ϵ, we need to specify the statistical properties and to perform averaging.

6.4 Statistical Averaging

We assume that at $t = 0$ the wave field is of RPA type, i.e. the complex amplitudes $b_k^{(0)}$ (which are equal to $b_k(0) = a_k(0)$) have random and independent magnitudes and phase factors, the later being uniformly distributed on $\mathbb{S}^1$.

Because the amplitudes and the phase factors of $b_k^{(0)}$ are statistically independent, we can perform the phase averaging and the amplitude averaging independently from each other. For example, in first order of ϵ in (6.26) we have

$$\epsilon\left\langle e^{\tilde{\lambda}_k J_k^{(0)}} \left\langle \tilde{\lambda}_k b^{(0)} b^{(1)*} + \text{c.c.}\right\rangle_\phi \right\rangle_J, \tag{6.28}$$

where the subscripts ϕ and J denote the phase and the amplitude averaging respectively. As explained in the previous chapter, the phase averaging has to be done first. For the order ϵ terms, this involves finding $\langle b^{(0)} b^{(1)*} \rangle_\phi$ which we find using expression (6.19):

$$\left\langle b_k^{(0)*} b_k^{(1)} \right\rangle_\phi = \sum_{1,2} V_{12k} \delta_{12}^k \left\langle b_k^{(0)*} b_1^{(0)} b_2^{(0)} \right\rangle_\phi \Delta_T(\omega_{12}^k). \qquad (6.29)$$

We now remember the Wick's contraction rule (see end of Sect. 5.6) and note the odd number of $b^{(0)}$'s in the average on the RHS of (6.29). Thus, we see that the average is zero, and the contribution to $\mathscr{Z}(T)$ from order ϵ^1 is zero.

We continue with the terms of order ϵ^2 in (6.27) and obtain:

$$\mathscr{Z}_k(T) - \mathscr{Z}_k(0) = \epsilon^2 \left\langle e^{\tilde{\lambda}_k J_k^{(0)}} \tilde{\lambda}_k \right.$$

$$\left. \times \left\{ \left| b_k^{(1)} \right|^2 + \left(b_k^{(0)*} b_k^{(2)} + \text{c.c.} \right) + \frac{\tilde{\lambda}_k}{2} \left(\left[\left(b_k^{(0)} b_k^{(1)*} \right)^2 + \text{c.c.} \right] + 2 \left| b_k^{(0)} \right|^2 \left| b_k^{(1)} \right|^2 \right) \right\} \right\rangle. $$
$$(6.30)$$

where we took into account that the ϵ^0-order for the generating function gives $\mathscr{Z}(0)$. Now we need to find the phase averages of $\left| b_k^{(1)} \right|^2$, $b_k^{(0)*} b_k^{(2)}$ and $\left(b_k^{(0)} b_k^{(1)*} \right)^2$. For the first term we write

$$\left\langle \left| b_k^{(1)} \right|^2 \right\rangle_\phi = \sum_{1,2,3,4} V_{12k} V_{34k}^* \left\langle b_1^{(0)} b_2^{(0)} b_3^{(0)*} b_4^{(0)*} \right\rangle_\phi \delta_{12}^k \delta_{34}^k \Delta_T(\omega_{12}^k) \Delta_T^*(\omega_{34}^k). \quad (6.31)$$

Now we need to use the Wick's contraction rule of Sect. 5.6 and seek combinations of wavevectors that will lead to a nonzero result. We pair the wavevectors as follows:

- $\mathbf{k}_1 = \mathbf{k}_3$ and $\mathbf{k}_2 = \mathbf{k}_4$, and
- $\mathbf{k}_1 = \mathbf{k}_4$ and $\mathbf{k}_2 = \mathbf{k}_3$.
- Since the original variables $\psi(\mathbf{x})$ are real-valued, we have $b_k^* = b_{-k}$, so another option would be $\mathbf{k}_1 = -\mathbf{k}_2$ and $\mathbf{k}_3 = -\mathbf{k}_4$.

We can immediately group the first two cases as they lead to the same result due to the $1 \leftrightarrow 2$ symmetry in (6.31). Also, it is easy to see that the third situation gives zero, for both δ's will be zero since $\mathbf{k} \neq 0$. Thus, we only consider the first option and multiply it by 2. We write

$$\left\langle \left| b_k^{(1)} \right|^2 \right\rangle_\phi = 2 \sum_{1,2} |V_{12k}|^2 J_1^{(0)} J_2^{(0)} \delta_{12}^k \left| \Delta_T(\omega_{12}^k) \right|^2. \qquad (6.32)$$

For the second term in (6.30) we use (6.23) and write

$$\langle b_k^{(0)*} b_k^{(2)} \rangle_\phi$$

$$= -2 \sum_{1,2,3,4} \text{sign}(k_x k_{1x}) V_{12k} V_{341} \langle b_k^{(0)*} b_2^{(0)} b_3^{(0)} b_4^{(0)} \rangle_\phi \delta_{34}^1 \delta_{12}^k E(\omega_{34}^1, \omega_{12}^k). \quad (6.33)$$

Again, we look for the wavevector combinations which lead to a nonzero phase average:

- $\mathbf{k} = \mathbf{k}_3$ and $\mathbf{k}_2 = -\mathbf{k}_4$.
- Similar situation as the first, with $\mathbf{k}_3$ and $\mathbf{k}_4$ swapped (symmetry).
- Similar situation as the first, with $\mathbf{k}_2$ and $\mathbf{k}_3$ swapped.

We can rule out the third situation, because when $\mathbf{k} = \mathbf{k}_2$ we get $\delta_{12}^k = \delta(\mathbf{k}_1)$, which only gives a solution for $k_1 = 0$, but then $V_{12k} = 0$. We group the first two options, because of the $3 \leftrightarrow 4$ symmetry, and write

$$\langle b_k^{(0)*} b_k^{(2)} \rangle_\phi = -4 \sum_{1,2} \text{sign}(k_x k_{1x}) \, V_{12k} V_{k,-2,1} J_k^{(0)} J_2^{(0)} \delta_{12}^k \, E(\omega_k^{12}, \omega_{12}^k), \quad (6.34)$$

where

$$V_{k,-2,1} = V_{12k}^*$$

and

$$\omega_k^{12} = -\omega_{12}^k,$$

so we write

$$\langle b_k^{(0)*} b_k^{(2)} \rangle_\phi = -4 \sum_{1,2} \text{sign}(k_x k_{1x}) |V_{12k}|^2 J_k^{(0)} J_2^{(0)} \delta_{12}^k \, E(-\omega_{12}^k, \omega_{12}^k). \quad (6.35)$$

Now you are asked to consider term $\left(b_k^{(0)} b_k^{(1)*} \right)^2$ as formulated in the following exercise.

Exercise 6.3 *Substitute $b_k^{(1)}$ from (6.19) into term $\left(b_k^{(0)} b_k^{(1)*} \right)^2$ and show that averaging it over phases gives zero. For this, you will need to apply the Wick's contraction rule.*

Now, we will perform the amplitude averaging using (5.30) and (5.31). For the contribution to $\mathcal{L}_k(T) - \mathcal{L}_k(0)$ arising from (6.32) we write

$$\epsilon^2 \left\langle \left(\tilde{\lambda}_k + \tilde{\lambda}_k^2 J_k \right) e^{\tilde{\lambda}_k J_k} \left\langle \left| b_k^{(1)} \right|^2 \right\rangle_\phi \right\rangle_J$$

$$= 2\epsilon^2 \left\langle \left(\tilde{\lambda}_k + \tilde{\lambda}_k^2 J_k \right) e^{\tilde{\lambda}_k J_k} \sum_{1,2} |V_{12k}|^2 \delta_{12}^k |\Delta_T(\omega_{12}^k)|^2 J_1 J_2 \right\rangle_J, \quad (6.36)$$

where from now on we are going to omit the superscripts (0) for brevity. For the
contribution arising from (6.35) we have

$$
\epsilon^2 \tilde{\lambda}_k \left\langle e^{\tilde{\lambda}_k J_k} \left\langle b_k^{(0)*} b_k^{(2)} + \text{c.c.} \right\rangle_\phi \right\rangle_J
$$
$$
= -8\epsilon^2 \tilde{\lambda}_k \sum_{1,2} \text{sign}(k_x k_{1x}) |V_{12k}|^2 \delta_{12}^k \, \Re\big[E(-\omega_{12}^k, \omega_{12}^k)\big] \left\langle J_2 J_k e^{\tilde{\lambda}_k J_k} \right\rangle_J, \quad (6.37)
$$

where $\Re$ denotes "the real part of". In both (6.36) and (6.37), we can exclude the
cases $\mathbf{k} = \mathbf{k}_1$ and $\mathbf{k} = \mathbf{k}_2$ because otherwise one of the wavevectors in the interaction coefficient is zero which would make this coefficient zero. We can also
ignore the case $\mathbf{k}_1 = \mathbf{k}_2$ because, as we will see later, this point is not on the
resonant curve (for now roughly: the term $\big|\Delta_T(\omega_{12}^k)\big|^2$ will be negligible in this
case). Thus we assume $\mathbf{k} \neq \mathbf{k}_1 \neq \mathbf{k}_2$, in which case we can write

$$
\left\langle J_1 J_2 e^{\tilde{\lambda}_k J_k} \right\rangle_J = \langle J_1 \rangle \langle J_2 \rangle \left\langle e^{\tilde{\lambda}_k J_k} \right\rangle,
$$
$$
\left\langle J_1 J_2 J_k e^{\tilde{\lambda}_k J_k} \right\rangle_J = \langle J_1 \rangle \langle J_2 \rangle \left\langle J_k e^{\tilde{\lambda}_k J_k} \right\rangle,
$$
$$
\left\langle J_2 J_k e^{\tilde{\lambda}_k J_k} \right\rangle_J = \langle J_2 \rangle \left\langle J_k e^{\tilde{\lambda}_k J_k} \right\rangle.
$$

Furthermore, we know from (5.17) and (5.18) that

$$
\langle J_k \rangle = \left(\frac{2\pi}{L}\right)^2 n_k,
$$
$$
\left\langle e^{\tilde{\lambda}_k J_k} \right\rangle = \mathscr{Z}_k,
$$
$$
\left\langle J_k e^{\tilde{\lambda}_k J_k} \right\rangle = \frac{\partial}{\partial \tilde{\lambda}_k} \left\langle e^{\tilde{\lambda}_k J_k} \right\rangle = \frac{\partial}{\partial \tilde{\lambda}_k} \mathscr{Z}_k.
$$

We combine these with (6.37), (6.36) and substitute $\tilde{\lambda}_k = \lambda_k \left(\frac{L}{2\pi}\right)^2$; this gives:

$$
\mathscr{Z}_k(T) - \mathscr{Z}_k(0)
$$
$$
= \left(\frac{2\pi}{L}\right)^2 2\epsilon^2 \left(\lambda_k \mathscr{Z}_k + \lambda_k^2 \frac{\partial}{\partial \lambda_k} \mathscr{Z}_k\right) \sum_{1,2} |V_{12k}|^2 \delta_{12}^k \big|\Delta_T(\omega_{12}^k)\big|^2 n_1 n_2
$$
$$
- 8\left(\frac{2\pi}{L}\right)^2 \epsilon^2 \lambda_k \sum_{1,2} \text{sign}(k_x k_{1x}) |V_{12k}|^2 \delta_{12}^k \, \Re\big[E(-\omega_{12}^k, \omega_{12}^k)\big] n_2 \frac{\partial}{\partial \lambda_k} \mathscr{Z}_k.
$$
$$
(6.38)
$$

6.5 Large-Box and Weak-Nonlinearity Limits

It is very important that in WT the large-box limit is taken before the weak-nonlinearity limit. Physically, this means that there is a vast number of quasi-resonances

each of whom is as important as the exact wave resonances; see Remark 6.5.1 later in this section. The opposite limit when the weakly nonlinear limit is taken before the large box limit, as well as the finite-box effects in general, will be discussed in Chap. 10.

6.5.1 Taking $\mathbf{L} \to \infty$

We now take the large-box limit, $L \to \infty$, following the standard rules of correspondence between the sums and the integrals:

$$\sum_{1,2} \to \int d\mathbf{k}_1 d\mathbf{k}_2 \left(\frac{L}{2\pi}\right)^4,$$

$$\text{Kronecker-}\delta \to \left(\frac{2\pi}{L}\right)^2 \times \text{Dirac-}\delta,$$

and we write for (6.38):

$$
\begin{aligned}
\mathscr{L}_k(T) &- \mathscr{L}_k(0) \\
&= 2\epsilon^2 \left(\lambda_k \mathscr{L}_k + \lambda_k^2 \frac{\partial}{\partial \lambda_k} \mathscr{L}_k\right) \int d\mathbf{k}_1\, d\mathbf{k}_2 |V_{12k}|^2 \delta_{12}^k |\Delta_T(\omega_{12}^k)|^2 n_1 n_2 \\
&\quad - 8\epsilon^2 \lambda_k \int d\mathbf{k}_1\, d\mathbf{k}_2\, \text{sign}(k_x k_{1x}) |V_{12k}|^2 \delta_{12}^k \Re[E] n_2 \frac{\partial}{\partial \lambda_k} \mathscr{L}_k.
\end{aligned}
\tag{6.39}
$$

6.5.2 Taking $\epsilon \to 0$

Recall that we set

$$\frac{2\pi}{\omega_k} \ll T \ll \frac{2\pi}{\epsilon^2 \omega_k},$$

and, in particular, one can take

$$T \sim \frac{2\pi}{\epsilon \omega_k},$$

so that

$$\lim_{\epsilon \to 0} T = \infty.$$

Thus, in (6.39) we need to take $T \to \infty$ limit in functions $|\Delta_T|^2$ and $\Re[E]$. We use (6.20) and write ($x = \omega_{12}^k$):

$$\Delta_T(x) = \int_0^T e^{ixt}dt = \frac{e^{ixT} - 1}{ix}, \tag{6.40}$$

so that

$$|\Delta_T(x)|^2 = \frac{\left|e^{\frac{ixT}{2}} - e^{-\frac{ixT}{2}}\right|^2}{x^2} = \frac{4\sin^2\frac{xT}{2}}{x^2} \xrightarrow{T\to\infty} 2\pi T\delta(x). \tag{6.41}$$

Exercise 6.4 *Explain the final step in* (6.41) *and find a similar* $T \to \infty$ *expression for* (6.24) *as follows:*

- *Sketch a graph for* $|\Delta_T|^2$ *and note what happens as* $T\to\infty$.
- *Show that the area under the curve of* $|\Delta_T|^2$ *is* $2\pi T$, *and thus obtain* (6.41).
- *Now recall* (6.24). *Show that as* $T\to\infty$,

$$\Re[E(-x,x)] \to \pi T\delta(x). \tag{6.42}$$

Using these results, we obtain

$$\frac{\mathscr{L}_k(T) - \mathscr{L}_k(0)}{T}$$
$$= 2\epsilon^2\left(\lambda_k\mathscr{L}_k + \lambda_k^2\frac{\partial}{\partial\lambda_k}\mathscr{L}_k\right)\int d\mathbf{k}_1 d\mathbf{k}_2 |V_{12k}|^2\delta_{12}^k 2\pi\delta(\omega_{12}^k)n_1 n_2$$
$$- 8\epsilon^2\lambda_k\int d\mathbf{k}_1 d\mathbf{k}_2 \operatorname{sign}(k_x k_{1x})|V_{12k}|^2\delta_{12}^k\pi\delta(\omega_{12}^k)n_2\frac{\partial}{\partial\lambda_k}\mathscr{L}_k. \tag{6.43}$$

Because we have $T \ll \frac{2\pi}{\epsilon^2\omega_k}$, time T is small compared to the characteristic evolution time of the averaged quantities such as $\mathscr{L}_k$. Thus, one can replace the difference $\mathscr{L}_k(T) - \mathscr{L}_k(0)$ by its expression in terms of the time derivative as follows,

$$\frac{\mathscr{L}_k(T) - \mathscr{L}_k(0)}{T} \approx \dot{\mathscr{L}}_k. \tag{6.44}$$

Note however, that this step involve more than just a Taylor approximation for a slowly varying function, but also a filtering-out of rapid oscillations on the linear time-scale. Indeed, as will be discussed in Exercise 6.7, the instantaneous time derivative can become of the same order or even greater than the rate of change described by (6.44), but such rapid changes are oscillatory and they drop out from the long-time cumulative change; see Fig. 7.2.

Finally, we can write the evolution equation for the generating function,

$$\dot{\mathscr{L}}_k = \lambda_k\eta_k\mathscr{L}_k + \left(\lambda_k^2\eta_k - \lambda_k\gamma_k\right)\frac{\partial}{\partial\lambda_k}\mathscr{L}_k, \tag{6.45}$$

with

$$\eta_k = 4\pi\epsilon^2 \int d\mathbf{k}_1 d\mathbf{k}_2 |V_{12k}|^2 \delta_{12}^k \delta(\omega_{12}^k) n_1 n_2, \qquad (6.46)$$

and

$$\gamma_k = 8\pi\epsilon^2 \int d\mathbf{k}_1 d\mathbf{k}_2 \mathrm{sign}(k_x k_{1x}) |V_{12k}|^2 \delta_{12}^k \delta(\omega_{12}^k) n_2. \qquad (6.47)$$

Remark 6.5.1 Three-wave resonant interaction is contained in both η_k and γ_k, via

$$\delta_{12}^k \Rightarrow \mathbf{k} = \mathbf{k}_1 + \mathbf{k}_2, \qquad (6.48)$$

$$\delta(\omega_{12}^k) \Rightarrow \omega_k = \omega_1 + \omega_2. \qquad (6.49)$$

Note, however, that such resonances are broadened by nonlinearity, i.e. the Fourier transform of the interaction term with respect to time will have a finite width in the frequency space $\Gamma \sim \gamma_k$. Thus, interaction takes place not only among the waves which are in exact frequency resonance (6.48), (6.49) but also among quasi-resonant waves with

$$|\omega(\mathbf{k}) - \omega(\mathbf{k}_1) - \omega(\mathbf{k}_2)| \lesssim \Gamma. \qquad (6.50)$$

Moreover, since the large box limit was taken before the weakly nonlinear limit, we see that the number of such quasi-resonances is vast (infinite in the infinite-box limit), and the exact resonances in this case do not play any special role.

6.6   The PDF

We recall from Sect. 5.4 that we can find the PDF via the inverse Laplace transform of the generating function,

$$\mathscr{P}(s) = \frac{1}{2\pi i} \int\limits_{\zeta+i\infty}^{\zeta+i\infty} e^{-\lambda s} \mathscr{Z}(\lambda) \, d\lambda,$$

where $\zeta \in \mathbb{R}^+$ is chosen in such a way that the integration contour is to the right of all singularities of $\mathscr{Z}(\lambda)$ in the complex λ-plane. Applying the inverse Laplace transform to (6.45), we have

$$\dot{\mathscr{P}}_k + \frac{\partial}{\partial s_k} \mathscr{F}_k = 0, \qquad (6.51)$$

where $\mathscr{F}_k$ is a flux in the wave-amplitude space:

$$\mathscr{F}_k = -s_k \left(\gamma_k \mathscr{P}_k + \eta_k \frac{\partial}{\partial s_k} \mathscr{P}_k \right). \tag{6.52}$$

The fact that (6.51) has a continuity equation form is natural, since it ensures the probability normalization condition

$$\int_0^\infty \mathscr{P}_k ds_k = 1. \tag{6.53}$$

Exercise 6.5 *Derive (6.51) and (6.52) by applying the inverse Laplace transform to (6.45).*

6.7 Kinetic Equation

Finally, let us find the kinetic equation for the wave spectrum

$$n_k = \left(\frac{L}{2\pi}\right)^2 \langle J_k \rangle = \left(\frac{L}{2\pi}\right)^2 \int_0^\infty s_k \mathscr{P} ds_k.$$

Using (6.51) and (6.52), we have

$$\begin{aligned} \dot{n}_k &= -\int_0^\infty s_k \frac{\partial}{\partial s_k} \mathscr{F}_k ds_k = \int_0^\infty \mathscr{F}_k ds_k \\ &= -\int_0^\infty \left(s_k \gamma_k \mathscr{P}_k + s_k \eta_k \frac{\partial}{\partial s_k} \mathscr{P}_k \right) ds_k. \end{aligned} \tag{6.54}$$

Integrating, the second term by parts and using (6.53), we get the *kinetic equation:*

$$\dot{n}_k = \eta_k - \gamma_k n_k. \tag{6.55}$$

Exercise 6.6 *Now multiply (6.51) by s_k^p (with $p \in \mathbb{N}$) and integrate over s_k to show that the higher-order moments $M_k^{(p)} = \left(\frac{L}{2\pi}\right)^{2p} \langle J_k^p \rangle$ satisfy the following equations,*

$$\dot{M}_k^{(p)} = -p\gamma_k M_k^{(p)} + p^2 \eta_k M_k^{(p-1)}. \tag{6.56}$$

For $p = 1$, this gives the kinetic equation (6.55).

Substituting η_k and γ_k from (6.46) and (6.47), we rewrite the kinetic equation in the following form,

$$\dot{n}_k = 4\pi\epsilon^2 \int d\mathbf{k}_1 d\mathbf{k}_2 |V_{12k}|^2 \delta_{12}^k \delta(\omega_{12}^k)$$

$$\times [n_1 n_2 - n_k n_1 \operatorname{sign}(k_x k_{2x}) - n_k n_2 \operatorname{sign}(k_x k_{1x})]. \qquad (6.57)$$

Similar equations can be derived for many other systems which are described by real-valued amplitudes in the x-space. As we will see later, it is also generalizable to the case of waves with complex-valued amplitudes.

Exercise 6.7 *At this point, let us try to understand better what we have done, and rewind back to the original dynamical equation in form (6.15), i.e.*

$$i\dot{b}_k = \epsilon \operatorname{sign}(k_x) \sum_{1,2} V_{12k} b_1 b_2 \delta_{12}^k e^{i\omega_{12}^k t}.$$

Multiplying this equation by b_k^* and adding the result to its complex conjugate and averaging, we have

$$i\dot{n}_k = \epsilon \operatorname{sign}(k_x) \sum_{1,2} V_{12k} \langle b_1 b_2 b_k^* \rangle \delta_{12}^k e^{i\omega_{12}^k t} + cc.$$

Now, assume as before that all b_k's have random phases at $t = 0$. Then the RHS of the above equation is zero, i.e. the rate of change of the spectrum is zero, in apparent contradiction with the kinetic equation we obtained before. How to resolve this "paradox"?

6.7.1 Symmetrical Form of the Kinetic Equation

Exercise 6.8 *Show that if field $\psi(\mathbf{x})$ is real-valued then its Fourier transform satisfies $\hat{\psi}_{-k} = \hat{\psi}_k^*$. This implies $n_{-k} = n_k$.*

Let us assume that the original wave variable $\psi(\mathbf{x})$ is real. Because in this case $\hat{\psi}_{-k} = \hat{\psi}_k^*$ and $n_{-k} = n_k$, we only need to consider half of the k-space. Let us consider the "positive" k-space, i.e. $k_x, k_{1x}, k_{2x} \geq 0$, and call it k^+-space. We can now rewrite the kinetic equation (6.57) in a symmetric form,

$$\dot{n}_k = \int (\mathscr{R}_{12k} - \mathscr{R}_{k12} - \mathscr{R}_{2k1}) d\mathbf{k}_1 d\mathbf{k}_2, \qquad (6.58)$$

where the integration is over the k^+-space (in both $\mathbf{k}_1$ and $\mathbf{k}_2$) and

$$\mathscr{R}_{12k} = 2\pi\epsilon^2 |V_{12k}|^2 \delta_{12}^k \delta(\omega_{12}^k)(n_1 n_2 - n_2 n_k - n_k n_1). \qquad (6.59)$$

This form of kinetic equation is somewhat more general compared to (6.57), as we will see in the next section.

6.8 Generalization to Complex Wavefields

6.8.1 Hamiltonian Wave Equations

Often, it is useful to start with a Hamiltonian form or wave equation, especially for the cases where the wave variable is not real in the x-space, or when the wave variable is real but the order of the wave equation in t is greater than one (e.g. for sound the wave equation is second-order in t). For example, for three-wave systems, this has been done for the capillary (surface tension) waves and sound waves.[3]

For three-wave systems, the (diagonalized) Hamiltonian wave equations look as follows,

$$i\dot{a}_k = \frac{\delta H}{\delta a_k^*},\qquad(6.60)$$

with

$$H = H_2 + H_3 + \cdots,\qquad(6.61)$$

$$H_2 = \sum_k \omega_k |a_k|^2,\qquad(6.62)$$

$$H_3 = \sum_{1,2,3}\left(V_{12}^3 a_1 a_2 a_3^* \delta_{12}^3 + \text{c.c.}\right) + \left(U_{123}\delta_{123}a_1 a_2 a_3 + \text{c.c.}\right),\qquad(6.63)$$

with the second term in (6.63) completely symmetric with respect to the three indices.

6.8.1.1 Rewriting Petviashvilli in Hamiltonian Form

Here we will show that our master example, the Petviashvili equation (6.11), can also be written in the Hamiltonian form (6.60) and (6.61), but for that we must use only half of the $\mathbf{k}$-space, i.e. the k^+-space introduced above. From (6.61) we have:

---

[3] It should also be applicable to MHD Alfvén waves and inertial waves, although this has not been done yet. In these particular applications WT was derived in terms of the other convenient variables, Elsässer variables [8] and the helical decomposition respectively. In both cases, Hamiltonian variables can be introduced via so-called Clebsch-Weber variables, see e.g. [9–12] for MHD.

$$i\dot{a}_k = \frac{\delta H}{\delta a_k^*} = \omega_k a_k + \text{sign}(k_x) \sum_{1,2} V_{12k} a_1 a_2 \delta_{12}^k$$

$$= \omega_k a_k + \sum_{k_{1x}, k_{2x} > 0} (V_{12k} a_1 a_2 \delta_{12}^k + V_{-12k} a_1^* a_2 \delta_2^{k1}$$

$$+ V_{1-2k} a_1 a_2^* \delta_1^{k2} + V_{-1-2k} a_1^* a_2^* \delta_{k12}), \tag{6.64}$$

where the last term is actually zero because δ_{k12} is always zero if $k_x > 0$ and k_{1x}, $k_{2x} \geq 0$. This means that in this case that there will be no U-terms in (6.63) if the phase space is k^+.

We find H_3 by functional integration over a_k^*:

$$H_3 = \sum_{k,1,2} (V_{123} a_1 a_2 a_3^* \delta_{12}^3 + V_{-123} a_1^* a_2 a_3^* \delta_2^{31}). \tag{6.65}$$

Note that the second term in (6.65) arises from the two symmetric terms in (6.64) and integrating over two variables a_j^*. This term is also the complex conjugate of the first term in the summation because $V_{-123} = V_{123}^*$. The latter condition is obviously true for Petviashvili, but it must also be true for any Hamiltonian system with the waveaction variables a_k having real (inverse-Fourier) images in the x-space. This makes Hamiltonian (6.64) real. Thus, redefining $V_{123} \rightarrow V_{12}^3$, we arrive at a system corresponding to (6.60)–(6.63) without U-terms in (6.63). We emphasize again that now our phase space is k^+ and all summations are over k_x, k_{1x}, $k_{2x} \geq 0$.

It is not hard to see that the resulting Hamiltonian is nothing but the integral (6.3) when written in the physical space representation.

6.8.1.2 Deriving WT from Hamiltonian Equations

Now consider the general case where the original wave variable in the x-space could be complex. For the complex wave fields, the system is described by (6.60)–(6.63) with respective sums taken over the the full k-space.

The steps of WT derivations remain as before, except the particular expressions for $b^{(1)}$ and $b^{(2)}$, leading to differences in $\langle |b^{(1)}|^2 \rangle_\phi$ and $\langle (b^{(0)*} b^{(2)}) \rangle_\phi$, which in turn lead to somewhat different expressions for η_k and γ_k. Apart from that the equations for the generating function and for the PDF will remain the same.

We leave these derivations for the following two exercises. First of all, let us get rid of an "unnecessary" term:

Exercise 6.9 *Assume that $\omega_k + \omega_1 + \omega_2$ can never become zero, and show how to eliminate the terms with U_{123} from the interaction Hamiltonian (6.63) using a quasi-identity canonical transformation of form*

$$b_k = a_k + \sum_{1,2} W_{12k} a_1^* a_2^* \delta_{12k}, \qquad (6.66)$$

i.e. find the coefficients W_{12k} by ignoring the cubic nonlinear terms in the required equation. How to prove that this transformation is canonical?

Now to the WT derivations:

Exercise 6.10 *Follow the scheme of WT derivations starting with the Hamiltonian equations (6.60)–(6.63) (with respective sums taken over the full **k**-space). Show that this leads to the same equations for the one-point generating function (6.45), the one-point PDF (6.53) and (6.53), and the kinetic equation for the wave spectrum (6.55), but with different expressions for η_k and γ_k:*

$$\eta_k = 4\pi\epsilon^2 \int \left(|V_{12}^k|^2 \delta_{12}^k \delta(\omega_{12}^k) + 2|V_{k1}^2|^2 \delta_{k1}^2 \delta(\omega_{k1}^2) \right) n_1 n_2 \, d\mathbf{k}_1 d\mathbf{k}_2, \qquad (6.67)$$

$$\gamma_k = 8\pi\epsilon^2 \int \left(|V_{12}^k|^2 \delta_{12}^k \delta(\omega_{12}^k) n_2 + |V_{k1}^2|^2 \delta_{k1}^2 \delta(\omega_{k1}^2)(n_1 - n_2) \right) d\mathbf{k}_1 d\mathbf{k}_2. \qquad (6.68)$$

Substituting (6.67) and (6.68) into (6.55) we get the following form of the kinetic equation,

$$\dot{n}_k = \int (\mathcal{R}_{12k} - \mathcal{R}_{k12} - \mathcal{R}_{2k1}) d\mathbf{k}_1 d\mathbf{k}_2, \qquad (6.69)$$

where the integration is over the entire **k**-space (in both $\mathbf{k}_1$ and $\mathbf{k}_2$) and

$$\mathcal{R}_{12k} = 2\pi\epsilon^2 |V_{12k}|^2 \delta_{12}^k \delta(\omega_{12}^k)(n_1 n_2 - n_2 n_k - n_k n_1). \qquad (6.70)$$

Note that the (6.69) is the same as the (6.58) but now the integration is over the entire **k**-space rather than its half as in (6.58).

6.9 Four-Wave and Higher-Order Systems

In the previous sections, we only considered the three-wave systems. However, with little extra work we can extend the analysis to the four-wave and the higher-order systems.

6.9.1 Four-Wave Systems

Let us start with the four-wave systems assuming that they are described by the following Hamiltonian

$$H = \sum_k \omega_k a_k a_k^* + \frac{1}{2} \sum_{1,2,3,4} W_{34}^{12} \delta_{34}^{12} a_1 a_2 a_3^* a_4^*, \qquad (6.71)$$

with W_{34}^{12} being a four-wave interaction coefficient. This type of Hamiltonian arises either when the nonlinearity of the original wave equation is cubic, like in the NLS equation (5.1) (in which case $W_{34}^{12} = 1$), or when the nonlinearity is quadratic in the natural variables (e.g. surface elevation and velocity potential for the water waves) but the dispersion relation does not permit three-wave resonances. In the later case, one can perform a nonlinear change of variables and bring the Hamiltonian to the form (6.71) (i.e. without a cubic Hamiltonian and, respectively, without quadratic terms in the wave equation). This change of variables is very well explained in book [13] (Appendix A3) and we will not reproduce it here. An important example here is the set of deep water gravity waves, and in the Appendix to this Part we provide, for reference, an expression for the Hamiltonian interaction coefficient obtained by such a canonical transformation [14, 15].

Similarly, if the lowest order of non-empty wave resonances is N then one can get rid of all terms of orders $\leq N - 1$ in the Hamiltonian by a nonlinear change of variables. When resonances are absent in all orders, constructing this change of variables amounts to integrating the system [16].

The k-space wave equation corresponding to the Hamiltonian (6.71) is

$$i\dot{a}_k = \tilde{\omega}_k a_k + \sum_{k_1,k_2 \neq k_3} W_{3k}^{12} \delta_{3k}^{12} a_1 a_2 a_3^*, \qquad (6.72)$$

where

$$\tilde{\omega}_k = \omega_k + \omega_{NL}, \quad \text{where} \quad \omega_{NL} = 2 \sum_{k_1} W_{1k}^{1k} |a_1|^2 \qquad (6.73)$$

is a nonlinear frequency shift. Note that we extracted the diagonal terms from the sum in (6.72), and wrote these terms separately as $\omega_{NL} a_k$. These terms constitute the leading nonlinear effect on the wave dynamics, but they do not change the wave intensity and, therefore, do not lead to transfers of energy between different k-modes.

Exercise 6.11 *Consider the phase of the amplitude* a_k *defined as* $\varphi_k = \Im(\ln a_k)$. *Show that in the leading order of small nonlinearity*

$$\langle \dot{\varphi}_k \rangle_\phi = -\tilde{\omega}_k = -\omega_k - \omega_{NL}. \qquad (6.74)$$

Let us now define the interaction representation variables b_k as

$$b_k = e^{i\omega_k t + i \int \omega_{NL} dt} a_k / \epsilon. \qquad (6.75)$$

Note the difference with the way we have defined the interaction representation before for the three-wave case: now we remove fast oscillations including the ones due to the nonlinear frequency correction. This step is very important. Should we not have done this, the subsequent derivation procedure would not be self-consistent.

Exercise 6.12 *Define the interaction representation variables b_k as before, without ω_{NL} and try to follow the WT derivation steps. At which point do you get inconsistency? What physical interpretation can you suggest?*

In terms of b_k we have

$$\dot{ib}_k = \epsilon^2 \sum_{k_1,k_2 \neq k_3} W_{3k}^{12} \delta_{3k}^{12} e^{i \int \omega_{12}^{3k} dt} b_1 b_2 b_3^*, \tag{6.76}$$

where $\omega_{12}^{3k} = \omega_k + \omega_{NL\,k} + \omega_3 + \omega_{NL\,3} - \omega_1 - \omega_{NL\,1} - \omega_2 - \omega_{NL\,2}$. (Note that for NLS the nonlinear frequency would drop out from this expression because it is **k**-independent).

As before, the next step is to consider solution $b_k(T)$ at an intermediate (between linear and nonlinear) time $2\pi/\omega_k \ll T \ll 2\pi/\epsilon^4 \omega_k$ (we do not have to demand $\omega_{NL} T \gg 1!$). We seek this solution as a series $b_k(T) = b_k^{(0)} + \epsilon^2 b_k^{(1)} + \epsilon^4 b_k^{(2)} + \cdots$ obtained by recursive substitution into (6.76). This gives $b_k^{(0)} = b_k|_{T=0}$ and

$$b_k^{(1)}(T) = -i \sum_{k_1,k_2 \neq k_3} W_{3k}^{12} \delta_{3k}^{12} b_1^{(0)} b_2^{(0)} b_3^{(0)*} \Delta_T(\omega_{12}^{3k}). \tag{6.77}$$

where function Δ_T is defined in (6.40). For the second iteration we get

$$b_k^{(2)}(T) = \sum_{1 \neq 3,4;5 \neq 6,2} \left(W_{15}^{34} W_{62}^{k5} \delta_{15}^{34} \delta_{62}^{k5} b_1^{(0)} b_6^{(0)} b_2^{(0)} b_3^{(0)*} b_4^{(0)*} E(\omega_{15}^{34}, \omega_{62}^{k5}) \right.$$
$$\left. -2 W_{34}^{16} W_{62}^{k5} \delta_{34}^{16} \delta_{62}^{k5} b_1^{(0)*} b_5^{(0)*} b_3^{(0)} b_4^{(0)} b_2^{(0)} E(\omega_{34}^{16}, \omega_{62}^{k5}) \right), \tag{6.78}$$

where $E(x,y) = \int_0^T \Delta_t(x) e^{iyt} dt$ (like in (6.24)).

The next step, again like before, is to find the phase averages of $|b_k^{(1)}|^2$ and of $b_k^{(2)} b_k^{(0)*}$, but now using the new expressions (6.77) and (6.78). This is to be followed by substitution into the expression (6.30), amplitude averaging and taking the large-box and the small-ϵ limits. These steps are very similar to what we did before and we leave them for the reader as an exercise:

Exercise 6.13 *Follow the steps outlined in the previous paragraph. Show that this leads to the same equations for the one-point generating function (6.45), the one-point PDF ((6.53) and (6.52)), and the kinetic equation (6.55), but now with new expressions for η_k and γ_k:*

$$\eta_k = 4\pi \epsilon^4 \int |W_{12}^{k3}|^2 \delta_{12}^{k3} \delta(\omega_{12}^{k3}) n_1 n_2 n_3 \, d\mathbf{k_1} d\mathbf{k_2} d\mathbf{k_3}, \tag{6.79}$$

$$\gamma_k = 4\pi\epsilon^4 \int |W_{12}^{k3}|^2 \delta_{12}^{k3} \delta(\omega_{12}^{k3})[n_3(n_1 + n_2) - n_1 n_2]\, d\mathbf{k}_1 d\mathbf{k}_2 d\mathbf{k}_3. \tag{6.80}$$

With these new expressions for η_k and γ_k, the kinetic equation (6.55) becomes

$$\dot{n}_k = 4\pi\epsilon^4 \int |W_{12}^{k3}|^2 \delta_{12}^{k3} \delta(\omega_{12}^{k3}) n_1 n_2 n_3 n_k \left[\frac{1}{n_k} + \frac{1}{n_3} - \frac{1}{n_1} - \frac{1}{n_2}\right] d\mathbf{k}_1 d\mathbf{k}_2 d\mathbf{k}_3. \tag{6.81}$$

6.9.2 Systems with Higher-Order Wave Resonances

If the reader has followed carefully the previous sections, he or she has enough knowledge and experience for dealing with systems characterized by wave resonances with orders higher than four. Basically, most work has already been done before, and the only thing to be done in each particular case is to calculate the first and the second order contributions in the interaction representation variables, $b_k^{(1)}$ and of $b_k^{(2)}$, and to find the phase averages of $|b_k^{(1)}|^2$ and of $b_k^{(2)} b_k^{(0)*}$. At times, things can get quite lengthy, particularly when the lower orders of nonlinearity need to be removed by nonlinear canonical transformations due to a great number of terms to be computed in these transformations.

Higher-order wave resonances typically arise in one-dimensional systems, because restricting the wavenumber configurations to 1D often leads to losses of the three-wave and four-wave interactions. Below we will consider examples of such 1D wave systems.

6.9.2.1 A Five-Wave System: 1D Water Gravity Waves

A very good example of five-wave turbulence is the 1D system of surface gravity waves on deep water [17]. In 2D, the lowest order resonant process for the the deep water gravity waves is four. However, the four-wave interaction coefficient turns out to become identically equal to zero for 1D configurations, i.e. when all four wavevectors are in the same direction. Thus, the leading process in 1D is five-wave, and the interaction Hamiltonians with the orders less than five are to be removed via a rather involved change of variables, as it was done in [17]. As a result, one gets the following Hamiltonian [17],

$$H = \sum_k \omega_k a_k a_k^* + \frac{1}{12} \sum_{1,2,3,4,5} W_{45}^{123} \delta_{123}^{45} a_1^* a_2^* a_3^* a_4 a_5 + c.c., \tag{6.82}$$

where (this is an amazing result of such a lengthy derivation!) the five-wave interaction coefficient looks remarkably simple:

$$W_{45}^{123} = \frac{2}{g^{1/4}\pi^{3/2}} \left| \frac{k_1 k_2 k_3}{k_4 k_5} \right|^{1/4} \frac{k_1 k_2 k_3 k_4 k_5}{\max(k_1, k_2, k_3)}, \tag{6.83}$$

where g is the gravity constant. The corresponding equation for the wave spectrum is [17]:

$$\dot{n}_k = \pi\epsilon^6 \int \left(\frac{1}{3} f_{k4}^{123} - \frac{1}{2} f_{34}^{k12} \right) d\mathbf{k}_1 d\mathbf{k}_2 d\mathbf{k}_3 d\mathbf{k}_4, \tag{6.84}$$

where

$$f_{45}^{123} = |W_{45}^{123}|^2 \delta_{45}^{123} \delta(\omega_{45}^{123}) n_1 n_2 n_3 n_4 n_5 \left[\frac{1}{n_4} + \frac{1}{n_5} - \frac{1}{n_1} - \frac{1}{n_2} - \frac{1}{n_3} \right]. \tag{6.85}$$

Starting with Hamiltonian (6.82), you could follow the WT derivation steps. As we said, this amounts to finding new expressions for $b_k^{(1)}$ and $b_k^{(2)}$, leading to new $|b_k^{(1)}|^2$ and $b_k^{(2)} b_k^{(0)*}$, and with the remaining steps being identical to what we did for the three-wave system. This will result in the same equations for the one-point generating function (6.45), the one-point PDF ((6.53) and (6.52)), but with new expressions for η_k and γ_k. This exercise could be a good thing to do, e.g. as a part of a course project, since it has not been done before in published literature (e.g. in [17]).

However, here you are asked to write these results without derivation. Namely:

Exercise 6.14 *Based on the known kinetic equation (6.84), and keeping in mind the expected form (6.55), please write out the expressions for the coefficients η_k and γ_k corresponding to the five-wave system (6.82). (This amounts to obtaining equations for the one-point generating function and for the one-point PDF).*

6.9.2.2 Six-Wave 1D Wave Systems

Let us consider one-dimensional six-wave systems described by the following Hamiltonian,

$$H = \sum_k \omega_k a_k a_k^* + \frac{1}{3} \sum_{1,2,3,4,5,6} W_{456}^{123} \delta_{123}^{456} a_1^* a_2^* a_3^* a_4 a_5 a_6. \tag{6.86}$$

Typically, systems like this arise in $U(1)$-symmetric equations which allow only even-order Hamiltoninans and when the dispersion relation $\omega \sim k^\alpha$ has index $\alpha > 1$,—because there is no four-wave resonances in 1D for $\alpha < 1$ (see Exercise 3.1). Often for the four-wave and the five-wave systems, bringing the Hamiltonian to the form (6.86) requires removing the lower-order interaction terms by a non-linear canonical transformation.

Examples include it 1D optical turbulence described by equation [18, 19]

$$i\frac{\partial\psi}{\partial t} = -\frac{\partial^2\psi}{\partial x^2} - \frac{1}{2\alpha}\psi|\psi|^2 - \frac{1}{2\alpha^2}\psi\frac{\partial^2|\psi|^2}{\partial x^2},\tag{6.87}$$

where $\psi(x, t)$ is the complex amplitude of the input beam propagating along "time axis" $\hat{t}$. Here, x is the coordinate across the beam, and $\alpha \gg 1$ is a real positive parameter. Ignoring the sub-leading nonlinear term $\frac{1}{2\alpha^2}\psi\frac{\partial^2|\psi|^2}{\partial x^2}$ one recovers the NLS equation. Since we are now in 1D, the NLS equation is integrable, which implies that there is no resonant wave interactions at any order,[4] which in turn means that there is no cascades over scales if the nonlinearity is weak. For larger nonlinearities, the system's dynamics is described by a set of solitons which can freely pass through each other and which preserve their identity infinitely long if there is no forcing or dissipation in the system. Adding forcing or/and dissipation would of course break integrability and would presumably create some transfer over the scales. However, such solitonic turbulence in forced/dissipated systems is largely unstudied subject and we will not discuss it. Instead, we will concentrate on the case when the integrability is broken by modifications of the nonlinear interactions, i.e. by the second nonlinear term in (6.87). This term makes six-wave resonant interactions possible and, after a nonlinear canonical transformation, one gets a Hamiltonian of type (6.86) with a constant six-wave interaction coefficient [18, 19]

$$W^{123}_{456} \equiv \frac{3}{16\alpha^3}.\tag{6.88}$$

Because in the leading order the 1D optical wave turbulence is integrable, one can expect that solitons should be seen along with random weakly nonlinear waves. Such coexisting coherent and incoherent components and their mutual interactions and transformations one into another are the typical manifestations of the WT cycle. Coexisting waves and solitons and their mutual interactions were indeed observed in numerical simulations reported in [18, 19]. See further discussion of this system in Sect. 15.10.

Another example of a six-wave 1D turbulence arises in quantum turbulence: it is a system of interacting it Kelvin waves propagating along quantized vortex lines. Like the optical WT, the Kelvin wave system is integrable in the leading order given by Local Induction Approximation (LIA).[5] The simplest models that allow Kelvin wave cascades are truncated-LIA model [21] and Local Nonlinear Model [22]; in both cases the expression for the interaction coefficient is:

---

[4] I.e. either the resonant conditions on the wavenumbers and frequencies do not have non-trivial solutions, or the interaction coefficients are zero in corresponding orders.

[5] In fact, the Local Induction Approximation model becomes 1D NLS after so-called Hasimoto transformation [20].

$$W_{456}^{123} \sim k_1 k_2 k_3 k_4 k_5 k_6.$$

The kinetic equation for the six-wave system (6.86) is:

$$\dot{n}_k = 12\pi\epsilon^8 \int |W_{456}^{123}|^2 \, \delta_{456}^{123} \, \delta(\omega_{456}^{123})$$

$$\times \, n_k n_1 n_2 n_3 n_4 n_5 \left(\frac{1}{n_k} + \frac{1}{n_1} + \frac{1}{n_2} - \frac{1}{n_3} - \frac{1}{n_4} - \frac{1}{n_5} \right) dk_1 dk_2 dk_3 dk_4 dk_5. \quad (6.89)$$

Exercise 6.15 *Derive WT for the six-wave Hamiltonian (6.86), particularly the coefficients η_k and γ_k for the evolution equations for the one-mode generating function, (6.45), and the one-mode PDF, (6.53) and (6.52). Use the result to derive the kinetic equation (6.89).*

6.10 Evolution of Multi-Mode Statistics

In this section, we will extend our WT description to multi-mode statistical objects, particularly the N-point generating function and the N-point PDF. This extension is useful if one is interested in checking validity of the underlying assumptions in WT, in particular, whether RPA property of wavefields can survive over the nonlinear evolution time. It is also useful for studying the question whether inter-mode amplitude correlations can develop, and if so, what is their role in the WT life cycle. Finally, the multi-mode description can hopefully provide a bridge to the areas of dynamical systems and statistical mechanics and to developing WT into a valid and rigorous part of mathematics. These issues we listed in the grand "mahayana" scheme of WT depicted in Fig. 6.1.

On the other hand, if the reader is not interested in these "esoteric" issues being happy with the one-mode descriptions and with taking RPA for granted (which is totally sufficient for most existing applications) he or she can safely skip this section.

The derivation for the multi-mode statistical objects will inevitably become longer, but fundamentally most derivation steps are very similar to what we have done for the one-mode objects. Namely, one has to take the wave equation in the interaction representation and find the solution for interaction representation amplitudes at an intermediate (between the linear and the nonlinear) time T in the form of expansion in the nonlinearity parameter ϵ. These steps are identical to what we have done before, and they result in the same expressions for $b_k^{(1)}$ and $b_k^{(2)}$. After that, one has to use the resulting $b_k^{(1)}$ and $b_k^{(2)}$ in finding an expression for the N-mode generating function $\mathcal{Z}^{(N)}\{\lambda, \mu\}(T)$. Let us start with this step.

6.10.1 Weak Nonlinearity Expansion of the Generating Function

Let us first obtain an asymptotic weak-nonlinearity expansion for the generating function $\mathscr{L}^{(N)}\{\lambda,\mu\}$ defined for the interaction representation variable via expression (5.15). Taylor expanding the amplitude and the phase parts in $\mathscr{L}^{(N)}$ we have,

$$e^{\tilde{\lambda}_l J_l} = e^{\tilde{\lambda}_l |b_l^{(0)} + \epsilon b_l^{(1)} + \epsilon^2 b_l^{(2)}|^2} = e^{\tilde{\lambda}_l J_l^{(0)}}(1 + \epsilon \alpha_{1l} + \epsilon^2 \alpha_{2l}), \tag{6.90}$$

and

$$\phi_l^{\mu_l} = \left(\frac{b_l^{(0)} + \epsilon b_l^{(1)} + \epsilon^2 b_l^{(2)}}{b_l^{(0)*} + \epsilon b_l^{(1)*} + \epsilon^2 b_l^{(2)*}}\right)^{\frac{\mu_l}{2}} = \phi_l^{(0)\mu_l}(1 + \epsilon \beta_{1l} + \epsilon^2 \beta_{2l}), \tag{6.91}$$

where, as before, we have neglected the ϵ^3—and the higher orders and where

$$\alpha_{1l} = \tilde{\lambda}_l(b_l^{(1)} b_l^{(0)*} + b_l^{(1)*} b_l^{(0)}), \tag{6.92}$$

$$\alpha_{2l} = (\tilde{\lambda}_l + \tilde{\lambda}_l^2 J_l^{(0)})|b_l^{(1)}|^2 + \tilde{\lambda}_l(b_l^{(2)} b_l^{(0)*} + b_l^{(2)*} b_l^{(0)})$$
$$+ \frac{\tilde{\lambda}_l^2}{2}[(b_l^{(1)} b_l^{(0)*})^2 + (b_l^{(1)*} b_l^{(0)})^2], \tag{6.93}$$

$$\beta_{1l} = \frac{\mu_l}{2J_l^{(0)}}(b_l^{(1)} b_l^{(0)*} - b_l^{(1)*} b_l^{(0)}), \tag{6.94}$$

$$\beta_{2l} = \frac{\mu_l}{2J_l^{(0)}}(b_l^{(0)*} b_l^{(2)} - b_l^{(2)*} b_l^{(0)})$$
$$+ \frac{\mu_l}{4}\left[\left(\frac{\mu_l}{2} - 1\right)\left(\frac{b_l^{(1)}}{b_l^{(0)}}\right)^2 + \left(\frac{\mu_l}{2} + 1\right)\left(\frac{b_l^{(1)*}}{b_l^{(0)*}}\right)^2\right] - \frac{\mu_l^2 |b_l^{(1)}|^2}{4J_l^{(0)}}. \tag{6.95}$$

Substituting (6.90) and (6.91) into the expression (5.15), we have

$$\mathscr{L}^{(N)}\{\lambda,\mu,T\} = \mathscr{L}^{(N)}\{\lambda,\mu,0\} + X\{\lambda,\mu,T\} + X^*\{\lambda,-\mu,T\} \tag{6.96}$$

with

$$X\{\lambda,\mu,T\} = (2\pi)^{2N}\left\langle \prod_{l \in B_N} e^{\tilde{\lambda}_l J_l^{(0)}}[\epsilon G_1 + \epsilon^2(G_2 + G_3 + G_4 + G_5)]\right\rangle_J, \tag{6.97}$$

where

$$G_1 = \left\langle \prod_{l \in B_N} \phi_l^{(0)\mu_l} \sum_j \left(\tilde{\lambda}_j + \frac{\mu_j}{2J_j^{(0)}}\right) b_j^{(1)*} b_j^{(0)}\right\rangle_\phi, \tag{6.98}$$

$$G_2 = \frac{1}{2} \left\langle \prod_{l \in B_N} \phi_l^{(0)\mu_l} \sum_j \left(\tilde{\lambda}_j + \tilde{\lambda}_j^2 J_j^{(0)} - \frac{\mu_j^2}{2J_j^{(0)}} \right) |b_j^{(1)}|^2 \right\rangle_\phi , \qquad (6.99)$$

$$G_3 = \left\langle \prod_{l \in B_N} \phi_l^{(0)\mu_l} \sum_j \left(\tilde{\lambda}_j + \frac{\mu_j}{2J_j^{(0)}} \right) b_j^{(2)} b_j^{(0)*} \right\rangle_\phi , \qquad (6.100)$$

$$G_4 = \left\langle \prod_{l \in B_N} \phi_l^{(0)\mu_l} \sum_j \left[\frac{\tilde{\lambda}_j^2}{2} + \frac{\mu_j}{4J_j^{(0)2}} \left(\frac{\mu_j}{2} - 1 \right) + \frac{\tilde{\lambda}_j \mu_j}{2J_j^{(0)}} \right] (b_j^{(1)} b_j^{(0)*})^2 \right\rangle_\phi , \qquad (6.101)$$

$$G_5 = \frac{1}{2} \left\langle \prod_{l \in B_N} \phi_l^{(0)\mu_l} \sum_{j \neq m} \tilde{\lambda}_j \tilde{\lambda}_m (b_j^{(1)} b_j^{(0)*} + b_j^{(1)*} b_j^{(0)}) b_m^{(1)} b_m^{(0)*} \right.$$

$$\left. + \left(\tilde{\lambda}_j + \frac{\mu_j}{4J_j^{(0)}} \right) \frac{\mu_m}{J_j^{(0)}} (b_m^{(1)} b_m^{(0)*} - b_m^{(1)*} b_m^{(0)}) b_j^{(1)} b_j^{(0)*} \right\rangle_\phi . \qquad (6.102)$$

Here, as before, $\langle \cdot \rangle_J$ and $\langle \cdot \rangle_\phi$ denote the averaging over the initial amplitudes and the initial phases which can be done independently because the initial fields are assumed to be RP (not necessarily RPA!). Note that so far our calculation for $\mathscr{Z}^{(N)}$ is the same for the three-wave and the four-wave systems, as well as for the systems with higher-order wave resonances. Now we have to substitute expressions for $b_k^{(1)}$ and $b_k^{(2)}$ which are different for the three-wave and the four-wave cases and given by (6.19), (6.23) and (6.77), (6.78) respectively.

6.10.2 Statistical Averaging and Graphs

Let us take the initial fields $a_k(0) = a_k^{(0)}$ to be of RP type (no assumption about the amplitudes yet). We will perform averaging over the statistics of the initial fields in order to obtain evolution equations, first for the multi-mode generating function, and then for the multi-mode PDF.

For simplicity, we will again use our main example, the Petviashvili model. We will only consider here contributions from G_1 and G_2 which will allow the reader to understand the basic method. Calculation of the rest of the terms, G_3, G_4 and G_5, follows the same principles, and it will be left as an exercise. First, the linear in ϵ terms are represented by G_1 which, after using (6.19), becomes

$$G_1 = -\left\langle \prod_l \phi_l^{\mu_l} \sum_{j,m,n} (\tilde{\lambda}_j + \frac{\mu_j}{2J_j}) i \, \text{sign}(k_{jx}) V_{mnj}^* b_m^* b_n^* \Delta^*(\omega_{mn}^j) \delta_{mn}^j b_j \right\rangle_\phi . \qquad (6.103)$$

Hereafter we omit, for brevity, the super-script (0) because no other super-scripts will appear from now on.

Let us introduce some graphical notations for a simple classification of different contributions to this and to other formulae that will follow. Combination $V_{mnj}\delta^j_{mn}$ will be marked by a vertex joining three lines with in-coming j and out-going m and n directions. Complex conjugate $V^*_{mnj}\delta^j_{mn}$ will be drawn by the same vertex but with the opposite in-coming and out-going directions. Respectively, $b^*_{m,n}$ and b_j will be indicated by dashed lines pointing to and away from the vertex. Thus, the expression inside the averaging bracket in (6.103) can be schematically represented as follows,

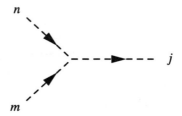

Let us average over all the phase factors in the set $\{\phi\}$ taking into account the statistical independence and the uniform distribution of these variables, which yield the Wick's contraction rule as explained in Sect. 5.6. The difference with the standard Wick, however, is that there exists the possibility of not only internal (with respect to the sum) matchings but also external ones with ϕ's in the pre-factor $\prod_l \phi^{\mu_l}_l$.

Thus, non-zero contributions can only arise for terms in which all ϕ's cancel out either via internal mutual couplings within the sum or via their external couplings to the ϕ's in the l-product. The internal couplings will be indicated by joining the dashed lines into loops whereas the external matchings will be shown as dashed lines pinned by blobs at the ends. The number of blobs in a particular graph will be called the *valence* of this graph.

Note that there will be no contribution from the internal couplings between the incoming and the out-going lines of the same vertex because, due to the δ-symbol, one of the wavenumbers is zero in this case, which means that $V = 0$. For G_1 we have

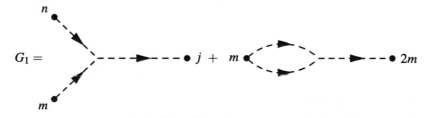

which corresponds to the following expression,

$$G_1 = -\sum_{j\neq m\neq n}(\tilde{\lambda}_j + \frac{\mu_j}{2J_j})i\,\text{sign}(k_{jx})V^*_{mnj}\sqrt{J_m J_n J_j}\Delta^*(\omega^j_{mn})$$

$$\times \delta^j_{mn}\delta(\mu_m - 1)\delta(\mu_n - 1)\delta(\mu_j + 1)\prod_{l\neq j,m,n}\delta(\mu_l)$$

$$-\sum_m(\lambda_{2m} + \frac{\mu_{2m}}{2J_{2m}})i\,\text{sign}(k_{2mx})V^*_{m,m,2m}J_m\sqrt{J_{2m}}$$

$$\times \Delta^*(\omega^{2m}_{mm})\delta(\mu_m - 2)\delta(\mu_{2m} + 1)\prod_{l\neq m,2m}\delta(\mu_l). \tag{6.104}$$

Because of the δ-symbols involving μ's, it takes very special combinations of the arguments μ in $\mathcal{X}^{(N)}\{\mu\}$ for the terms in the above expressions to be non-zero. For example, a particular term in the first sum of (6.104) may be non-zero if two μ's in the set $\{\mu\}$ are equal to 1, one of the μ's is -1 and the rest of them are 0. But in this case there is only one other term in this sum (corresponding to $m \leftrightarrow n$) that may be non-zero too. Similarly, there is at most one non-zero term in the second sum of (6.104): it corresponds to the cases when one of the μ's is 2, one μ (with twice the index of the first one) is -1, and the rest of the μ's are zero. In other words, *each external pinning of the dashed line removes summation in one index* and, since all the indices are pinned in the above diagrams, we are left with no summation at all in G_1 i.e. the number of terms in G_1 is $O(1)$ with respect to large N. We will see later that the dominant contributions have $O(N^2)$ terms. Although these terms come in the ϵ^2-order, they will be much greater that the ϵ^1-terms because the limit $N \to \infty$ must always be taken before $\epsilon \to 0$.

Let us consider G_2 which comes in the ϵ^2-order. Substituting (6.19) into (6.99) , we have

$$G_2 = \frac{1}{2}\langle \prod_l \phi_l^{(0)\mu_l} \sum_{j,m,n,\kappa,\nu}(\tilde{\lambda}_j + \tilde{\lambda}^2_j J_j - \frac{\mu^2_j}{2J_j})$$

$$\times V_{mnj}b_m b_n \Delta(\omega^j_{mn})\delta^j_{mn} V^*_{\kappa\nu j}b^*_\kappa b^*_\nu \Delta^*(\omega^j_{\kappa\nu})\delta^j_{\kappa\nu}\rangle_\phi$$

$$= \langle B\rangle_\phi, \tag{6.105}$$

where

$$B = \tag{6.106}$$

Here, the graphical notations for the interaction coefficients V and the amplitude b are the same as explained above and the dotted line with index j indicates that there is a summation over j but there is no amplitude b_j in the corresponding expression.

Let us now perform the phase averaging which corresponds to the internal and external couplings of the dashed lines. We have

$$(6.107)$$

where

$$= \frac{1}{2} \sum_{j \neq m \neq n \neq \kappa \neq v} (\lambda_j + \lambda_j^2 J_j - \frac{\mu_j^2}{2J_j}) V_{mnj} V_{\kappa v j}^*$$

$$\times \Delta(\omega_{mn}^j) \Delta^*(\omega_{\kappa v}^j) \delta_{mn}^j \delta_{\kappa v}^j \sqrt{J_m J_n J_\kappa J_v}$$

$$\times \delta(\mu_m + 1) \delta(\mu_n + 1) \delta(\mu_\kappa - 1) \delta(\mu_v - 1) \prod_{l \neq m,n,\kappa,v} \delta(\mu_l),$$

$$= \frac{1}{2} \sum_{m \neq n \neq v} (\tilde{\lambda}_{2v} + \tilde{\lambda}_{2v}^2 J_{2v} - \frac{\mu_{2v}^2}{2J_{2v}}) V_{m,n,2v} \delta_{mn}^{2v} V_{v,v,2v}^*$$

$$\times \Delta(\omega_{mn}^{2v}) \Delta^*(\omega_{vv}^{2v}) \sqrt{J_m J_n J_v}$$

$$\times \delta(\mu_m + 1) \delta(\mu_n + 1) \delta(\mu_v - 2) \prod_{l \neq m,n,v} \delta(\mu_l),$$

$$= \prod_l \delta(\mu_l) \sum_{j,m,n} (\tilde{\lambda}_j + \tilde{\lambda}_j^2 J_j) |V_{mnj}|^2 |\Delta_{mn}^j|^2 \delta_{mn}^j J_m J_n.$$

We have not written out the third term in (6.107) because it is just a complex conjugate of the second one. All the non-zero valence diagrams in (6.107) are $O(1)$ with respect to large N because all of the summations are lost due to the external couplings (similarly as in G_1 above). On the other hand, the zero-valence diagram contains two purely-internal couplings and is therefore $O(N^2)$. This is because the number of indices over which the summation survives is equal to the number of purely internal couplings. Thus, the zero-valent graph is dominant and we can write

$$G_2 = \prod_l \delta(\mu_l) \sum_{j,m,n} (\tilde{\lambda}_j + \tilde{\lambda}_j^2 J_j)|V_{mnj}|^2 |\Delta(\omega_{mn}^j)|^2 \delta_{mn}^j J_m J_n [1 + O(1/N^2)]. \quad (6.108)$$

Above, we considered in detail the different terms involved in G_2 so that the reader could clearly see that the dominant contributions come from the zero-valent graphs because they have more summation indices involved. This appears to be the general rule which allows one to simplify the calculation by discarding a significant number of graphs with non-zero valence. After this observation finding the rest of the terms, G_3–G_5, becomes straightforward and we leave it as an exercise.

Exercise 6.16 *Show that*

$$G_3 = -4 \prod_l \delta(\mu_l) \sum_{j,m,n} \tilde{\lambda}_j \, \text{sign}(k_{jx} k_{mx})|V_{mnj}|^2 E(-\omega_{mn}^j, \omega_{mn}^j) \delta_{mn}^j J_j J_n [1 + O(1/N)],$$

$$(6.109)$$

$$G_4 = G_2 \times O(1/N^2), \text{ (i.e. } \ll G_2 \text{ for large } N), \quad (6.110)$$

$$G_5 = -2 \prod_l \delta(\mu_l) \sum_{j,m,n} \tilde{\lambda}_j \tilde{\lambda}_m \, \text{sign}(k_{jx} k_{mx})|V_{mnj}|^2 \delta_{mn}^j |\Delta_{mn}^j|^2 J_j J_m J_n [1 + O(1/N)].$$

$$(6.111)$$

6.10.3 Equation for $\mathscr{Z}^{(N)}$

Now we can observe that all contributions to the evolution of $\mathscr{Z}^{(N)}$ (namely G_1–G_5, see the previous section and Exercise 6.16) contain the factor $\prod_l \delta(\mu_l)$ which means that the phase factors $\{\phi\}$ remain a set of statistically independent random variables uniformly distributed on S^1. This is true with accuracy $O(\epsilon^2)$ (assuming that the N-limit is taken first, i.e. $1/N \ll \epsilon^2$) and this proves persistence of the RP property. This result is interesting because it has been obtained without any assumptions on the statistics of the amplitudes $\{J\}$ and, therefore, it is valid beyond the RPA approach. It may appear useful in the future for study of fields with random phases but with correlated amplitudes.

Let us now derive an evolution equation for the generating functional. Using our results for G_1–G_5 in (6.97) and (6.96) we have

$$
\mathscr{Z}^{(N)}(T) - \mathscr{Z}^{(N)}(0) = 2\epsilon^2 \sum_{j,m,n} |V_{mnj}|^2 \delta^j_{mn} \left[|\Delta(\omega^j_{mn})|^2 (\tilde{\lambda}_j + \tilde{\lambda}^2_j \frac{\partial}{\partial \tilde{\lambda}_j}) \frac{\partial^2 \mathscr{Z}^{(N)}(0)}{\partial \tilde{\lambda}_m \partial \tilde{\lambda}_n} \right.
$$

$$
- 4\tilde{\lambda}_j \mathrm{sign}(k_{jx}k_{mx}) \, \mathrm{Re}[E(-\omega^j_{mn}, \omega^j_{mn})] \frac{\partial^2 \mathscr{Z}^{(N)}(0)}{\partial \tilde{\lambda}_j \partial \tilde{\lambda}_n}
$$

$$
\left. -2\tilde{\lambda}_j \tilde{\lambda}_m \mathrm{sign}(k_{jx}k_{mx}) |\Delta(\omega^j_{mn})|^2 \frac{\partial^3 \mathscr{Z}^{(N)}(0)}{\partial \tilde{\lambda}_j \partial \tilde{\lambda}_m \partial \tilde{\lambda}_n} \right]. \tag{6.112}
$$

Here, partial derivatives with respect to λ's have appeared because of the J-factors. This expression is valid up to $O(\epsilon^4)$ and $O(\epsilon^2/N)$ corrections. Note that we still have not used any assumption about the statistics of J's. This is a linear equation: as usual in statistics we have traded nonlinearity for higher dimensions.

Let us now take the limit $N \to \infty$ followed by $T \sim 1/\epsilon \to \infty$ (we repeat that this order of the limits is essential). Taking into account (6.41) and (6.42), using the same rules for replacing the sums by integrals and discrete δ's with continuous δ's as we did before (when considering the one-mode statistics), and replacing $[\mathscr{Z}^{(N)}(T) - \mathscr{Z}^{(N)}(0)]/T$ by $\dot{Z}$ we have

$$
\dot{\mathscr{Z}}^{(N)} = 4\pi\epsilon^2 \int |V_{mnj}|^2 \delta^j_{mn} \delta(\omega^j_{mn}) \left[(\lambda_j + \lambda^2_j \frac{\delta}{\delta \lambda_j}) \frac{\delta^2 \mathscr{Z}^{(N)}}{\delta \lambda_m \delta \lambda_n} \right.
$$

$$
\left. -2\lambda_j \mathrm{sign}(k_{jx}k_{mx}) \left(\frac{\delta^2 \mathscr{Z}^{(N)}}{\delta \lambda_j \delta \lambda_n} + \lambda_m \frac{\delta^3 \mathscr{Z}^{(N)}}{\delta \lambda_j \delta \lambda_m \delta \lambda_n} \right) \right] d\mathbf{k}_j d\mathbf{k}_m d\mathbf{k}_n. \tag{6.113}
$$

Here, the variational derivatives have appeared instead of the partial derivatives because of the $N \to \infty$ limit.

6.10.4 Equation for the PDF

Taking the inverse Laplace transform of (6.113) we have the following equation for the PDF,

$$
\dot{\mathscr{P}} = -\int \frac{\delta \mathscr{F}_j}{\delta s_j} \, d\mathbf{k}_j, \tag{6.114}
$$

where $\mathscr{F} = \{\mathscr{F}_1, \mathscr{F}_2, \ldots, \mathscr{F}_N\}$ is a probability flux in the N-dimensional space of the amplitudes $\{s_1, s_2, \ldots, s_N\}$, whose components are given by the following expression,

$$\mathcal{F}_j = -4\pi\epsilon^2 \int |V_{mnj}|^2 \delta(\omega_{mn}^j)\delta_{mn}^j s_j s_n s_m$$

$$\times \left[\frac{\delta\mathcal{P}}{\delta s_j} - 2\,\text{sign}(k_{jx}k_{mx})\frac{\delta\mathcal{P}}{\delta s_m}\right] d\mathbf{k}_m d\mathbf{k}_n. \tag{6.115}$$

Hereafter for brevity we omit superscript (N) whenever it does not cause confusion.

Equation (6.114) can be written in a symmetric form,

$$\dot{\mathcal{P}} = 8\pi\epsilon^2 \int\limits_{k_{jx},k_{mx},k_{nx}>0} |V_{mnj}|^2 \delta(\omega_{mn}^j)\delta_{m+n}^j$$

$$\times \left[\frac{\delta}{\delta s}\right]_3 \left(s_j s_m s_n \left[\frac{\delta}{\delta s}\right]_3 \mathcal{P}\right) d\mathbf{k}_j d\mathbf{k}_m d\mathbf{k}_n, \tag{6.116}$$

where

$$\left[\frac{\delta}{\delta s}\right]_3 = \frac{\delta}{\delta s_j} - \frac{\delta}{\delta s_m} - \frac{\delta}{\delta s_n}. \tag{6.117}$$

Written in this form, the PDF equation is easily generalizable to the general case of three-wave systems for which the phase space is the full $\mathbf{k}$-space (i.e. not just a half of the $\mathbf{k}$-space as in the considered Petviashvili example). For this, we simply extend the integration to the whole of the $\mathbf{k}$-space and replace the interaction coefficient with one less symmetric with respect to the indices,

$$\dot{\mathcal{P}} = 8\pi\epsilon^2 \int |V_{mn}^j|^2 \delta(\omega_{mn}^j)\delta_{m+n}^j \left[\frac{\delta}{\delta s}\right]_3 \left(s_j s_m s_n \left[\frac{\delta}{\delta s}\right]_3 \mathcal{P}\right) d\mathbf{k}_j d\mathbf{k}_m d\mathbf{k}_n. \tag{6.118}$$

An equation of this kind was first found by Peierls in the context of the physics of anharmonic crystals [23] and later discussed by Brout and Prigogine [24] and Zaslavski and Sagdeev [25].

Here we should again emphasize the importance of the taken order of limits: first $N \to \infty$ and $\epsilon \to 0$ second. Physically this means that the frequency resonance is broad enough to cover a great many modes. Some authors, e.g. [23–25], leave the sum notation in the PDF equation even after the $\epsilon \to 0$ limit is taken, giving $\delta(\omega_{jm}^n)$. One has to be careful interpreting such formulae since formally the RHS is null in most of the cases because there may be no exact resonances between the discrete $\mathbf{k}$-modes (as is the case, e.g. for the capillary waves). In real finite-size physical systems, this condition means that the wave amplitudes, although small, should not be too small so that the frequency broadening is sufficient to allow the resonant interactions. Our functional integral notation is meant to indicate that the $N \to \infty$ limit has already been taken.

Having found equations for the N-mode joint PDF for the RP fields, one can now recover the previously obtained equation for the one-mode PDF if one further assumes independence of the wave intensities J_j (which amounts to assuming RPA). We leave this as an exercise.

Exercise 6.17 *Assume that the initial wave fields are RPA (i.e. that the initial N-mode PDF is a product of the one-mode PDFs and is independent of the angle variables) and obtain the evolution (6.51) for the one-mode PDF via integrating the N-mode equation (6.116) over all the amplitude variables but one.*

6.11 Generalization to the Four-Wave and the Higher-Order Systems

In this section we are going to find an equation for the joint N-mode PDF for the four-wave case, and then further generalize the result to the wave systems with higher-order resonances. The calculations are similar to those of the previous section, namely one can use the same derivation up to formulae (6.96), (6.97) with the G_1–G_5 terms (6.98)–(6.102), but now for $b_k^{(1)}$ and $b_k^{(2)}$ one should use expressions (6.77) and (6.78) corresponding to the four-wave systems. Obviously, one must also replace $\epsilon^n \to \epsilon^{2n}$ (because the expansion is in powers of ϵ^2 in the four-wave case). Further, for the same reason as for the three-wave case, all μ-dependent terms will not contribute to the final result. Indeed, any term with nonzero μ would correspond to a nonzero-valence graph containing less summations and therefore subdominant when N is large. We formulate this step as an exercise.

Exercise 6.18 *Using formulae (6.77), (6.78) in expressions (6.98)–(6.102) and leaving only zero-valence graphs (putting $\mu_j = 0$) and using the Wick's contraction rule show that in the leading (for large N) order:*

$$G_1 = 0,$$

$$G_2 = \sum_{j\alpha\mu\nu} \tilde{\lambda}_j (1 + \tilde{\lambda}_j J_j) |W_{\mu\nu}^{j\alpha}|^2 |\Delta_{\mu\nu}^{j\alpha}|^2 \delta_{\mu\nu}^{j\alpha} J_\mu J_\alpha J_\nu,$$

$$G_3 = 2 \sum_{j\alpha\mu\nu} \tilde{\lambda}_j |W_{\mu\nu}^{j\alpha}|^2 \delta_{\mu\nu}^{j\alpha} J_j J_\nu (J_\mu - 2J_\alpha) E(-\omega_{\mu\nu}^{j\alpha}, \omega_{\mu\nu}^{j\alpha}),$$

$$G_4 = 0,$$

$$G_5 = \sum_{j \neq k} \tilde{\lambda}_j \tilde{\lambda}_k \left[-2|W_{k\nu}^{j\alpha}|^2 |\Delta_{k\nu}^{j\alpha}|^2 \delta_{k\nu}^{j\alpha} J_\alpha + |W_{\mu\nu}^{jk}|^2 |\Delta_{\mu\nu}^{jk}|^2 \delta_{\mu\nu}^{jk} J_\mu \right] J_j J_k J_\nu.$$

By putting everything together in (6.96) and (6.97) and taking the limit $N \to \infty$ limit followed by $T \to \infty$ we finally obtain

$$\dot{\mathscr{L}} = 4\pi\epsilon^4 \int |W_{nm}^{jl}|^2 \delta(\omega_{nm}^{jl}) \delta_{nm}^{jl} \lambda_j$$

$$\times \frac{\partial}{\partial \lambda_j} \frac{\partial}{\partial \lambda_l} \frac{\partial}{\partial \lambda_n} \frac{\partial}{\partial \lambda_m} \left[(\lambda_j + \lambda_l - \lambda_m - \lambda_n) \mathscr{L} \right] d\mathbf{k}_j d\mathbf{k}_l d\mathbf{k}_m d\mathbf{k}_n. \qquad (6.119)$$

By applying to this equation the inverse Laplace transform, we get an equation for the N-mode PDF:

$$\dot{\mathscr{P}} = \pi \epsilon^4 \int |W_{nm}^{jl}|^2 \delta(\omega_{nm}^{jl}) \delta_{nm}^{jl}$$

$$\times \left[\frac{\delta}{\delta s}\right]_4 \left(s_j s_l s_m s_n \left[\frac{\delta}{\delta s}\right]_4 \mathscr{P}\right) d\mathbf{k}_j d\mathbf{k}_l d\mathbf{k}_m d\mathbf{k}_n, \qquad (6.120)$$

where

$$\left[\frac{\delta}{\delta s}\right]_4 = \frac{\delta}{\delta s_j} + \frac{\delta}{\delta s_l} - \frac{\delta}{\delta s_m} - \frac{\delta}{\delta s_n}. \qquad (6.121)$$

This equation can be easily written in the continuity equation form (6.114) with the flux given in this case by

$$\mathscr{F}_j = -4 s_j \pi \epsilon^4 \int |W_{nm}^{jl}|^2 \delta(\omega_{nm}^{jl}) \delta_{nm}^{jl} s_l s_m s_n \left[\frac{\delta}{\delta s}\right]_4 \mathscr{P} \, d\mathbf{k}_l d\mathbf{k}_m d\mathbf{k}_n. \qquad (6.122)$$

Comparing the results for the three-wave and the four-wave cases, it becomes rather obvious how to generalize the result to higher-order wave systems. Namely, let us consider a p-wave Hamiltonian

$$H = \sum_k \omega_k a_k a_k^* + \sum_{\substack{j_1,\ldots,j_q; \\ l_1,\ldots,l_r}} W_{j_1,\ldots,j_q}^{l_1,\ldots,l_r} \delta_{j_1,\ldots,j_q}^{l_1,\ldots,l_r} a_{j_1} \ldots, a_{j_q} a_{l_1}^* \ldots, a_{l_r}^* + c.c., \qquad (6.123)$$

which represents a $q \to r$ resonant wave process with $q + r = p$. Then the equation for the N-mode PDF will be

$$\dot{\mathscr{P}} = C \, \epsilon^{2(p-2)} \int W_{j_1,\ldots,j_r}^{l_1,\ldots,l_r} \delta(\omega_{j_1,\ldots,j_q}^{l_1,\ldots,l_r}) \, \delta_{j_1,\ldots,j_q}^{l_1,\ldots,l_r}$$

$$\times \left[\frac{\delta}{\delta s}\right]_q^r \left(s_{j_1} \ldots s_{j_q} s_{l_1} \ldots s_{l_r} \left[\frac{\delta}{\delta s}\right]_q^r \mathscr{P}\right) d\mathbf{k}_{j_1} \ldots d\mathbf{k}_{j_q} d\mathbf{k}_{l_1} \ldots d\mathbf{k}_{l_r}, \qquad (6.124)$$

where C is a dimensionless constant and

$$\left[\frac{\delta}{\delta s}\right]_q^r = \frac{\delta}{\delta s_{l_1}} + \cdots + \frac{\delta}{\delta s_{l_r}} - \frac{\delta}{\delta s_{j_1}} - \cdots - \frac{\delta}{\delta s_{j_q}}. \qquad (6.125)$$

We have now obtained evolution equations for all the essential statistical objects in WT. Next we will consider most important properties and solutions of these equations.

References

1. Choi, Y., Lvov, Y., Nazarenko, S.V.: Probability densities and preservation of randomness in wave turbulence. Phys. Lett. A **332**(3–4), 230 (2004)
2. Choi, Y., Lvov, Y., Nazarenko, S.V., Pokorni, B.: Anomalous probability of large amplitudes in wave turbulence. Phys. Lett. A **339**(3–5), 361 (2004)

3. Choi, Y., Lvov, Y., Nazarenko, S.V.: Joint statistics of amplitudes and phases in wave turbulence. Physica D **201**, 121 (2005)
4. Choi, Y., Lvov, Y.V., Nazarenko, S.: Wave turbulence. In: Recent Developments in Fluid Dynamics, vol. 5. Transworld Research Network. Kepala, India (2004)
5. Jakobsen, P., Newell, A.: Invariant measures and entropy production in wave turbulence. J. Stat. Mech. 2004, L10002 (2004)
6. Choi, Y., Jo, G.G., Kim, H.I., Nazarenko, S.: Aspects of two-mode probability density function in weak wave turbulence. J. Phys. Soc. Jpn. **78**(8), 084403 (2009)
7. Petviashvili, V.I.: Nonlinear waves and solitons. In: Leontovich, M.A. (ed.) Reviews of Plasma Physics, vol. 9, p. 59. Consultants Bureau, New York (1986)
8. Galtier, S.: Weak inertial-wave turbulence theory. Phys. Rev. E **68**, 015301 (2003)
9. Zakharov, V.E., Kuznetsov, E.A.: Variational principle and canonical variables in magnetohydrodynamics. Sov. Phys. Dokl. **15**, 913–914 (1971)
10. Zakharov, V.E., Kuznetsov, E.A.: DAN SSSR **194**, 1288 (1970)
11. Zakharov, V.E., Kuznetsov, E.A.: Hamiltonian formalism for nonlinear waves. Usp. Fiz. Nauk **167**(11), 1137–1168 (1997)
12. Kuznetsov, E.A.: Weak magnetohydrodynamic turbulence of a magnetized plasma. J. Exp. Theor. Phys. **93**(5), 1052–1064 (2001)
13. Zakharov, V.E., L'vov, V.S., Falkovich, G.: Kolmogorov spectra of turbulence series in nonlinear dynamics. Springer, New York (1992)
14. Krasitskii, V.P.: On the canonical transformation of the theory of weakly nonlinear waves with nondecay dispersion law. Sov. Phys. JETP **98**, 1644–1655 (1990)
15. Zakharov, V.E.: Statistical theory of surface waves on fluid of finite depth. Eur. J. Mech. B Fluids **18**, 327–344 (1999)
16. Zakharov V.E., Schulman E.I.: Integrability of Nonlinear Systems and Perturbation Theory: In What is Integrability? Springer Series Nonlinear Dynamics, pp. 185–250. Springer, Berlin (1991)
17. Dyachenko, A.I., Lvov, Y.V., Zakharov, V.E.: Five-wave interaction on the surface of deep fluid. Physica D **87**, 233 (1995)
18. Bortolozzo, U., Laurie, J., Nazarenko, S., Residori, S.: Optical wave turbulence and the condensation of light. J. Opt. Soc. Am. B **26**(12), 2280–2284 (2009)
19. Bortolozzo, U., Laurie, J., Nazarenko, S., Residori, S.: One-dimensional optical wave turbulence, submitted to Physica D (long paper) (2009)
20. Hasimoto, H.: J. Fluid Mech. **51**, 477 (1972)
21. Boffetta, G., Celani, A., Dezzani, D., Laurie, J., Nazarenko, S.: Modeling kelvin wave cascades in superfluid helium. J. Low Temp. Phys. **156**(3–6), 193–214 (2009)
22. Laurie, J., L'vov, V.S., Nazarenko, S., Rudenko, O.: Interaction of Kelvin waves and non-locality of the energy transfer in superfluids. Phys. Rev. B **81**(10), Article Number: 104526 (2010)
23. Peierls, R.: Ann. Phys. **3**, 1055 (1929)
24. Brout, R., Prigogine, I.: Physica **22**, 621–636 (1956)
25. Zaslavskii, G.M., Sagdeev, R.Z.: Sov. Phys. JETP **25**, 718 (1967)

Chapter 7
Solutions to Exercises

7.1 One-Mode Generating Function for Gaussian Fields: Exercise 5.1

Let us substitute Rayleigh distribution (corresponding to a Gaussian field) into the expression for $\mathscr{Z}_j^{(a)}$ (5.17) and find the resulting integral,

$$\mathscr{Z}_j^{(a)}(\lambda_j) = \left\langle e^{\left(\frac{L}{2\pi}\right)^d J_j \lambda_j} \right\rangle = \int_0^\infty \mathscr{P}_j^{(a)}(s_j) e^{\left(\frac{L}{2\pi}\right)^d s_j \lambda_j} ds_j$$

$$= \frac{1}{\langle J_j \rangle} \int_0^\infty e^{-\frac{s_j}{\langle J_j \rangle} + \left(\frac{L}{2\pi}\right)^d s_j \lambda_j} ds_j = \frac{1}{1 - n_j \lambda_j}. \tag{7.1}$$

7.2 One-Mode Moments for Gaussian Fields: Exercise 5.2

Let us substitute Rayleigh distribution (corresponding to a Gaussian field) into the expression for $M_k^{(p)}$ (5.21) and find the resulting integral,

$$M_k^{(p)} = \left(\frac{L}{2\pi}\right)^{pd} \int_0^\infty s_k^p \mathscr{P}^{(a)}(s_k) ds_k$$

$$= \left(\frac{L}{2\pi}\right)^{pd} \frac{1}{\langle J_j \rangle} \int_0^\infty s_k^p e^{-\frac{s_j}{\langle J_j \rangle}} ds_j = p! n_k^p.$$

S. Nazarenko, *Wave Turbulence*, Lecture Notes in Physics, 825,
DOI: 10.1007/978-3-642-15942-8_7, © Springer-Verlag Berlin Heidelberg 2011

7.3 Six-Order Multi-Point Moment: Exercise 5.3

First of all, let us use the Wick contractions in a way similar to what was done to obtain (5.29). For the six-order multi-mode correlator we have:

$$
\begin{aligned}
\left\langle \hat{\psi}_1 \hat{\psi}_2 \hat{\psi}_3 \hat{\psi}_4^* \hat{\psi}_5^* \hat{\psi}_6^* \right\rangle &= \langle J_1 J_2 J_3 \rangle_J \big[\delta_4^1 \delta_5^2 \delta_6^3 + \delta_4^2 \delta_5^3 \delta_6^1 + \delta_4^3 \delta_5^1 \delta_6^2 + \delta_4^1 \delta_5^3 \delta_6^2 + \delta_4^3 \delta_5^2 \delta_6^1 \\
&\quad + \delta_4^2 \delta_5^1 \delta_6^3 - (\delta_2^1 + \delta_3^1 + \delta_3^2)\big(\delta_4^1 \delta_5^2 \delta_6^3 + \delta_4^2 \delta_5^3 \delta_6^1 + \delta_4^3 \delta_5^1 \delta_6^2\big) \\
&\quad + 4\delta_2^1 \delta_3^1 \delta_4^1 \delta_5^1 \delta_6^1 \big].
\end{aligned}
$$

The last line here is to prevent double counting of the cases where the Wick-contracted pairs have coinciding wavenumbers (e.g. coefficient 4 in the last term reflects the fact that there should be just one term corresponding to the case where all the wavenumbers coincide whereas the first line gives 6 of these terms and the first term in the second line "takes away" 9 of these terms).

So far we only used randomness of the phases only. Now let us use statistical independence of the amplitudes. We must differentiate between the cases when all wavenumbers $\mathbf{k}_1$, $\mathbf{k}_2$ and $\mathbf{k}_3$ are different and the cases when some of the wavenumbers (or all) coincide:

$$
\langle J_1 J_2 J_3 \rangle_J = \begin{cases}
\langle J_1 \rangle \langle J_2 \rangle \langle J_3 \rangle & \text{if } \mathbf{k}_1 \neq \mathbf{k}_2 \neq \mathbf{k}_3, \\
\langle J_1^2 \rangle \langle J_3 \rangle & \text{if } \mathbf{k}_1 = \mathbf{k}_2 \neq \mathbf{k}_3, \\
\langle J_2^2 \rangle \langle J_1 \rangle & \text{if } \mathbf{k}_2 = \mathbf{k}_3 \neq \mathbf{k}_1, \\
\langle J_3^2 \rangle \langle J_2 \rangle & \text{if } \mathbf{k}_3 = \mathbf{k}_1 \neq \mathbf{k}_2, \\
\langle J_1^3 \rangle & \text{if } \mathbf{k}_1 = \mathbf{k}_2 = \mathbf{k}_3.
\end{cases}
\tag{7.2}
$$

We can finally write the six-order correlator as

$$
\begin{aligned}
\left\langle \hat{\psi}_1 \hat{\psi}_2 \hat{\psi}_3 \hat{\psi}_4^* \hat{\psi}_5^* \hat{\psi}_6^* \right\rangle &= \langle J_1 \rangle \langle J_2 \rangle \langle J_3 \rangle \big(\delta_4^1 \delta_5^2 \delta_6^3 + \delta_4^2 \delta_5^3 \delta_6^1 + \delta_4^3 \delta_5^1 \delta_6^2 \\
&\quad + \delta_4^3 \delta_5^2 \delta_6^1 + \delta_4^2 \delta_5^1 \delta_6^3 \big) \\
&\quad + \langle J_3 \rangle \Big(\langle J_1^2 \rangle - 2\langle J_1 \rangle^2 \Big) \delta_2^1 \big(\delta_4^1 \delta_5^2 \delta_6^3 + \delta_4^2 \delta_5^3 \delta_6^1 + \delta_4^3 \delta_5^1 \delta_6^2 \big) \\
&\quad + \langle J_2 \rangle \Big(\langle J_3^2 \rangle - 2\langle J_3 \rangle^2 \Big) \delta_3^1 \big(\delta_4^1 \delta_5^2 \delta_6^3 + \delta_4^2 \delta_5^3 \delta_6^1 + \delta_4^3 \delta_5^1 \delta_6^2 \big) \\
&\quad + \langle J_1 \rangle \Big(\langle J_2^2 \rangle - 2\langle J_2 \rangle^2 \Big) \delta_3^2 \big(\delta_4^1 \delta_5^2 \delta_6^3 + \delta_4^2 \delta_5^3 \delta_6^1 + \delta_4^3 \delta_5^1 \delta_6^2 \big) \\
&\quad + \Big(\langle J_1^3 \rangle - 9\langle J_1^2 \rangle \langle J_1 \rangle + 12\langle J_1 \rangle^3 \Big) \delta_2^1 \delta_3^1 \delta_4^1 \delta_5^1 \delta_6^1.
\end{aligned}
$$

Here, the last four lines represent the cumulant part which is absent for Gaussian fields (check that!).

7.4 Fourth-Order Structure Function: Exercise 5.4

Let us start with the definition of the fourth-order structure function, and use the Wick's contraction:

$$\mathscr{S}^{(4)}(r) = \left\langle |\psi(\mathbf{x},t) - \psi(\mathbf{x}-\mathbf{r},t)|^4 \right\rangle = \left\langle \left| \sum_l \hat{\psi}_l \left(e^{i\mathbf{k}_l \cdot \mathbf{x}} - e^{i\mathbf{k}_l \cdot (\mathbf{x}-\mathbf{r})} \right) \right|^4 \right\rangle$$

$$= \sum_{1,2,3,4} \left\langle \overline{\hat{\psi}_1 \hat{\psi}_2 \hat{\psi}_3^* \hat{\psi}_4^*} \right\rangle (e^{i\mathbf{k}_1 \cdot \mathbf{x}} - e^{i\mathbf{k}_1 \cdot (\mathbf{x}-\mathbf{r})})(e^{i\mathbf{k}_2 \cdot \mathbf{x}} - e^{i\mathbf{k}_2 \cdot (\mathbf{x}-\mathbf{r})})$$

$$\times (e^{-i\mathbf{k}_3 \cdot \mathbf{x}} - e^{-i\mathbf{k}_3 \cdot (\mathbf{x}-\mathbf{r})})(e^{-i\mathbf{k}_4 \cdot \mathbf{x}} - e^{-i\mathbf{k}_4 \cdot (\mathbf{x}-\mathbf{r})})$$

(The two ways of contraction give the same result due to symmetry $1 \rightleftarrows 2$: hence factor 2)

$$= 2 \sum_{\mathbf{k}_1 \neq \mathbf{k}_2} \langle J_1 J_2 \rangle \left| e^{i\mathbf{k}_1 \cdot \mathbf{x}} - e^{i\mathbf{k}_1 \cdot (\mathbf{x}-\mathbf{r})} \right|^2 \left| e^{i\mathbf{k}_2 \cdot \mathbf{x}} - e^{i\mathbf{k}_2 \cdot (\mathbf{x}-\mathbf{r})} \right|^2 + \sum_{\mathbf{k}} \langle J_k^2 \rangle \left| e^{i\mathbf{k} \cdot \mathbf{x}} - e^{i\mathbf{k} \cdot (\mathbf{x}-\mathbf{r})} \right|^4$$

$$= 32 \sum_{\mathbf{k}_1 \neq \mathbf{k}_2} \langle J_1 J_2 \rangle \left| \sin \frac{\mathbf{k}_1 \cdot \mathbf{r}}{2} \right|^2 \left| \sin \frac{\mathbf{k}_2 \cdot \mathbf{r}}{2} \right|^2 + 16 \sum_{\mathbf{k}} \langle J_k^2 \rangle \left| \sin \frac{\mathbf{k} \cdot \mathbf{r}}{2} \right|^4$$

$$= 32 \left(\sum_{\mathbf{k}} \langle J_k \rangle \left| \sin \frac{\mathbf{k} \cdot \mathbf{r}}{2} \right|^2 \right)^2 + 16 \sum_{\mathbf{k}} \left(\langle J_k^2 \rangle - 2\langle J_k \rangle^2 \right) \left| \sin \frac{\mathbf{k} \cdot \mathbf{r}}{2} \right|^4$$

$$= 2 \left[\mathscr{S}^{(2)}(r) \right]^2 + 16 \left(\frac{2\pi}{L} \right)^{2d} \sum_{\mathbf{k}} \left(M_k^{(2)} - 2n_k^2 \right) \left| \sin \frac{\mathbf{k} \cdot \mathbf{r}}{2} \right|^4,$$

as required.

7.5 Invariants of the Petviashvilli Equation: Exercise 6.1

Differentiating (6.2) and substituting (6.1) we have

$$\rho^2 \dot{E} = \int \psi \dot{\psi} d\mathbf{x} = - \int \nabla \psi \cdot \partial_x \nabla \psi d\mathbf{x} - \frac{1}{3} \int \partial_x \psi^3 d\mathbf{x}, \qquad (7.3)$$

where we integrated by parts in the first term on the RHS. Further, since $\nabla \psi \cdot \partial_x \nabla \psi = \partial_x (\nabla \psi)^2$, this term integrates to zero; and so does the last term. Thus, $\dot{E} = 0$ and we obtain the conservation law (6.2).

Differentiating (6.3) and substituting (6.1) we have

$$
\begin{aligned}
\dot{\Omega} &= \frac{1}{2} \int \left(-2\Delta\psi + \psi^2 \right) \psi_t d\mathbf{x} \\
&= \frac{1}{2} \int \left(-2\Delta\psi + \psi^2 \right) \left(\Delta\partial_x\psi - \frac{1}{2}\partial_x\psi^2 \right) d\mathbf{x} \\
&= -\frac{1}{2} \int \partial_x \left(\Delta\psi - \frac{1}{2}\psi^2 \right)^2 d\mathbf{x} = 0,
\end{aligned}
\tag{7.4}
$$

where we integrated by parts in the first term on the RHS of line 1. Thus, Ω is conserved.

7.6 Charney-Hassegawa-Mima Model: Exercise 6.2

Writing (6.13) in Fourier space, we have

$$
\left(1 + \rho^2 k^2 \right) \dot{\hat{\psi}}_k - i\beta\rho^2 k_x \hat{\psi}_k = \rho^2 \sum_{\mathbf{k}_1 + \mathbf{k}_2 = \mathbf{k}} (k_{1x}k_{2y} - k_{1y}k_{2x}) k_1^2 \hat{\psi}_1 \hat{\psi}_2.
\tag{7.5}
$$

Introducing the waveaction variable as

$$
a_k = \frac{1 + \rho^2 k^2}{\sqrt{\beta k_x}} \hat{\psi}_k,
\tag{7.6}
$$

we re-write (7.5) as

$$
\begin{aligned}
\dot{a}_k + i\omega_k a_k &= \frac{\rho^2 \beta^{1/2}}{2} \sum_{\mathbf{k}_1 + \mathbf{k}_2 =} (k_x k_{2y} - k_y k_{2x}) \left(\frac{k_{1x}k_{2x}}{k_x} \right)^{1/2} \frac{a_1 a_2}{1 + \rho^2 k_2^2} \\
&= \frac{\rho^2 \beta^{1/2}}{2} \sum_{\mathbf{k}_1 + \mathbf{k}_2 =} (k_{1x}k_{2x}k_x)^{1/2} \left(\frac{k_{2y}}{1 + \rho^2 k_2^2} - \frac{k_y k_{2x}/k_x}{1 + \rho^2 k_2^2} \right) a_1 a_2 \\
&= \frac{\rho^2 \beta^{1/2}}{2} \sum_{\mathbf{k}_1 + \mathbf{k}_2 = \mathbf{k}} (k_{1x}k_{2x}k_x)^{1/2} \left(\frac{k_{2y}}{1 + \rho^2 k_2^2} - \frac{k_y k_{2x}/k_x}{1 + \rho^2 k_2^2} + \frac{k_{1y}}{1 + \rho^2 k_1^2} - \frac{k_y k_{1x}/k_x}{1 + \rho^2 k_1^2} \right) a_1 a_2 \\
&= \frac{\rho^2 \beta^{1/2}}{2} \sum_{\mathbf{k}_1 + \mathbf{k}_2 = \mathbf{k}} (k_{1x}k_{2x}k_x)^{1/2} \left(\frac{k_{2y}}{1 + \rho^2 k_2^2} + \frac{k_{1y}}{1 + \rho^2 k_1^2} - \frac{k_y}{1 + \rho^2 k_1^2} \right) a_1 a_2
\end{aligned}
$$

with

$$
\omega_k = -\frac{\beta\rho^2 k_x}{1 + \rho^2 k^2}.
\tag{7.7}
$$

In the last line of this derivation we used the frequency resonance condition $\omega_1 + \omega_2 = \omega_k$. Thus we have obtained (6.11) with

$$V_{12k} = \frac{i\rho^2\beta^{1/2}}{2}\sqrt{|k_{1x}k_{2x}k_x|}\left(\frac{k_{1y}}{1+\rho^2k_1^2} + \frac{k_{2y}}{1+\rho^2k_2^2} - \frac{k_y}{1+\rho^2k^2}\right). \qquad (7.8)$$

We emphasize, however, that we have used the frequency resonance condition and, therefore, our equation is *valid only on the resonant manifold*. Details of a more general Hamiltonian formulation of the CHM equation can be found in [1].

7.7 $T\to\infty$ limit: Exercise 6.4

Graph for $|\Delta_T(x)|^2$ is shown in Fig. 7.1. As $T \to \infty$, the graph is becoming taller, $\lim_{x\to 0}|\Delta_T(x)|^2 = T^2$, and narrower (width $\sim 1/T$). Thus, a typical Dirac-δ shape is emerging in the large-T limit.

The area under the curve of $|\Delta_T(x)|^2$ is given by

$$\int_{-\infty}^{+\infty} \frac{4\sin^2\frac{xT}{2}}{x^2}dx = 2T\int_{-\infty}^{+\infty}\frac{\sin^2 y}{y^2}dy = 2\pi T,$$

and thus we have obtained (6.41).

From (6.24) we have:

$$E(-x,x) = \int_0^T \frac{e^{-ixt}-1}{-ix}e^{ixt}dt = \frac{iT}{x} + \frac{1-e^{ixT}}{x^2}.$$

Taking the real part, we have

$$\Re[E(-x,x)] = \frac{1-\cos(xT)}{x^2} = \frac{2\sin^2\frac{xT}{2}}{x^2},$$

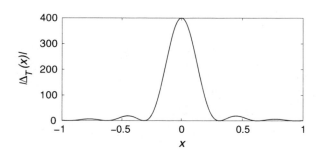

Fig. 7.1 Function $4\sin^2(xT/2)/x^2$ for $T = 20$

i.e.

$$\Re[E(-x,x)] = \frac{1}{2}|\Delta_T(x)|^2 \to \pi T \delta(x),$$

as required.

7.8 Slow and Fast Timescales in the Wave Amplitude Evolution: Exercise 6.7

As illustrated in Fig. 7.2, the wave amplitude evolves in time at two timescales: fast (with characteristic time of the linear wave period) and slow (at the nonlinear characteristic time, i.e. $\tau_{NL} \sim 1/(\epsilon^2 \omega)$). The fluctuations are ϵ-smaller compared to the time mean (time average over many wave periods) but the instantaneous time derivative is $1/\epsilon$ greater than the mean time derivative. When the phase averaging is performed, e.g. to obtain the spectrum $n \sim \langle J \rangle$, the fluctuations become smaller, $\sim \epsilon^2$, and yet the instantaneous time derivative remains at least as large as the rate of cumulative (wave-period averaged) change. The time derivative introduced in our method via T in fact corresponds only to wave-period averaged change, because defining the time derivative via the time difference with the increment T is effectively a filtering operation (it filters out all the timescales less than T).

Therefore, all our statistical objects are defined via averaging not only over the ensemble of random phases, but also in time over many wave periods. Thus, it is not surprising that the time derivative of the spectrum at $t = 0$ without such time averaging is zero (but non-zero at $t > 0$), whereas the time derivative after the time averaging is finite. However, the difference between the time averaged and non-averaged statistical objects remain small (in contrast to their time derivatives) because the time derivative associated with the rapid timescale is oscillatory and bounded.

Further, it is precisely the rapidly oscillating part of the wave amplitude that is responsible for deviations from the random-phase statistics and, therefore, non-

Fig. 7.2 Function $J(t) = \sin(t) + \epsilon \sin(t/\epsilon^2)$ illustrating typical two-scale behavior of the wave amplitude. Fluctuations are ϵ-smaller compared to the time average (over many wave periods) but the instantaneous time derivative is $1/\epsilon$ greater than the mean time derivative

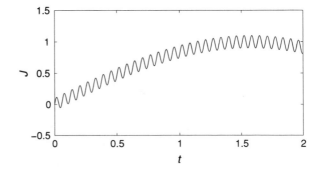

zero values of the third moments which are necessary for the nonlinear evolution. Note however, that these deviations from the random-phase statistics are small and oscillatory, i.e. the time averaged PDF's are RP.

7.9 Eliminating U_{123}: Exercise 6.9

Differentiating (6.66) with respect to time, we get

$$i\dot{b}_k = i\dot{a}_k + i\sum_{1,2} W_{12k}\left(\dot{a}_1^* a_2^* + a_1^* \dot{a}_2^*\right)\delta_{12k}.$$

Substituting $\dot{a}_k$ from the dynamical equation (6.60), and ignoring the cubic non-linearity, we have

$$i\dot{b}_k = \omega_k b_k + \sum_{1,2}\left[3U_{12k}^* - W_{12k}(\omega_k + \omega_1 + \omega_2)\right]b_1^* b_2^* \delta_{12k}$$
$$+ \sum_{1,2}\left[V_{12}^k b_1 b_2 \delta_{12}^k + 2V_{1k}^{3*} b_1^* b_3 \delta_{1k}^3\right]. \tag{7.9}$$

Thus, by choosing

$$W_{12k} = \frac{3U_{12k}^*}{(\omega_k + \omega_1 + \omega_2)}, \tag{7.10}$$

which is possible since $\omega_k + \omega_1 + \omega_2 \neq 0$, we eliminate the terms with U_{123}, and the dynamical equation (7.9) acquires a standard form,

$$i\dot{b}_k = \omega_k b_k + \sum_{1,2}\left[V_{12}^k b_1 b_2 \delta_{12}^k + 2V_{1k}^{3*} b_1^* b_3 \delta_{1k}^3\right] = \frac{\delta H}{\delta b_k^*}. \tag{7.11}$$

For the Hamiltonian we have:

$$H = \sum_k \omega_k |b_k|^2 + \sum_{1,2,3}\left(V_{12}^3 b_1 b_2 b_3^* \delta_{12}^3 + \text{c.c.}\right). \tag{7.12}$$

In fact, up to the higher-order terms (h.o.t.) this is the same Hamiltonian as the original Hamiltonian (6.61). But this means that, up to h.o.t., transformation (6.66) is canonical.

7.10 Nonlinear Phase Evolution: Exercise 6.11

Let us time-differentiate the phase $\langle \varphi_k \rangle_\phi = \langle \Im(\ln a_k) \rangle_\phi$ and use the evolution equation (6.72). In the leading order of small nonlinearity

$$\frac{d}{dt}\langle\varphi_k\rangle_\phi = \langle\Im(\dot{a}_k/a_k)\rangle_\phi = -\omega_k - \Re\sum_{\mathbf{k}_1,\mathbf{k}_2\neq\mathbf{k}_3} W_{3k}^{12}\delta_{3k}^{12}\langle a_1 a_2 a_3^*/a_k\rangle_\phi$$

$$= -\omega_k - \Re\sum_{\mathbf{k}_1,\mathbf{k}_2\neq\mathbf{k}_3} W_{3k}^{12}\delta_{3k}^{12}\frac{J_1 J_2}{J_k}\langle\phi_1\phi_2\phi_3^*\phi_k^*\rangle_\phi = -\omega_k - 2\sum_{\mathbf{k}_1} W_{1k}^{1k} J_1$$

$$= -\omega_k - \omega_{NL}, \tag{7.13}$$

where we used the Wick's contraction rule when doing the random phase averaging.

7.11 Inconsistency of WT Expansions Without Frequency Re-Normalization: Exercise 6.12

If not eliminated by the frequency re-normalization, the diagonal terms in the interaction term (the ones with coinciding indices in the multi-index sums) yield terms $\sim T^2$. For example, the phase averages of $\left|b_k^{(1)}\right|^2$ would have an additional term $\omega_{NL}^2 T^2 n_k$. Importantly, the T^2-terms arising from $\left|b_k^{(1)}\right|^2$ and from $b_k^{(2)}b_k^{(0)*}$ do not cancel each other neither in the equations for the generating function or the PDF, nor in the kinetic equation for the energy spectrum (please check this!).

Since $\omega_k T \gg 1$, the T^2-terms (rather than the T-terms) will be dominant in this case. Thus, the respective evolution equations cannot be represented as T-independent PDE's for averaged objects (the generation function, the PDF's or the wave spectrum). Namely, no longer one can absorb the T-dependence into the time-filtered time derivatives of these objects, as it was done before for the T-terms. Thus, the time separation procedure appears to be inconsistent in this case.

The inconsistency is related with the fact that without the frequency correction the interactions amplitudes oscillate with frequency ω_{NL} (just like the original complex wave amplitudes a_k rotate with frequency ω_k). Thus, it one can only assume their approximate time independence in the iterative procedure if $\omega_{NL} T \ll 1$. On the other hand, such T would be insufficient for filtering out fast oscillations and thereby finding time derivatives corresponding to slow cumulative evolution of the statistical objects (the generation function, the PDF's or the wave spectrum). This is because these objects would have a high-derivative component oscillating with frequency ω_{NL}. Compare these oscillations with the ones considered in Exercise 6.7 and shown in Fig. 7.2.

7.12 Finding G_3-G_5: Exercise 6.16

When computing G_2 we established that the leading contributions arise from the zero-valence graphs which correspond to the internal pairings (of the phase factors) only. This is true for the other terms too. Thus, considering only the zero-valence contributions in G_3-G_5 (given by (6.100)–(6.102)) we get zero for G_4 and

$$G_3 = -\prod_l \delta(\mu_l) \sum_j \left(\tilde{\lambda}_j + \frac{\mu_j}{2J_j^{(0)}}\right) \left\langle b_j^{(2)} b_j^{(0)*} \right\rangle_\phi, \qquad (7.14)$$

$$G_5 = \frac{1}{2}\prod_l \delta(\mu_l) \sum_{j\neq m} \tilde{\lambda}_j \tilde{\lambda}_m \left\langle \left(b_j^{(1)} b_j^{(0)*} + b_j^{(1)*} b_j^{(0)}\right) b_m^{(1)} b_m^{(0)*} \right\rangle_\phi. \qquad (7.15)$$

Thus in (7.14) we can use (6.35) which immediately gives (6.109). In (7.15), one should use (6.19) and use the Wick's contraction rule. The second term in the sum of (7.15) will give zero, whereas the first one results in (6.111).

7.13 Appendix: Interaction Coefficient for the Deep Water Surface Waves

Dispersion relation for the deep water surface waves, $\omega = \sqrt{gk}$, does not allow three-wave resonances. Thus the leading process is four-wave, and by a quasi-identity canonical transformation the cubic interaction Hamiltonian (corresponding to quadratic nonlinear terms in the dynamical equation) can be removed from to the description, which transforms the Hamiltonian to the form (6.71). This task appears to be quite cumbersome, and the correct canonical transformation and the resulting Hamiltonian was obtained only in 1990 (24 years after finding the Zakharov-Filonenko spectrum for the gravity WT!) by Krasitskii [2]. Krasitskii's derivation is quite complicated, and a more efficient procedure was described in ZLF book [3] in Appendix 3 for general four-wave systems. This method uses a trick of finding the canonical transformation as a infinitesimal-time evolution operator with an auxiliary Hamiltonian. Interested reader should read this place in ZLF, as it gives a very clear and detailed explanation of the method. The only remark we add here is that the resulting expression (A3.7) in ZLF can be significantly simplified by removing the non-resonant part (the second and the third lines in (A3.7)) by an appropriate choice of function $\tilde{W}$.

The most compact and, therefore, practically useful expression for the interaction coefficient for the deep water surface waves was obtained by Zakharov in [4]. Here, we reproduce it for reference purposes:

$$W_{12}^{34} = -\frac{1}{32\pi^2(k_1k_2k_3k_4)^{1/4}}\{-12k_1k_2k_3k_4$$

$$-\frac{2}{g^2}(\omega_1+\omega_2)^2[\omega_3\omega_4((\mathbf{k}_1\cdot\mathbf{k}_2)-k_1k_2)+\omega_1\omega_2((\mathbf{k}_3\cdot\mathbf{k}_4)-k_3k_4)]$$

$$-\frac{2}{g^2}(\omega_1-\omega_3)^2[\omega_2\omega_4((\mathbf{k}_1\cdot\mathbf{k}_3)+k_1k_3)+\omega_1\omega_3((\mathbf{k}_2\cdot\mathbf{k}_4)+k_2k_4)]$$

$$-\frac{2}{g^2}(\omega_1-\omega_4)^2[\omega_2\omega_3((\mathbf{k}_1\cdot\mathbf{k}_4)+k_1k_4)+\omega_1\omega_4((\mathbf{k}_2\cdot\mathbf{k}_3)+k_2k_3)]$$

$$+((\mathbf{k}_1\cdot\mathbf{k}_2)+k_1k_2)((\mathbf{k}_3\cdot\mathbf{k}_4)+k_3k_4)$$

$$+(-(\mathbf{k}_1\cdot\mathbf{k}_3)+k_1k_3)(-(\mathbf{k}_2\cdot\mathbf{k}_4)+k_2k_4)$$

$$+(-(\mathbf{k}_1\cdot\mathbf{k}_4)+k_1k_4)(-(\mathbf{k}_2\cdot\mathbf{k}_3)+k_2k_3)$$

$$+4(\omega_1+\omega_2)^2\frac{((\mathbf{k}_1\cdot\mathbf{k}_2)-k_1k_2)((\mathbf{k}_3\cdot\mathbf{k}_4)-k_3k_4)}{\omega_{1+2}^2-(\omega_1+\omega_2)^2}$$

$$+4(\omega_1-\omega_3)^2\frac{((\mathbf{k}_1\cdot\mathbf{k}_3)+k_1k_3)((\mathbf{k}_2\cdot\mathbf{k}_4)+k_2k_4)}{\omega_{1-3}^2-(\omega_1-\omega_3)^2}$$

$$+4(\omega_1-\omega_4)^2\frac{((\mathbf{k}_1\cdot\mathbf{k}_4)+k_1k_4)((\mathbf{k}_2\cdot\mathbf{k}_3)+k_2k_3)}{\omega_{1-4}^2-(\omega_1-\omega_4)^2}\}, \qquad (7.16)$$

where we have used a shorthand notation $\omega_{1+2}\equiv\omega(\mathbf{k}_1+\mathbf{k}_2)$, etc.

From (7.16), one can see that the four-wave interaction coefficient W_{12}^{34} is zero if all four wavevectors are collinear. Thus, in 1D the gravity wave system is five-wave [5], see Hamiltonian (6.82).

References

1. Zakharov, V.E., Piterbarg, L.I.: Sov. Phys. Dokl. **32**, 560 (1987)
2. Krasitskii, V.P.: On the canonical transformation of the theory of weakly nonlinear waves with nondecay dispersion law. Sov. Phys. JETP **98**, 1644–1655 (1990)
3. Zakharov, V.E., L'vov, V.S., Falkovich, G.: Kolmogorov Spectra of Turbulence. Series in Nonlinear Dynamics. Springer (1992)
4. Zakharov, V.E.: Statistical theory of surface waves on fluid of finite depth. Eur. J. Mech. B/Fluids **18**, 327–344 (1999)
5. Dyachenko, A.I., Lvov, Y.V., Zakharov, V.E.: Five-wave interaction on the surface of deep fluid. Phys. D, **87**, 233 (1995)

Part III
Wave Turbulence Predictions

In this part we will discuss some key properties and solutions of the Wave Turbulence closures. We will talk about conservation of energy, waveaction and other invariants. We will discuss the cascade directions of such invariants, as well as Kolmogorov–Zakharov spectra corresponding to constant k-space fluxes. We will study locality and stability of WT spectra, finite size effects—discrete and mesoscopic WT, intermittent WT states with fluxes over the wave amplitude variable and interplay between the wavenumber and the amplitude fluxes in the Wave Turbulence cycle.

Chapter 8
Conserved Quantities in Wave Turbulence and their Cascades

8.1 Conserved Quantities in Wave Turbulence

Conservation laws represent the most general and important features of the wave systems, and therefore we will start with considering these laws. We will start with the *energy* and the *momentum* which are conserved for all WT systems. Then we will consider the *waveaction* invariant which is conserved for the even-order resonant wave interactions (four-wave, six-wave, etc). Furthermore, we will consider conditions for the presence of extra invariants. After this, we will consider our master example, the Petviashvili model, and will discuss relations between the energy and the momentum in the WT closure and the energy and the potential enstrophy invariants of the original Petviashvili equation. We will also find an extra invariant present in this system—*zonostrophy*.

8.1.1 Energy and Momentum

Energy and momentum are conserved for the WT closure at all levels, from the one-point statistics obtained within the RPA approach (random phases and amplitudes), to the N-point statistics requiring RP only (random phases but not necessarily random amplitudes). For simplicity, let us start with the one-point statistics.

The energy and the momentum *physical-space densities* are defined in WT as follows,

$$E = \int \omega_k n_k d\mathbf{k} \tag{8.1}$$

and

$$\mathbf{M} = \int \mathbf{k} n_k d\mathbf{k}. \tag{8.2}$$

S. Nazarenko, *Wave Turbulence*, Lecture Notes in Physics, 825,
DOI: 10.1007/978-3-642-15942-8_8, © Springer-Verlag Berlin Heidelberg 2011

For brevity, we will call these quantities simply *energy* and *momentum* and omit mention that they are actually physical-space densities.

Generally, one can write for a conserved quantity Φ with density ρ_k:

$$\Phi = \int \rho_k n_k d\mathbf{k}. \tag{8.3}$$

For example, for energy we have $\rho_k = \omega_k$, and for the j-th component of the momentum we have $\rho_k = k_j$.

8.1.2 Three-Wave Systems

Let us consider a general three-wave system with a kinetic equation given by (6.69), where the integration is either over half of the $\mathbf{k}$-space (as for the Petviashvili waves and similar systems described by a real physical-space amplitude) or over the full $\mathbf{k}$-space (as for wavefields with a complex physical-space amplitude). We find for $\dot{\Phi}$:

$$\dot{\Phi} = \int \rho_k \dot{n}_k d\mathbf{k} = \int \int \int (\rho_k \mathscr{R}_{12k} - \rho_k \mathscr{R}_{k12} - \rho_k \mathscr{R}_{2k1}) d\mathbf{k}_1 d\mathbf{k}_2 d\mathbf{k}$$
$$= \int \int \int \mathscr{R}_{12k}(\rho_k - \rho_1 - \rho_2) d\mathbf{k}_1 d\mathbf{k}_2 d\mathbf{k}, \tag{8.4}$$

where we have made a change of integration variables $\mathbf{k}_1, \mathbf{k}_2, \mathbf{k} \to \mathbf{k}, \mathbf{k}_1, \mathbf{k}_2$ and $\mathbf{k}_1, \mathbf{k}_2, \mathbf{k} \to \mathbf{k}_2, \mathbf{k}, \mathbf{k}_1$ in the second and the third terms in the integrand respectively.

Thus, if

$$\rho_k - \rho_1 - \rho_2 = 0 \tag{8.5}$$

on the resonant manifold given by the frequency and the wavevector resonant conditions,

$$\omega_k - \omega_1 - \omega_2 = 0 \tag{8.6}$$

and

$$\mathbf{k} - \mathbf{k}_1 - \mathbf{k}_2 = 0, \tag{8.7}$$

then Φ is conserved,

$$\Phi = \int \rho_k n_k d\mathbf{k} = \text{const.} \tag{8.8}$$

For the energy and the momentum, the resonant condition (8.5) is obviously satisfied, which proves conservation of these quantities,

$$E = \int \omega_k n_k d\mathbf{k} = \text{const} \tag{8.9}$$

and

$$\mathbf{M} = \int \mathbf{k} n_k d\mathbf{k} = \text{const.} \tag{8.10}$$

Usually, the energy and the momentum are the only two invariants in WT of three-wave systems. The only known (so far) example of the three-wave systems with an additional invariant is the Rossby/drift wave turbulence. The Petviashvili model is a special case of such systems, and we will consider this example and discuss the extra invariant in the next section.

8.1.2.1 Petviashvili Model

Recall that the Petviashvili equation (6.1) (Sect. 6.2) deals with a real physical-space wave amplitude and we only need half of the Fourier space, $k_x \geq 0$. In this situation, the x-component of the momentum,

$$M_x = \int_{k_x \geq 0} k_x n_k dk \geq 0, \tag{8.11}$$

is a positive invariant, but the y-component of the momentum,

$$M_y = \int_{k_x \geq 0} k_y n_k dk, \tag{8.12}$$

can be either positive or negative.

Similarly, the energy is also positive,

$$E = \int_{k_x \geq 0} \omega_k n_k d\mathbf{k} = \int_{k_x \geq 0} k_x k^2 n_k d\mathbf{k} = \text{const.} \tag{8.13}$$

A very distinct and exceptional property of the Rossby/drift wave systems (of which the Petviashvili model is a special case) is the presence of a *single* extra WT invariant called *zonostrophy* (because it is related to the formation of zonal jets, as we will see later) [1–4]. For the Petviashvili model, the zonostrophy density is [1–3]:

$$\rho_k = \frac{k_x k^2}{k_y^2 - 3k_x^2}. \tag{8.14}$$

Exercise 8.1 *Prove that the quantity (8.14) satisfies the condition (8.5) on the resonant manifold given by the frequency and the wavevector resonant conditions, (8.6) and (8.7), and therefore represents a density of a WT invariant.*

Like for the y-component of the momentum, the zonostrophy invariant is not sign-definite, i.e. it can be either positive or negative. However, for an important range of *zonal scales* $k_y > k_x\sqrt{3}$ the zonostrophy is positive, and as such it can be useful for finding the turbulent cascade directions, as we will see later.

For reference, we will also give an expression for the zonostrophy density for the most general form of the Rossby/drift systems with frequency

$$\omega_k = -\frac{\beta\rho^2 k_x}{1 + \rho^2 k^2}, \tag{8.15}$$

where ρ is a constant called the Rossby deformation radius (in the geophysical context) or the ion Larmor radius (in the plasma context) and β is a constant related to the medium's inhomogeneity.

The zonostrophy density in this case is [4]:

$$\rho_k = \arctan\frac{k_y + k_x\sqrt{3}}{\rho k^2} - \arctan\frac{k_y - k_x\sqrt{3}}{\rho k^2}. \tag{8.16}$$

This expression is not scale-invariant.

A particularly important special case which is most frequently considered in literature is the limit of small-scale Rossby/drift turbulence, $\rho k \to \infty$. The most relevant form of the extra invariant's density is to be obtained by subtracting an energy dominated part [1]:

$$\zeta_k = -\lim_{\rho\to\infty}\frac{5\rho^5}{8\sqrt{3}}(\rho_k - 2\sqrt{3}\omega/\beta\rho) = \frac{k_x^3}{k^{10}}(k_x^2 + 5k_y^2). \tag{8.17}$$

It is remarkable that this density is scale-invariant and positive (without any further assumptions of anisotropy), and, therefore, it plays an important quantity which allows to find the turbulence cascade directions, as was done in this small-scale limit in [5].

8.1.3 Four-Wave Systems

Now let us consider the four-wave systems with wave spectra described by the kinetic equation (6.81). Then for $\dot{\Phi}$ we have

$$\dot{\Phi} = \int \rho_k \dot{n}_k d\mathbf{k} = 4\pi\epsilon^4 \int \rho_k |W_{12}^{k3}|^2 \delta_{12}^{k3}\delta(\omega_{12}^{k3})n_1 n_2 n_3 n_k \left[\frac{1}{n_k} + \frac{1}{n_3} - \frac{1}{n_1} - \frac{1}{n_2}\right] d\mathbf{k}_1 d\mathbf{k}_2 d\mathbf{k}_3$$

$$= \int (\rho_k + \rho_3 - \rho_1 - \rho_2)|W_{12}^{k3}|^2\delta_{12}^{k3}\delta(\omega_{12}^{k3})n_1 n_2 n_3 n_k d\mathbf{k}_1 d\mathbf{k}_2 d\mathbf{k}_3 d\mathbf{k},$$

$$\tag{8.18}$$

where we changed variables $\mathbf{k} \rightleftarrows \mathbf{k}_3$, $\mathbf{k} \rightleftarrows \mathbf{k}_1$ and $\mathbf{k} \rightleftarrows \mathbf{k}_2$ in the second, the third and the fourth terms in the integrand.

Thus, if

$$\rho_k + \rho_3 - \rho_1 - \rho_2 = 0 \tag{8.19}$$

on the resonant manifold given by the frequency and the wavenumber resonant conditions,

$$\omega_k + \omega_3 - \omega_1 - \omega_2 = 0 \tag{8.20}$$

and

$$\mathbf{k} + \mathbf{k}_3 - \mathbf{k}_1 - \mathbf{k}_2 = 0, \tag{8.21}$$

then Φ is conserved.

Again, this immediately proves conservation of the energy and momentum. However, we now can also take $\rho_k \equiv 1$ which obviously satisfies the condition (8.19) (everywhere in the $\mathbf{k}$-space including the resonant manifold). This corresponds to conservation of the waveaction integral

$$\mathcal{N} = \int \omega_k n_k d\mathbf{k} = \text{const}, \tag{8.22}$$

which is also called the *number of particles* (or quasi-particles, depending on a particular application).

At this point, from our experience with the three-wave and the four-wave systems, it should be clear that the energy and the momentum are conserved for WT systems with any order of the resonant wave interactions, whereas the waveaction integral is conserved for all even-order resonant processes of the kind $2p \to 2p, \ (p \in \mathbb{N})$.

8.1.4 Conservation Laws in the Multi-Particle Statistics

Let us now consider the N-mode joint PDF, the equation for which was obtained in a more general setting (RP) compared to the one-mode description (RPA). Then, in place of (8.3) the general form of the conservation law is:

$$\Phi = \int \rho_k \langle J_k \rangle d\mathbf{k} = \int d\mathbf{k} \rho_k \int s_k \mathcal{P}^{(N,a)} \{s_k\} \mathcal{D} s_k. \tag{8.23}$$

For transparency of derivations, let us "undo" the infinite-box limit and consider the discrete version of this expression,

$$\Phi = \sum_{\mathbf{k} \in B_N} \rho_k \int_0^\infty \int_0^\infty \cdots \int_0^\infty s_k \mathcal{P}^{(N,a)} ds_1 ds_2 \ldots ds_N. \tag{8.24}$$

We will first deal with the three-wave case. Correspondingly, let us also consider the discrete version of the equation (6.118) for the N-mode PDF,

$$\dot{\mathscr{P}} = 8\pi\epsilon^2 \sum_{j,m,n} |V_{mn}^j|^2 \delta(\omega_{mn}^j) \delta_{m+n}^j \left[\frac{\delta}{\delta s}\right]_3 \left(s_j s_m s_n \left[\frac{\delta}{\delta s}\right]_3 \mathscr{P}\right). \qquad (8.25)$$

Of course, as we mentioned earlier, the frequency delta-function strictly speaking is undefined in this formulation, and we should loosely think of a "broadened" delta-function which covers many Fourier modes. Once the main steps in the discrete settings are done, it should become clear how to do it rigorously manipulating directly with the functional integrals in the continuous case.

Let us differentiate Φ in the expression (8.24) and use the evolution equation (8.26); we have:

$$\dot{\Phi} = \sum_{k\in B_N} \rho_k \int_0^\infty \int_0^\infty \cdots \int_0^\infty s_k \dot{\mathscr{P}} ds_1 ds_2 ... ds_N$$

$$= 8\pi\epsilon^2 \int_0^\infty \int_0^\infty \cdots \int_0^\infty \rho_k s_k \sum_{k,j,m,n} |V_{mn}^j|^2 \delta(\omega_{mn}^j) \delta_{m+n}^j \left[\frac{\delta}{\delta s}\right]_3 \left(s_j s_m s_n \left[\frac{\delta}{\delta s}\right]_3 \mathscr{P}\right) ds_1 ds_2 ... ds_N$$

$$= 8\pi\epsilon^2 \int_0^\infty \int_0^\infty \cdots \int_0^\infty \sum_{j,m,n} (\rho_m + \rho_n - \rho_j)|V_{mn}^j|^2 \delta(\omega_{mn}^j) \delta_{m+n}^j s_j s_m s_n \left[\frac{\delta}{\delta s}\right]_3 \mathscr{P} ds_1 ds_2 ... ds_N.$$

Thus we see that the N-mode PDF equation has the same invariants as the one-mode kinetic equation, namely *the quantity Φ defined in (8.23) is an invariant of the N-mode PDF equation (6.118) if its density ρ_k satisfies the resonant condition (8.5) on the manifold of the frequency and the wavenumber resonances defined by the conditions (8.6) and (8.7).*

Remark 8.1 Conditions for conservation. Let us repeat that the N-mode PDF equation (6.118) is obtained under the RP condition which is less restrictive than the RPA condition leading to the one-mode description. Therefore, the invariants we have discussed here are conserved in a broader class of statistical WT systems than the ones described by the wave kinetic equation (6.69).

For example, one can consider a set of waves which initially have random phases but strongly correlated amplitudes. Then the amplitude de-correlation will occur at the time-scale of the N-mode PDF equation (6.118) (which is the same as the time-scale of the kinetic equation, i.e. $1/\epsilon^2$ in the three-wave case). In this case, the kinetic equation (6.69) will be inapplicable, but the N-mode PDF equation (6.118) will hold, and therefore the invariants discussed above will be conserved. Later we will consider solutions of the N-mode PDF equation (6.118), and we will particularly discuss the regimes where the amplitudes are correlated and the one-mode description is inapplicable.

Exercise 8.2 *Consider the N-mode PDF equation corresponding to the four-wave case, (6.122). Show that this equation conserves a quantity of form (8.23) if its* *density ρ_k satisfies the resonant condition (8.19) on the manifold of the frequency and the wavenumber resonances defined by the conditions (8.20) and (8.21). In particular, argue that the energy (with $\rho_k = \omega_k$), the momentum (with $\rho_k = \mathbf{k}$) and the waveaction (with $\rho_k = 1$) are conserved.*

8.1.5 Relation Between the Dynamical and the Statistical Invariants

The statistical invariants discussed above, obtained within the RPA or RP closures, are often related to respective invariants of the original nonlinear wave equation. Consider for example the de-focusing NLS model (5.1) which is a four-wave system. In this case, the energy invariant is given by Hamiltonian (6.71) with $W_{34}^{12} = 1$, i.e.

$$E = \int (|\nabla \psi|^2 + \tfrac{1}{2}|\psi|^4)d\mathbf{x} = H_2 + H_4 = \sum_k \omega_k \hat{\psi}_k \hat{\psi}_k^* + \frac{1}{2}\sum_{1,2,3,4} \delta_{34}^{12} \hat{\psi}_1 \hat{\psi}_2 \hat{\psi}_3^* \hat{\psi}_4^*$$

(8.26)

with $\omega_k = k^2$.

Let us consider the case of weak nonlinearity, so that in the Hamiltonian above $H_2 \gg H_4$. We write for the energy

$$E \approx H_2 = \int \omega_k n_k d\mathbf{k}.$$

(8.27)

Thus we see that the statistical energy invariant is approximately equal to the dynamical energy invariant under one of the assumptions of the WT closure, namely the weak nonlinearity. Note however that the other assumption (PR or RPA) is not necessary in this case.

The same is true for our other example, the Petviashvili model, in which case the dynamical counterpart of the statistical energy invariant is given in (6.3). It is interesting that the WT x-momentum invariant in this case *exactly* coincides with its dynamical counterpart given in (6.2).

On the other hand, the waveaction invariant of the dynamical NLS model,

$$\mathcal{N} = \int |\psi|^2 d\mathbf{x} = \int n_k d\mathbf{k},$$

(8.28)

coincides *exactly* with the waveaction invariant of the WT closure, i.e. without further weak nonlinearity or phase randomness assumptions. However, this fact is exceptional and in most four-wave (or six-wave) systems the waveaction invariant is conserved *approximately* in the original dynamical nonlinear system. The condition for such an approximate conservation is, again, weakness of nonlinearity, which is why such an invariant is often called *adiabatic*. Note that the role of the further statistical condition for the WT closure in conservation of the waveaction (e.g. for the water surface gravity waves) is not necessary.

Exercise 8.3 *Show that for the systems with Hamiltonian of type* (6.71) *the waveaction is conserved exactly.Why then do we say that the waveaction for the water surface gravity waves is conserved approximately?*

For some other WT invariants, approximate conservation in the original dynamical equation also typically requires weakness of nonlinearity, whereas the role of the phase randomness (RP) could be unclear. An example of such an invariant, zonostrophy, was considered above when we considered the Petviashvilli model, see (8.14).

8.2 Directions of Turbulent Cascades

Recall from Sects. 2.1.3.2 and 2.1.3.3 the dual cascade in 2D turbulence. The directions of the cascades of the respective two invariants, the energy and the enstrophy were determined via the Fjørtoft argument (which, as we saw, has several variations, e.g. for the one for forced and for decaying turbulence as in Sects. 2.1.3.2 and 2.1.3.3, respectively). Now we can use similar arguments for finding the cascade directions for the WT invariants discussed above.

The Fjørtoft argument is only meaningful if there are at least two quadratic (with respect to the wave amplitude) positive invariants—otherwise there is only one turbulent cascade transferring the energy from large to small scales (similarly to the 3D Navier–Stokes turbulence).

8.2.1 Dual Cascade in the NLS and Other Even-Wave Systems

The most straightforward application of the methods we have used for the 2D turbulence to WT can be found for the NLS example. Indeed, comparing the two NLS invariants (8.27) and (8.28) and the two 2D Navier–Stokes invariants (2.6) and (2.7), we see a clear analogy $E \rightarrow \Omega$ and $\mathcal{N} \rightarrow E$, see the table below.

	Invariant 1	Invariant 2
2D Navier–Stokes	$E = \int_0^\infty E^{(1D)}(k)dk$	$\Omega = \int_0^\infty k^2 E^{(1D)}(k)dk$
NLS	$\mathcal{N} = \int n_k d\mathbf{k}$	$E = \int k^2 n_k d\mathbf{k}$

In both systems, the ratios of the spectral densities of the invariants are k^2 and, therefore, the Fjørtoft argument for the NLS invariants $\mathcal{N}$ and E is word-to-word

Fig. 8.1 Dual cascade in the NLS and other even-wave systems with $\omega_k \sim k^\alpha$, $\alpha > 0$

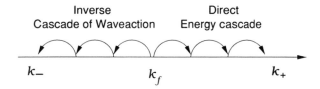

identical to the one given in Sects. 2.1.3.2 and 2.1.3.3 for the invariants E and Ω in 2D turbulence.

Thus we conclude that in the NLS wave turbulence, the energy E cascades *directly*, from small to large k's, whereas the "particles" cascade *inversely*, from large to small k's, see Fig. 8.1.

Further, the same argument applies almost without change to any even-wave systems which conserve the energy (8.27) and the "particles" (8.22). Indeed, according to the expressions (8.27) and (8.22), the ratio of the densities of invariants E and $\mathcal{N}$ is ω_k and, therefore, the Fjørtoft argument in this case leads to a conclusion that E must be transferred to the high-ω_k scales and $\mathcal{N}$ to the low-ω_k scales. In particular, for $\omega_k \propto k^\alpha$ with $\alpha > 0$ the cascade of E and $\mathcal{N}$ are directed to the high and the low k's respectively (recall Exercise 3.3), i.e. exactly like for the NLS system, as shown in Fig. 8.1. This is true, for example, for the water-surface gravity waves, for which $\alpha = 1/2$.

Remark 8.2.1 Fjørtoft argument does not use locality of the transfers of the invariants in the k-space. Thus, it will hold even if the dominant transfer is non-local, i.e. directly from very small to very large (or vice versa) scales, rather than a step-by-step transfer between similar-size scales as in a local turbulent cascade.

Remark 8.2.2 To reach a steady state, an effective large-scale dissipation must be present, for example a bottom friction (like in the 2D turbulence). On the other hand, often there is no effective dissipation at large scales. In this case, the inverse cascade leads to *condensation* at the largest scales of the system, which is similar to the Bose-Einstein condensation (BEC) phenomenon. For example, in the sea wave system such a nonlinear condensation at long waves is called the *swell* (or wave aging).

Remark 8.2.3 De-focusing NLS is often used as a model for BEC, and as mentioned in the previous remark, the connection between the inverse cascade and the condensation is more than just an analogy: it models a real condensation process in highly non-equilibrium (turbulent) setting. The forward energy cascade in this example can be linked to the so-called "evaporation cooling", when the high-energy particles "spill out" over the potential barrier of the magnetic trap retaining BEC.

Remark 8.2.4 Fjørtoft argument relies on the both invariants being quadratic with respect to the wave amplitude. For NLS, as well as for any other WT system, the energy E is quadratic only for weakly nonlinear waves, $H_2 \gg H_4$. The inverse cascade induced condensation may lead to gradual accumulation of energy until the breakdown of the weak nonlinearity assumption so that $H_4 \sim H_2$. The Fjørtoft argument does not apply in these conditions, and one of two things may happen:

- The condensation continues, even though it cannot be deduced from a Fjørtoft-like argument, and eventually a strong condensate state with $H_4 \gg H_2$ is achieved. After that the relevant description is the one introduced by Bogo-liubov, i.e. acoustic-like WT on the background of a strong coherent condensate mode which is uniform (or almost uniform) in the physical space.
- The condensation is arrested when $H_4 \sim H_2$ is reached. In addition, the k-space densities of H_4 and H_2 may become equal on the *scale-by-scale* basis, i.e. the system reaches the *Critical Balance* state described in Sect. 3.2.

In Chap. 15 we will consider the NLS system in greater detail, and will show that both of the above possibilities can be realized by choosing different types of dissipation (or absence of thereof) at low k's.

8.2.2 Cascade of Momentum and Other Non-Positive Invariants

Let us now include into consideration a quadratic invariant which is not sign-definite (i.e. can take both positive and negative values). The most important of such invariants are the components of the momentum (8.2) for isotropic WT systems (both three- and four-wave), or just the y-component of the momentum for the Petviashvili model (or other systems with reduced k-space).

With respect to such invariants, the Fjørtoft-like argument works only one-way: positive invariants restrict the k-space fluxes of non-positive invariants, but a non-positive invariant cannot restrict cascades of any other invariants (positive or non-positive ones).

Let us consider, for example, the y-component of the momentum, M_y, which has the k-space density k_y, and let the system be isotropic with frequency $\omega_k \propto k^\alpha$. Let us excite WT near a scale $\mathbf{k}_0$ and dissipate at scales remote from $\mathbf{k}_0$ (at large $k = |\mathbf{k}|$ or at small values of components of $\mathbf{k}$).

Suppose that, together with energy, some non-zero (positive or negative) momentum is generated at the forcing scale $\mathbf{k}_0$ (i.e. the forcing is anisotropic). The Fjørtoft *ad absurdum* argument tells us that the y-momentum cannot be dissipated at scales where $|k_y/k_{0y}| \ll k^\alpha/k_0^\alpha$ because otherwise at these scales there would be a huge amount of dissipated energy, a much greater amount than the energy produced by the forcing, which is impossible in steady-state turbulence. Thus the y-momentum is dissipated at the scales where $|k_y/k_{0y}| \gg k^\alpha/k_0^\alpha$. For $\alpha > 1$, this means that *the y-momentum is dissipated at the small k's*, $k \ll k_0$, and not too close to the k_x-axis (i.e. excluding $|k_y/k_{0y}| \ll |k_x|^\alpha/k_0^\alpha$). Similarly, for $\alpha > 1$, this means that *the x-momentum is dissipated at the small k's* and not too close to the k_y-axis.

Summarizing, *for $\alpha > 1$*, all components of the momentum $\mathbf{M}$ can be dissipated only at the large scales where $k \ll k_0$, which means that *the momentum cascade is inverse* in this case.

Now let us consider $\alpha < 1$. In this case the y-momentum can only be dissipated at the small scales excluding the scales close to the k_x-axis, i.e. excluding the sector

$$|k_y/k_{0y}| \ll |k_x|^\alpha/k_0^\alpha. \tag{8.29}$$

Respectively, the x-momentum can only be dissipated at the small scales excluding the scales close to the k_y-axis,

$$|k_x/k_{0x}| \ll |k_y|^\alpha/k_0^\alpha. \tag{8.30}$$

Also, for $\alpha < 1$ the system may be four-wave in which case it conserves the particles. It is easy to see that the presence of the particle invariant means that the **M**-components must dissipate not too close to the respective axis, i.e. excluding sector $|k_y| \ll |k_{0y}|$ for the y-momentum and sector $|k_x| \ll |k_{0x}|$ for the x-momentum. However, these inequalities are weaker than (8.29) and (8.30) for large k's, and, therefore, they do not impose extra restrictions on the the momentum cascades.

Thus we see that for *for* $\alpha < 1$, all components of the momentum **M** can be dissipated only at the small scales where $k \gg k_0$ (excluding the mentioned sectors close to the axes), which means that *the momentum cascade is direct.*

Remarks 8.2.5 We emphasize that the component M_y cannot impose any extra restrictions on the cascades of the energy or the particles. This is because a conclusion about dissipating a huge amount of M_y would not lead to a contradiction, because it could be compensated by an equally huge dissipation of M_y of the opposite sign elsewhere in the **k**-space (so that the total dissipated amount is finite and equal to the one produced at the forcing scale). The same is true for the other non-positive invariants: none of them can impose any extra restrictions on the cascades in the system via a Fjørtoft-type argument.

8.2.3 Triple Cascade in the Petviashvilli and Other Rossby/Drift Wave Systems

Now let us consider the Petviashvilli system. Recall that the Petviashvilli model was introduced as the simplest example which was supposed to exhibit all generic WT properties. However, we have already seen that the Petviashvilli model appears to represent a rather unique class of the Rossby/drift systems which possess a single extra quadratic invariant, zonostrophy. Thus, we are now going to consider the three invariants,

$$E = \int k_x k^2 n_k d\mathbf{k},$$

$$M_x = \Omega = \int k_x n_k d\mathbf{k}, \tag{8.31}$$

$$\Phi = \int \rho_k n_k d\mathbf{k},$$

with $\rho_k = k_x k^2/(k_y^2 - 3\ k_x^2)$. Note that Φ are not sign definite. however, let us further assume that we deal only with the nearly zonal scales such that $|k_y| \gg k_x$. In this case

$$\rho_k = \frac{k_x k^2}{k_y^2 - 3k_x^2} \approx k_x(1 + 4k_x^2/k_y^2),$$

so that the quantity

$$\tilde{\rho}_k = \frac{1}{4}(\rho_k - k_x) \approx k_x^3/k_y^2 \tag{8.32}$$

represents a **k**-space density of a *positive* invariant,

$$\tilde{\Phi} = \int \tilde{\rho}_k n_k d\mathbf{k}.$$

Now we have three positive quadratic invariants, E, Ω and $\tilde{\Phi}$, which is a very special property of the Rossby/drift wave systems. However, we are going to focus on this un-typical situation for a while because, as we will immediately see, it leads to a fascinating picture of a *triple cascade*. It is easy to see that three cascades cannot divide the **k**-space in an isotropic way (as was possible for the dual cascade), and this has an important practical consequence—generation of zonal flows.

Let us, following papers [2, 3], develop a Fjørtoft-type argument for this case. Suppose that WT is generated at scales around some $\mathbf{k}_0$ on the 2D **k**-plane, and dissipated at the scales which are separated from $\mathbf{k}_0$, namely at $|\mathbf{k}| \gg |\mathbf{k}_0|$ and at $k_x \ll k_{0x}$ and $|k_y| \ll |k_{0y}|$, see Fig. 8.2 (for simplicity we consider turbulence whose statistical properties are symmetric with respect to $k_y \to -k_y$, and therefore we show only one quadrant of the **k**-space). In Kolmogorov's spirit, the scales between the forcing and the dissipation are conservative, and thus we have a ring-shaped inertial range in this case.

Fig. 8.2 Triple cascade in the Petviashvilli model of Rossby/drift wave turbulence

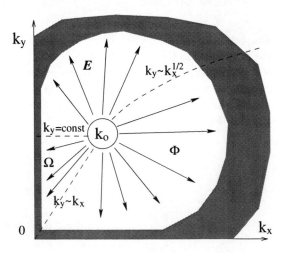

In fact, the Fjørtoft argument can be generalized and summarized in a very compact way:

Theorem 8.1 *Each positive quadratic invariant must dissipate in its own sector of the **k**-space in which its density is greater than the density of the other positive quadratic invariants.*

We have formulated this as a theorem, and indeed it can be proven to be asymptotically rigorous when the forcing/dissipation scale separation tends to infinity. We are not going to give rigorous bounds, but the idea is roughly that, if we assumed *ad absurdum* that a particular invariant was dissipated outside of thus prescribed sector, then another invariant, with the greater density at these scales, would have to be hugely over-dissipated, which is impossible in the steady state.

It is clear that the cascade sectors do not intersect and their unity covers the whole of the 2D **k**-plane. The boundary of the three sectors is given by equating the respective **k**-space densities, see Fig. 8.2:

1. $\omega_k \propto k_x$ for the boundary between the cascades of E and Ω. Substituting for ω_k, we have $k_y \sim$ const.
2. $\omega_k \propto \tilde{\rho}_k$ for the boundary between the cascades of E and $\tilde{\Phi}$. This gives $k_y \propto \sqrt{k_x}$.
3. $k_x \propto \tilde{\rho}_k$ for the boundary between the cascades of Ω and $\tilde{\Phi}$. This gives $k_y \propto k_x$.

Remark 8.2.6 The Fjørtoft argument deals with strong inequalities (i.e. $\gg$ and $\ll$ instead of $>$ and $<$). This means that the boundaries between the different cascade sectors in Fig. 8.2 are not sharp. In other words, cascade of a particular invariant can follow a boundary of its designated sector and even slightly cross this boundary. The Fjørtoft argument only prohibits crossing the designated boundaries strongly, so that the density of the respective invariant should never become *much less* than the density of any other invariant.

Remark 8.2.7 The Fjørtoft argument prescribes energy to cascade to lower k's and closer to the k_y-axis, which corresponds to the zonal scales. Thus, presence of the extra invariant makes the Rossby/drift turbulence anisotropic and leads to generation of large-scale zonal jets. This is a prototype process for the jet formation mechanism in atmospheres of giant planets, like Jupiter and Saturn.

Remark 8.2.8 For 2D turbulence, we have developed an alternative version of Fjørtoft argument in terms of the centroids, see exercise 2.1. Obviously, the same argument can be used for finding the boundary between the energy and the enstrophy cascades for the Rossby/drift turbulence. However, using the centroids for finding the $E - \Phi$ and the $\Phi - \Omega$ boundaries appears to be much harder, and it has not been done as yet.

Remark 8.2.9 Conservation of zonostrophy is a property of the dispersion relation for the wave frequency only. It holds for different types of the nonlinear interaction, in particular for the nonlinearity of the Charney–Hassegawa–Mima model (6.13).

Exercise 8.4 *For the small-scale Rossby/drift turbulence, the zonostrophy density is given by (8.17). Derive the cascade directions for this case* [5].

References

1. Nazarenko, S.V.: Physical realisability of anisotropic spectra of wave turbulence and structure of the Rossby/drift turbulence, PhD thesis, Landau Institute for Theoretical Physics AN USSR, Chernogolovka (1990)
2. Balk, A.M., Nazarenko, S.V., Zakharov, V.E.: On the structure of the Rossby/drift turbulence and zonal flows. In: Proceedings of the International Symposium "Generation of Large-Scale Structures in Continuous Media", June 11–20, 1990, Perm-Moscow, USSR, pp. 34–35
3. Balk, A.M., Nazarenko, S.V., Zakharov, V.E.: New invariant for drift turbulence. Phys. Lett. A **152**, 5, 6, 276 (1991)
4. Balk, A.M.: A new invariant for Rossbywave systems. Phys. Lett. A **155**(1), 20 (1991)
5. Nazarenko, S., Quinn, B.: Triple cascade behaviour in QG and drift turbulence and generation of zonal jets. Phys. Rev. Lett. **103**(11) (2009), Article Number: 118501; Extended version on arxiv.org: nlin.CD arXiv:0905.1314

Chapter 9
Steady State and Evolving Solutions for the Wave Spectrum

9.1 Thermodynamic Equilibrium States: Rayleigh-Jeans Spectra

First, we will consider the simplest steady states which have no fluxes of energy or other invariants and the wave system is in thermodynamic equilibrium. These states are described by Rayleigh-Jeans (RJ) spectra, which in general have a form

$$n_k = \frac{1}{\sum_j C_j \rho_k^{(j)}}, \tag{9.1}$$

where $\rho_k^{(j)}$'s are the densities of the motion integrals described in the previous section and C_j's are dimensional constants corresponding to the thermodynamic potentials of the respective invariants.

The RJ spectrum (9.1) corresponds to equipartition in the **k**-space of the invariant whose density is given by the expression in the denominator of (9.1) (and which is a linear combination of the main invariants in the system).

On the RJ spectrum (9.1), the integrand in the RHS of the kinetic equation is equal to zero pointwise, i.e. for each individual triad (or quartet, etc.). Thus, it corresponds to a non-dissipative system with "detailed balance" of the local transfer rates, so that the net flux over scales is null.

Exercise 9.1 *Check that the form of n_k in (9.1) is indeed a solution of the kinetic equation for any wave systems: three-wave, four-wave, five-wave, etc.*

For example, for the Petviashvili system the RJ spectrum is

$$n_k = \frac{T}{\omega_k + \lambda \rho_k + \mathbf{u} \cdot \mathbf{k}}, \tag{9.2}$$

S. Nazarenko, *Wave Turbulence*, Lecture Notes in Physics, 825,
DOI: 10.1007/978-3-642-15942-8_9, © Springer-Verlag Berlin Heidelberg 2011

where T, and ρ are dimensional constants which could be interpreted as "temperature" and "zonostrophy potential" and constant vector $\mathbf{u} = (u_x, u_y)$ is a "momentum potential".

For the even-order systems, the RJ spectrum is

$$n_k = \frac{T}{\omega_k + \mu + \mathbf{u} \cdot \mathbf{k}}, \tag{9.3}$$

where μ is a "chemical potential" constant which corresponds to conservation of the waveaction (particles). Here we assumed that there is no extra invariants besides the energy, the particles and the momentum.

Remark 9.1.1 In choosing the constants in the RJ spectrum, one has to ensure that the physical condition $n_k > 0$ is satisfied throughout the parts of the k-space occupied by the RJ spectrum, taking particular care for the contributions that may change sign.

Particularly, the momentum contribution $\mathbf{u} \cdot \mathbf{k}$ may change sign. Recall that for the Petviashvili system $k_x \geq 0$, so choosing $u_x > 0$ would ensure that $u_x k_x > 0$. However, k_y takes both positive and negative values, so whenever $u_y k_y < 0$ it would have to be compensated by the positive contributions in the denominator of (9.2).

RJ-states have been known since Peierls derived the very first wave kinetic equation (for phonons in anharmonic crystals) in 1929 [1]. These solutions are, however, of limited relevance to WT. They can only be realized in truncated in k-space systems, with a finite number of modes without forcing or dissipation, which are unrealistic. No RJ spectrum can be realized throughout the *infinite* k-space because this would mean that the x-space density of one (or several) motion integrals would be infinite. This is true, for example, for the invariant whose density is given by the expression in the denominator of (9.1), because its density is constant in the k-space and, therefore, the k-space integration would give infinity. Un-realizability of RJ spectra is related to so-called *Ultraviolet Catastrophe*, see discussion in Chap. 15.

However, a RJ spectrum could form in a *part* of the k-space, for example when a turbulent energy cascade cannot be effectively dissipated and the energy flux is stagnated at small scales forming a "thermalized" set of waves, which was called a *warm cascade* state [2]. This stagnation is often called a *bottleneck phenomenon* and it is observed, e.g., in the crossover scales between the hydrodynamic and the Kelvin wave scales in superfluid turbulence [3]. Bottlenecks also arise in WT of Bose-Einstein condensates, and they will be discussed in this context in Chap. 15.

9.2 Cascade States: Kolmogorov-Zakharov Spectra

The most important and relevant solutions in WT are those which correspond to k-space fluxes of the motion integrals. They are called Kolmogorov-Zakharov

(KZ) spectra because they are analogous to the famous Kolmogorov-Obukhov (K41) spectrum of hydrodynamic turbulence, and because they were first discovered in WT by Zakharov in 1965 [4]. In 1966–1967, Zakharov and Filonenko proved the existence of such solutions for the capillary waves and for the deep-water waves [5, 6].

9.2.1 Three-Wave Systems

To illustrate the ideas we will first consider isotropic media (e.g., capillary waves), and then we will return to our main example, the Petviashvili model, which is anisotropic. Let us consider a three-wave system and recall the kinetic equation (6.69) and (6.70),

$$\dot{n}_k = \int (\mathscr{R}_{12k} - \mathscr{R}_{k12} - \mathscr{R}_{2k1})d\mathbf{k}_1 d\mathbf{k}_2,$$

$$\mathscr{R}_{12k} = 2\pi\epsilon^2 |V_{12k}|^2 \delta_{12}^k \delta(\omega_{12}^k)(n_1 n_2 - n_2 n_k - n_k n_1).$$

Furthermore, let us assume that the dispersion relation has a simple power-law form,

$$\omega_k = \lambda k^\alpha, \tag{9.4}$$

with $k = |\mathbf{k}|$ and with λ and α constant. We also introduce the interaction coefficient

$$V_{12}^k = V(\mathbf{k}, \mathbf{k}_1, \mathbf{k}_2), \tag{9.5}$$

which only depends on the relative wave vector orientations with respect to each other, not the absolute orientations. Note that when explaining the KZ solutions, the ZLF book [7] assumed instead

$$V_{12}^k = V(k, k_1, k_2), \tag{9.6}$$

(see page 87 of [7]) which unfortunately rules out most known examples (even though it was done properly in later research papers, e.g. [8]).

Finally, we assume an isotropic spectrum, i.e. $n_k = n(k)$. We can now write the kinetic equation

$$\dot{n}_k = \int (\hat{\mathscr{R}}_{k12} - \hat{\mathscr{R}}_{12k} - \hat{\mathscr{R}}_{2k1})k_1^{d-1}k_2^{d-1}dk_1 dk_2, \tag{9.7}$$

with

$$\hat{\mathscr{R}}_{k12} = 2\pi\epsilon^2 \delta(\omega_{12}^k)(n_1 n_2 - n_2 n_k - n_k n_1)\Upsilon_{12}^k, \tag{9.8}$$

where

$$\Upsilon_{12}^k = \int |V_{12}^k|^2 \delta_{12}^k d\mathbb{S}_1^{d-1} d\mathbb{S}_2^{d-1}, \tag{9.9}$$

and $\mathbb{S}_k^{d-1}$ is the $(d-1)$-dimensional unit sphere in the d-dimensional **k**-space.

We note that for scale-invariant solutions, we require homogeneity of the interaction coefficient, i.e.

$$V(\mu\mathbf{k}, \mu\mathbf{k}_1, \mu\mathbf{k}_2) = \mu^\beta V_{12}^k, \tag{9.10}$$

for any $\mu > 0$ with constant β which is called the homogeneity degree. We may check the homogeneity of Υ :

$$\begin{aligned}\Upsilon(\mu k, \mu k_1, \mu k_2) &= \int |V(\mu\mathbf{k}, \mu\mathbf{k}_1, \mu\mathbf{k}_2)|^2 \delta(\mu k - \mu k_1 - \mu k_2) d\mathbb{S}_1^{d-1} d\mathbb{S}_2^{d-1} \\ &= \mu^{2\beta-d} \Upsilon_{12}^k.\end{aligned} \tag{9.11}$$

Let us now look for a steady state solution of the power-law form, $n_k = Ak^\nu$, with constant ν and A. Thus, we write

$$0 = \int \left(\hat{\mathscr{R}}_{k12} - \hat{\mathscr{R}}_{12k} - \hat{\mathscr{R}}_{2k1}\right) k_1^{d-1} k_2^{d-1} dk_1 dk_2. \tag{9.12}$$

We now focus on the term with $\hat{\mathscr{R}}_{12k}$ and perform a *Zakharov transformation* (ZT) of the integration variables k_1 and k_2:

$$k_1 = \frac{k^2}{\tilde{k}_1}, \quad k_2 = \frac{k\tilde{k}_2}{\tilde{k}_1}. \tag{9.13}$$

Remark 9.2.1 The transformation (9.13) is a non-identity transformation, so we will need to test the solutions by substitution in the original equation (9.7). Indeed, ZT takes the limit of 0 wavenumber to ∞ and vice versa, which may lead (and often does) to cancelation of a divergence with an oppositely signed divergence in the other terms, $\hat{\mathscr{R}}_{k12}$ and $\hat{\mathscr{R}}_{21k}$. If the original collision integral converges (e.g. as $k_1 \to 0$ and $k_1 \to \infty$) then the found spectrum is indeed a valid solution, in which case the spectrum is called local. If not, then the spectrum in question is a spurious solution; it is called a nonlocal spectrum. Furthermore, even a valid local spectrum may turn out unstable or evolutionally non-local, in which case we do not expect it to be realizable in practice. We will come back to the issues of locality and stability in Sect. 9.2.6.

Let us perform ZT in the different parts of $\hat{\mathscr{R}}_{12k}$. We write

$$\delta(\omega_{2k}^1) = \lambda^{-1}\delta(k_1^\alpha - k_2^\alpha - k^\alpha) \overset{ZT}{=} \lambda^{-1}\delta\left(\frac{k^{2\alpha}}{\tilde{k}_1^\alpha} - \frac{k^\alpha\tilde{k}_2^\alpha}{\tilde{k}_1^\alpha} - k^\alpha\right)$$

$$= \lambda^{-1}\frac{\tilde{k}_1^\alpha}{k^\alpha}\delta\left(k^\alpha - \tilde{k}_2^\alpha - \tilde{k}_1^\alpha\right) = \frac{\tilde{k}_1^\alpha}{k^\alpha}\delta(\tilde{\omega}_{12}^k). \tag{9.14}$$

For the term involving n_k's we write

$$(n_2 n_k - n_k n_1 - n_1 n_2) = A^2\left(k_2^\nu k^\nu - k^\nu k_1^\nu - k_1^\nu k_2^\nu\right)$$

$$\overset{ZT}{=} A^2 k^{2\nu}\tilde{k}_1^{-2\nu}\left(\tilde{k}_1^\nu\tilde{k}_2^\nu - k^\nu\tilde{k}_1^\nu - k^\nu\tilde{k}_2^\nu\right) \tag{9.15}$$

$$= \left(\frac{k}{\tilde{k}_1}\right)^{2\nu}(\tilde{n}_1\tilde{n}_2 - \tilde{n}_k\tilde{n}_1 - \tilde{n}_k\tilde{n}_2).$$

Further,

$$\Upsilon_{k2}^1 = \Upsilon(k_1, k, k_2) \overset{ZT}{=} \Upsilon\left(\frac{k^2}{\tilde{k}_1}, k, \frac{k\tilde{k}_2}{\tilde{k}_1}\right) = \left(\frac{k}{\tilde{k}_1}\right)^{2\beta-d}\tilde{\Upsilon}_{12}^k, \tag{9.16}$$

where we have used the homogeneity of Υ, as expressed in (9.33) with $\mu = \frac{k}{\tilde{k}_1}$. The final term to be transformed in Eq. (9.7) is

$$(k_1 k_2)^{d-1}dk_1 dk_2. \tag{9.17}$$

The Jacobian of the ZT is $\left(\frac{k}{\tilde{k}_1}\right)^3$, so that the term (9.17) becomes

$$\left(\frac{k^2}{\tilde{k}_1}\frac{k\tilde{k}_2}{\tilde{k}_1}\right)^{d-1}\left(\frac{k}{\tilde{k}_1}\right)^3 d\tilde{k}_1 d\tilde{k}_2 = (\tilde{k}_1\tilde{k}_2)^{d-1}\left(\frac{k}{\tilde{k}_1}\right)^{3d}d\tilde{k}_1 d\tilde{k}_2. \tag{9.18}$$

Combining all the terms (9.14), (9.15), (9.16), and (9.18), we find

$$\hat{\mathscr{R}}_{12k}(k_1 k_2)^{d-1}dk_1 dk_2 = 2\pi\epsilon^2\left(\frac{\tilde{k}_1}{k}\right)^\alpha \delta(\tilde{\omega}_{12}^k)A^2\left(\frac{k}{\tilde{k}_1}\right)^{2\nu}$$

$$\times (\tilde{n}_1\tilde{n}_2 - \tilde{n}_k\tilde{n}_1 - \tilde{n}_k\tilde{n}_2)\left(\frac{k}{\tilde{k}_1}\right)^{2\beta-d}\tilde{\Upsilon}_{12}^k(\tilde{k}_1\tilde{k}_2)^{d-1}\left(\frac{k}{\tilde{k}_1}\right)^{3d}d\tilde{k}_1 d\tilde{k}_2$$

$$= \left(\frac{\tilde{k}_1}{k}\right)^x \tilde{\mathscr{R}}_{k12}(\tilde{k}_1\tilde{k}_2)^{d-1}d\tilde{k}_1 d\tilde{k}_2, \tag{9.19}$$

where

$$x = \alpha - 2\nu - 2\beta - 2d. \tag{9.20}$$

We can perform a similar transformation on $\hat{\mathscr{R}}_{2k1}$, so that we can write our collision integral (9.12) as

$$0 = \int \hat{\mathscr{R}}_{k12}\left[1 - \left(\frac{k_1}{k}\right)^x - \left(\frac{k_2}{k}\right)^x\right](k_1 k_2)^{d-1} dk_1 dk_2. \qquad (9.21)$$

Finally, we recall the dispersion relation (9.4) and note that if $x = \alpha$, the collision integral (9.21) contains the factor

$$\delta(\omega_{12}^k)\left[1 - \left(\frac{k_1}{k}\right)^x - \left(\frac{k_2}{k}\right)^x\right] = \frac{1}{\lambda k^\alpha}\delta(k^\alpha - k_1^\alpha - k_2^\alpha)(k^\alpha - k_1^\alpha - k_2^\alpha) = 0.$$

Hence $x = \alpha$ corresponds to a steady state solution of the kinetic equation (9.7). The scaling solution in this situation is

$$n_k = A k^\nu = A k^{-\beta - d}, \qquad (9.22)$$

which is called the Kolmogorov-Zakharov (KZ) spectrum. As we will see in the next section, this spectrum corresponds to a state with a constant energy flux through k.

9.2.1.1 Energy Flux in the k-space

We recall the Eq. (8.27) for the energy density,

$$E = \int \omega_k n_k d\mathbf{k} = \int E_k^{(1D)} dk$$

$$\text{where} \quad E_k^{(1D)} = \mathbb{A}^{(d-1)} \omega_k n_k k^{d-1} \qquad (9.23)$$

is the 1D energy spectrum and $\mathbb{A}^{(d-1)}$ is the area of $\mathbb{S}^{d-1}$, the $(d-1)$-dimensional unit sphere (for example $\mathbb{A}^{(1)} = 2\pi$ and $\mathbb{A}^{(2)} = 4\pi^2$).

Let us now write the continuity equation for the energy

$$\partial_t E_k^{(1D)} + \partial_k \varepsilon = 0, \qquad (9.24)$$

where $\varepsilon = \varepsilon(k)$ is the energy flux which, taking into account (9.7), can be expressed as

$$\epsilon(k) = -\mathbb{A}^{(d-1)} \int_0^k dq \omega_q \int_0^\infty\int_0^\infty dk_1 dk_2 (\hat{\mathscr{R}}_{q12} - \hat{\mathscr{R}}_{12q} - \hat{\mathscr{R}}_{2q1})(k_1 k_2 q)^{d-1}. \qquad (9.25)$$

In steady state $\partial_k \varepsilon = 0$, so we expect a constant energy flux, $\varepsilon = $ const. Let us substitute $n_k \sim k^\nu$ into Eq. (9.25) and find the scaling of ε in k (by taking all the k-dependence away from the integral in (9.25) so that the remaining integral is a non-dimensional number). We note

$$\hat{\mathscr{R}}_{k12} \sim k^{-\alpha + 2\nu - d + 2\beta},$$

where power $-\alpha$ comes from the frequency delta-function, 2ν comes from the bracket containing n's, $-d$ comes from the wavenumber delta-function, and 2β comes from the squared interaction coefficient. Thus, we have for the constant flux solution

$$\epsilon(k) \sim k^{\alpha} k^{-\alpha+2\nu-d+2\beta} k^{3d} \sim k^0$$
$$\Rightarrow \nu = -\beta - d.$$

As we see, this is the KZ solution (9.22) and, therefore, we have just showed that the KZ solution corresponds to constant energy flux over k.

Remark 9.2.2 The correspondence between the KZ solution and the constant energy flux is a direct analogue to the original Kolmogorov spectrum in hydrodynamic turbulence, $E_k \sim k^{-5/3}$. In the latter, we have a constant energy flux from lower to higher wavenumbers k.

The direction of the energy flux can be obtained from the following simple argument suggested by Uriel Frisch in the context of Navier–Stokes turbulence [9, 10]. Here, we will adopt it to WT. Consider a spectrum $n_k \sim k^\nu$ and let us insert, at some moment of time, barriers dividing the k-space into many non-interacting shells, see Figs. 9.1 and 9.2. Then, in each of the shells WT will quickly come to thermodynamic equilibrium with RJ spectrum $n_k \sim 1 / \omega_k \sim k^{-\alpha}$. At it is clear from Figs. 9.1 and 9.2, the energy will shift to the higher k's in each shell if the original spectrum is steeper than the thermodynamic one, $\nu < -\alpha$, and it will shift toward the lower k's otherwise. Now repeatedly withdraw the barriers and put them back in and obtain a cascading system where the energy flux is direct if $\nu < -\alpha$ and inverse if $\nu > -\alpha$.

Remark 9.2.3 The above argument implies that in the KZ solution (9.22) the energy cascade is direct if $\beta + d > \alpha$ and inverse otherwise.

Fig. 9.1 Forward cascade of energy in three-wave systems with spectra which are steeper than the thermodynamic spectrum

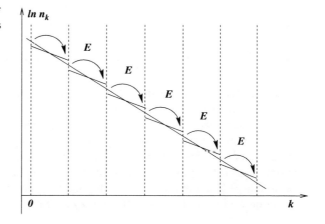

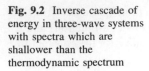

Fig. 9.2 Inverse cascade of energy in three-wave systems with spectra which are shallower than the thermodynamic spectrum

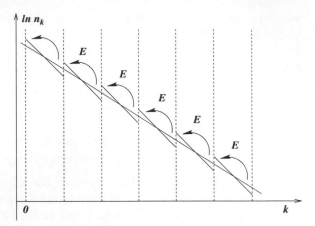

Exercise 9.2 *Prove, using results* (3.9) *and* (3.10), *that in three-wave WT systems with a single relevant dimensional parameter the energy flux in the KZ solutions is direct if* $\alpha < 5 + d$.

9.2.2 Four-Wave Systems

Let us recall the kinetic equation for the four-wave systems (6.81) and let us again assume that the medium and the wave spectrum are isotropic. We have

$$\dot{n}_k = 4\pi\varepsilon^4 \int |W_{12}^{k3}|^2 \delta_{12}^{k3} \delta(\omega_{12}^{k3}) n_1 n_2 n_3 n_k \left[\frac{1}{n_k} + \frac{1}{n_3} - \frac{1}{n_1} - \frac{1}{n_2}\right] d\mathbf{k}_1 d\mathbf{k}_2 d\mathbf{k}_3$$

$$= \int \mathfrak{T}_{12}^{k3} n_1 n_2 n_3 n_k \, (k_1 k_2 k_3)^{d-1} dk_1 dk_2 dk_3, \qquad (9.26)$$

where

$$\mathfrak{T}_{12}^{k3} = 4\pi\varepsilon^4 \left[\frac{1}{n_k} + \frac{1}{n_3} - \frac{1}{n_1} - \frac{1}{n_2}\right] \delta(\omega_{12}^{k3}) \int |W_{12}^{k3}|^2 \delta_{12}^{k3} \, d\mathbb{S}_1^{d-1} d\mathbb{S}_2^{d-1} d\mathbb{S}_3^{d-1}. \quad (9.27)$$

In the ZLF book [7], derivation of the KZ spectra proceeds by splitting the integration domain into four regions and by mapping three of these regions onto the fourth one. Here we will take a different route which is closer to the method we used for the three-wave case above. An extra advantage of this approach is that its generalizations to higher-order systems (five-wave, six wave, etc.) is rather straightforward. For this, let us first rewrite the kinetic equation (9.26) in a form resembling (9.12). First we note that, because the system is Hamiltonian, the interaction coefficient has the following symmetries,

$$W_{12}^{k3} = W_{12}^{3k} = W_{21}^{k3} = W_{k3}^{12*},\qquad(9.28)$$

using which we can rewrite (9.26) as

$$\dot{n}_k = \frac{1}{4}\int\left[\mathfrak{T}_{12}^{k3} + \mathfrak{T}_{12}^{3k} - \mathfrak{T}_{k2}^{13} - \mathfrak{T}_{1k}^{23}\right]n_1 n_2 n_3 n_k\,(k_1 k_2 k_3)^{d-1}dk_1 dk_2 dk_3,\qquad(9.29)$$

where the third and the fourth terms we obtained via the changes of indices of the integration variables as $3 \rightleftarrows 2$ and $3 \rightleftarrows 1$ respectively.

Now we look for a power-law steady state solution $n_k = Ak^\nu$ (with constant ν and A), and perform ZT in the second term of (9.29) as

$$k_1 = \frac{k\tilde{k}_1}{\tilde{k}_3},\qquad k_2 = \frac{k\tilde{k}_2}{\tilde{k}_3},\qquad k_3 = \frac{k^2}{\tilde{k}_3},\qquad(9.30)$$

in the third term as

$$k_1 = \frac{k^2}{\tilde{k}_1},\qquad k_2 = \frac{k\tilde{k}_2}{\tilde{k}_1},\qquad k_3 = \frac{k\tilde{k}_3}{\tilde{k}_1},\qquad(9.31)$$

and in the fourth term as

$$k_1 = \frac{k\tilde{k}_1}{\tilde{k}_2},\qquad k_2 = \frac{k^2}{\tilde{k}_2},\qquad k_3 = \frac{k\tilde{k}_3}{\tilde{k}_2}.\qquad(9.32)$$

Assuming homogeneity of the interaction coefficient,

$$W_{\mu k_1,\mu k_2}^{\mu k,\mu k_3} = \mu^\beta W_{k_1,k_2}^{k,k_3},\qquad(9.33)$$

and a power-law dispersion relation $\omega \sim k^\alpha$, we thus transform equation (9.29) into

$$\int \mathfrak{T}_{12}^{k3}\left[1 + \left(\frac{k_3}{k}\right)^x - \left(\frac{k_1}{k}\right)^x - \left(\frac{k_2}{k}\right)^x\right]n_1 n_2 n_3 n_k\,(k_1 k_2 k_3)^{d-1}dk_1 dk_2 dk_3 = 0,\qquad(9.34)$$

with

$$x = \alpha - 3\nu - 2\beta - 3d.\qquad(9.35)$$

Exercise 9.3 *Use homogeneity of the interaction coefficient, (9.33), and dispersion relation $\omega \sim k^\alpha$ to derive (9.34) and (9.35).*

Now we can find the KZ spectra, of which there will be two in this case. Like in the three-wave case, one solution will be given by $x = \alpha$:

$$\nu_E = -2\beta/3 - d.\qquad(9.36)$$

This solution corresponds to the direct energy cascade. The second KZ spectrum is given by $x = 0$:

$$\nu_N = -2\beta/3 - d + \alpha/3. \tag{9.37}$$

It corresponds to the inverse waveaction cascade.

Again, the KZ solutions are only meaningful if they are local, i.e. when the collision integral in the original kinetic equation converges.

Exercise 9.4 *Write expressions for the flux of energy and the flux of waveaction. Substitute into these expressions a power-law spectrum and find the scalings of the fluxes in k. Show that the KZ spectra (9.36) and (9.37) correspond to constant (k-independent) fluxes of the energy and the waveaction respectively.*

9.2.2.1 KZ Solutions With "Wrong" Flux Furections

Sometimes, the formal KZ solutions turn out with a "wrong" flux sign contradicting the Fjørtoft argument. Such solutions cannot be realized in any finite inertial range because, according to the Fjørtoft argument, it would be impossible to match them to the WT source and sink regions in a steady state. For example, the waveaction cascade KZ spectrum in the 2D NLS model is known to have a positive waveaction flux and, therefore, cannot be formed [7, 11] What would we observe instead in between of the forcing and the dissipation scales? Because the pure KZ solution does not suit, one should expect a "warm cascade" solution in which both the flux and thermal components are present. Such mixed solutions where discussed in [11, 12, 13] and their counterparts for strong 3D hydrodynamic turbulence were discussed in [2] (where the term "warm cascade" was coined) and for 2D hydrodynamic turbulence in [14]. It is natural to think that above a certain threshold value of temperature or/and chemical potential, the warm cascade solution may allow reversal of the "bad" flux, so that it starts pointing in the correct (by Fjørtoft argument) direction. Such a condition for the flux reversal should provide an expression for determining the systems temperature or/and chemical potential at the equilibrium state. In Chap. 15, we give examples of such expressions for turbulence in Bose-Einstein condensates and the quantum kinetics.

Finding the flux directions on the KZ solutions can be done directly by calculating the respective fluxes based on the kinetic equations, as was done, e.g., in [7, 11]. However, there exists a simple way to determine the flux directions based solely on the relative values of the exponents of the RJ and the KZ spectra, similar to the one presented in Chap. 2, Sect. 2.1.3.5 for the 2D turbulence. Indeed, let us consider a power-law spectrum $n_k = Ak^\nu$ (not necessarily a stationary one) and ask ourselves what will be the signs of the fluxes of E and $\mathcal{N}$. It is clear that the fluxes will both be positive if ν is very small (i.e. large negative) and they will be both negative if ν is very large, see Fig. 9.3. Indeed, if this was not the case then a localized initial spectrum would get narrower rather than spread out, and this would correspond to an unphysical "negative viscosity" behavior.

Further, the energy flux will cross zero for both of the RJ exponents and at the waveaction flux KZ exponent. In turn, the waveaction flux will cross zero for both of the RJ exponents and at the energy flux KZ exponent. Given the relative position of the spectrum exponents, this will uniquely determine the flux directions on the KZ solutions. Let us denote the stationary spectrum exponents of the thermodynamic waveaction and energy equipartitions as v_{TN} and v_{TE} respectively, and of the respective KZ-flux solutions as v_{FN} and v_{FE}. We have $v_{TN} = 0 > v_{TE} = -\alpha$, and the KZ-flux exponents are given by (9.36) and (9.37) from which it follows that $v_{FN} > v_{FE}$.

Most typical situation is when

$$v_{FE} < v_{FN} < v_{TE} < v_{TN}, \tag{9.38}$$

see Fig. 9.3. It is clear that in this case the flux directions are as they should be according to the Fjørtoft argument: the energy cascade is direct and the waveaction cascade is inverse. Note that according to the formula (3.11) in the systems with a single relevant dimensional parameter (like for the water surface gravity waves) we have $v_{FN} < v_{TE}$ when $\alpha < d/(2N - 4) + 5/2$, which ensures the ordering (9.38).

Troubles arise when the ordering (9.38) is violated, e.g. when one (or both) of the KZ exponents get in between of the RJ indices, as is the case for the 2D NLS; see Fig. 15.3 in Chap. 15. In this case the waveaction cascade spectrum has an index which is in between of the RJ indices, and as a result its waveaction flux is in the wrong direction (i.e. positive). In Chap. 15, we will also meet two other examples of systems in which pure KZ states have wrong fluxes: Boltzmann kinetic equation (Exercise 15.4) and 1D optical turbulence (Exercise 15.7).

9.2.3 Temporal Evolution Leading to KZ Spectra: Finite and Infinite Capacity Systems

The most important quantity which determines the character of temporal evolution in both forced-dissipated and freely decaying turbulence (weak or strong) is the *capacity* defined as follows.

Definition 9.1 Consider a KZ spectrum describing a constant flux of invariant Φ with density ρ_k. If the integral $\int \rho_k n_k \, d\mathbf{k}$ converges at the scales toward which Φ is

Fig. 9.3 The energy and the waveaction fluxes as a function of the spectral index in a typical four-wave case

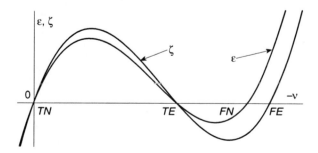

cascading then the KZ spectrum is said to have *finite capacity*. Otherwise, this KZ spectrum is said to have *infinite* capacity.

For example, for KZ spectra corresponding to a direct energy cascade, we need to examine convergence of the integral $\int \omega_k n_k \; d\mathbf{k} = \int E_k^{(1D)} dk$ at $k \to \infty$. If it converges (diverges) then the KZ spectrum has a finite (infinite) capacity. In particular, the famous Kolmgorov spectrum $E_k^{(1D)} \sim k^{-5/3}$ has a finite capacity. Similarly, for KZ spectra corresponding to an inverse waveaction cascade, we need to examine convergence of the integral $\int n_k \; d\mathbf{k} = \int \omega_k^{-1} E_k^{(1D)} dk$ at $k \to 0$. If it converges (diverges) then the KZ spectrum has a finite (infinite) capacity.

In the case of infinite capacity, it takes infinitely long to form the KZ spectrum in a forced-dissipated system in the limit of infinitely remote dissipative scales. In fact, even if we had only forcing and no dissipation, a solution would still exist for infinite time and the KZ spectrum would asymptotically form at any fixed k. The KZ spectrum forms right behind the propagating front of the non-stationary spectrum, see Fig. 9.4. In freely decaying systems, the KZ spectrum would not form at all, because any finite initial energy (or another cascading invariant) would be insufficient to form an infinite capacity tail.

In the case of finite capacity, it takes finite time t^* to form the KZ spectrum in a forced-dissipated system, no matter how far the dissipative scales are. In this case, the KZ spectrum can also form in the freely decaying WT, because the energy provided by the initial conditions would serve as a reservoir of energy which is able to keep the finite capacity tail "filled" for a long time. The energy (or another invariant's) flux in such KZ spectrum would, of course, adiabatically decrease in time in this case. The most prominent example here is the Kolmogorov spectrum of strong hydrodynamic turbulence, which is known to form out of initial conditions in absence of continuous forcing.

There is an important difference in the way the KZ spectra form in the finite and the infinite capacity systems. In contrast with the infinite capacity systems, in the finite capacity systems the KZ spectrum does not form right behind the propagating front. Instead, a steeper power-law spectrum appears behind the front, and

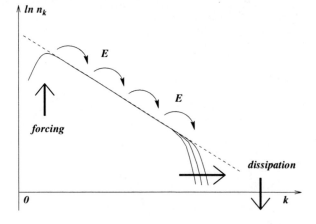

Fig. 9.4 Formation of stationary spectrum in direct cascades in forced-dissipated infinite-capacity systems. KZ spectrum forms behind the propagating front

the KZ spectrum starts forming only at finite time $t = t^*$ as a backscatter wave propagating from the dissipation scales to the forcing scales, see Fig. 9.5. This kind of behavior was observed for the direct energy cascade in the MHD wave turbulence [15] (see Chap. 12), and the inverse waveaction cascade in the NLS wave turbulence [16, 17] (see Chap. 14). In fact, this type of formation of constant-flux finite capacity spectra is more general and was observed beyond WT, particularly for the 3D Navier-Stokes turbulence within the Leith model [2] and for Burgers equation [18]. What determines the spectral index for $t < t^*$ is still unclear, except for the Burgers case, where it is known to be due to the pre-shock cubic-root singularity.

Time evolution of non-stationary WT spectra preceding the KZ steady states can be described by self-similar solutions,

$$n(k, t) = \tau^a f(\xi), \tag{9.39}$$

with self-similar variable $\xi = k\tau^b$, where $\tau = t$ for the infinite capacity systems and $\tau = t^* - t$ for the finite capacity systems. The self-similar solutions become asymptotically valid in the limit $\tau = t \to \infty$ for the infinite capacity systems and $\tau \to +0 \quad (t \to t^* -0)$ for the finite capacity systems.

Substitution of (9.39) to the kinetic equation (6.69) (three-wave case) or (6.81) (four-wave case), and requirement that the resulting equation depends only on ξ leads to a condition on the similarity constants a and b. For the three-wave case we have

$$(2\beta + d - \alpha)b = a + 1. \tag{9.40}$$

Exercise 9.5 *Derive a similar condition on the similarity constants a and b for the four-wave case.*

The second condition on a and b can be derived for the infinite capacity cases. In the forced-dissipated systems, this condition is given by the requirement that the relevant invariant grows linearly in time. This condition implies that WT gets

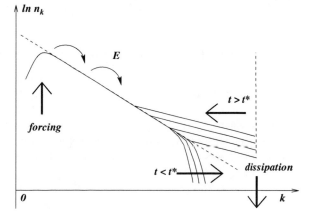

Fig. 9.5 Formation of stationary spectrum in direct cascades in finite-capacity systems. An anomalous (steeper than KZ) power-law spectrum forms behind the propagating front for $t < t^*$. The KZ spectrum forms at $t > t^*$ as a backscatter wave propagating from the dissipative scale toward lower k's

stabilized at the forcing scales so that the forcing rate is constant. For the energy cascade we have:

$$E = \int \omega_k n_k \, d\mathbf{k} \propto t \quad \Rightarrow \quad a - (\alpha + d)b = 1 \quad \text{(forced systems)}. \tag{9.41}$$

Conditions (9.39) and (9.41) allow us to find a and b, and show that *the KZ spectrum forms behind the front*, as expressed in the following exercise.

Exercise 9.6 *Find a and b from (9.39) and (9.41). Try a power-law self-similar solution $f(\xi) \sim \xi^x$ and show that index x coincides with the KZ exponent (9.22). Obtain similar results for the infinite capacity energy and waveaction cascades in four-wave systems.*

For freely decaying WT in the infinite capacity case, one should replace condition (9.41) with the respective conservation law. In the case of the energy cascade we have:

$$E = \int \omega_k \, n_k \, d\mathbf{k} = \text{const} \quad \Rightarrow \quad a - (\alpha + d)b = 0 \quad \text{(freely decaying systems)}. \tag{9.42}$$

Obviously, no KZ spectrum would form in this case.

For the finite capacity systems, treatments of the forced-dissipated and the freely decaying WT are similar because the role of forcing is negligible during the short time close to t^* when the self-similar evolution occurs. Unfortunately, one cannot use conditions of type (9.41) or (9.42) because the self-similar finite-capacity tail contains only a small fraction of the energy (or another relevant invariant) and most of the energy is contained in the part which is not self-similar. Thus, there remains a one-parametric family of solutions and it is presently unknown what condition or algorithm could be used for choosing one of these solutions as most relevant. Correspondingly, it remains unknown what determines the numerically observed anomalous (steeper than KZ) slope forming behind the propagating front for $t < t^*$.

9.2.4 KZ Spectra in Anisotropic Media

Let us now generalize the ZT method of finding the KZ spectra to anisotropic media. For simplicity we will think of systems with the same type of anisotropy as in the Petviashvilli model, i.e. 2D three-wave systems with $k_x \geq 0$. A similar approach for various types of 3D anisotropic systems was used in [19].

Let the dispersion relation have the following form,

$$\omega_k = |k_x|^{\alpha_x} |k_y|^{\alpha_y} = k^\alpha, \tag{9.43}$$

where we introduced a rather straightforward short-hand notation to avoid writing the wavenumber components. For the Petviashvili example (long Rossby/drift waves) in the quasi-zonal case, $|k_y| \gg |k_x|$, according to formula (6.9) we have:

$$\omega_k = k_x k^2 \approx k_x k_y^2, \tag{9.44}$$

which means $\alpha = (1, 2)$ (recall that for Petviashvili model $k_x \geq 0$).

We look for a solution of the kinetic equation (6.69) of the form

$$n_k = |k_x|^{\nu_x} |k_y|^{\nu_y} = \mathbf{k}^\nu, \tag{9.45}$$

The interaction coefficient and its homogeneity degree then become

$$V\left(\mu_x k_x, \mu_y k_y, \mu_x k_{1x}, \mu_y k_{1y}, \mu_x k_{2x}, \mu_y k_{2y}\right) = \mu^\beta V(k, k_1, k_2). \tag{9.46}$$

In the case of the Petviashvili model, the interaction coefficient is given by (6.12), $V_{12k} = \frac{1}{2}\sqrt{|k_x k_{1x} k_{2x}|}$ so that we have $\boldsymbol{\beta} = \left(\frac{3}{2}, 0\right)$.

Let us now generalize Zakharov transform (9.13) as follows,

$$
k_{1x} = \frac{k_x^2}{\tilde{k}_{1x}} \qquad k_{1y} = \frac{k_y^2}{\tilde{k}_{1y}},
$$

$$
k_{2x} = \frac{k_x \tilde{k}_{2x}}{\tilde{k}_{1x}} \qquad k_{2y} = \frac{k_y \tilde{k}_{2y}}{\tilde{k}_{1y}}. \tag{9.47}
$$

This form of ZT performed on the kinetic equation (6.69) and (6.70) results in

$$0 = \int \mathscr{R}_{k12}\left[1 - \left(\frac{k_1}{k}\right)^x - \left(\frac{k_2}{k}\right)^x, \right] dk_1 dk_2, \tag{9.48}$$

with the exponent

$$x = -2\beta + \alpha - 2\nu - 2d, \tag{9.49}$$

where $d = (1, 1)$.

Exercise 9.7 *Derive (9.48) and (9.49).*

From Eq. (9.48) we can immediately identify several solutions. First of all, $s = \alpha$ is a KZ solution because of the factor $\delta(\omega_{12}^k)$ in $\mathscr{R}$ (compare with the isotropic case considered before). This solution corresponds to the energy cascade and it has the exponent

$$\nu_E^{KZ} = -\beta - d. \tag{9.50}$$

For the Petviashvili system we have

$$\nu_E^{KZ} = \left(-\frac{5}{2}, -1\right). \tag{9.51}$$

Secondly, because of the factor $\delta(k_x - k_{1x} - k_{2x})$, and because $k_x, k_{1x}, k_{2x} \geq 0$ in $\mathscr{R}$, we can also write down $s = (1,0)$ as a solution, with resulting exponent

$$v_\Omega^{KZ} = -\beta - d + \frac{1}{2}(\alpha - e_x), \tag{9.52}$$

with $e_x = (1,0)$. For the Petviashvili system we have

$$v_\Omega^{KZ} = \left(-\frac{5}{2}, 0\right). \tag{9.53}$$

This KZ solution corresponds to the cascade of the potential vorticity Ω (i.e. the x-component of the momentum M_x).

Remark 9.2.4 Note that there is no KZ solution corresponding to the y-momentum cascade because M_y is not sign definite. Technically speaking, one cannot use $\delta(k_y - k_{1y} - k_{2y})$ to find a new solution because factors $\left(\frac{k_1}{k}\right)^x$ and $\left(\frac{k_2}{k}\right)^x$ contain $|k_y|$ and not k_y.

Finally, if there exists an additional positive invariant Φ with a density ρ_k of form

$$\rho_k = k^\gamma, \tag{9.54}$$

then we also have the KZ solution $s = \gamma$, or

$$v_\Phi^{KZ} = -\beta - d + \frac{1}{2}(\alpha - \gamma). \tag{9.55}$$

Respectively, this spectrum corresponds to a cascade of invariant Φ.

For the Petviashvili system with nearly-zonal waves the extra invariant, zonostrophy, has density $\tilde{\rho}_k$ given in formula (8.32). Thus in this case $\gamma = (3, -2)$ and

$$v_\Phi^{KZ} = \left(-\frac{7}{2}, 1\right). \tag{9.56}$$

9.2.5   Other Power-Law Spectra in Anisotropic Media

We can immediately write three other power-law spectra which are limiting cases of the general RJ expression (9.2):

$$v_E^{RJ} = -\alpha, \tag{9.57}$$

which describes equipartition of the energy E,

$$v_\Omega^{RJ} = -e_x, \tag{9.58}$$

which describes equipartition of the potential enstrophy Ω, and

$$v_\Phi^{RJ} = -\gamma, \tag{9.59}$$

which describes equipartition of the invariant Φ.

For the Petviashvilli model we have respectively

$$v_E^{RJ} = (-1, -2), \quad v_\Omega^{RJ} = (-1, 0), \quad v_\Phi^{RJ} = (-3, 2). \tag{9.60}$$

It is easy to see that there are plenty more power-law stationary solutions in such anisotropic media. Indeed, substituting a power-law spectrum (9.45) into the steady state kinetic equation gives us the whole family of solutions which could be thought of as a 1D set (curve or curves) on the (v_x, v_y)-plane [19]:

$$f(v_x, v_y) = 0. \tag{9.61}$$

Here, the function f is given by the RHS of equation (9.48).

For the Petviashvilli system (and for the other Rossby/drift wave systems), this family of solutions was found numerically in [19]. The respective curves on the (v_x, v_y)-plane are shown in Fig. 9.6. Obviously these curves pass through the KZ and the RJ exponents. The dashed line corresponds to nonlocal spectra. The locality and the stability properties will be discussed in the next section.

Fig. 9.6 Family of the stationary power-law solutions on the (v_x, v_y)-plane for the Petviashvilli model. The region of locality is marked by the dotted lines area *abcd* for the stationary locality and *abef* for the evolutionary locality with respect to odd in k_y perturbations. Dashed line corresponds to nonlocal solutions. The RJ and the KZ indices are marked by the hollow and by the solid dots respectively

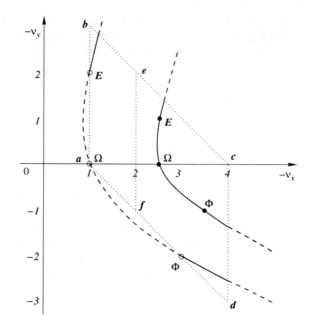

9.2.6 Locality and Stability

Locality and stability theory for the isotropic KZ spectra was developed by Balk and Zakharov and described in detail in several papers and reviews [20–23]. It was extended to anisotropic systems by Balk and Nazarenko in [19]. A significant part of the ZLF book [7] was devoted to describing this theory, and we therefore will not repeat it here. However, we will present (without derivation) a very short overview of the main results on the stability and locality. An emphasis will be on the anisotropic case: firstly for completeness of description of our master example, the Petviashvili model, and also because we will present a few extra results which were not in [19].

9.2.6.1 Stationary Locality

Locality and stability properties affect realizability of KZ and of other power-law spectra $n_k^{(0)} = Ak^\nu$ in real physical wave systems. As we have already mentioned, the very first check to be done is to find whether the original (i.e. before the ZT) integral in the kinetic equation, $\mathscr{I}(n_k)$, converges on the spectrum in question, $\mathscr{I}(n_k^{(0)}) < \infty$. If yes then the spectrum is a valid mathematical solution of the kinetic equation, and it is called a *local spectrum*. Often this situation is called *stationary locality* to distinguish it from *evolutional locality* described below. If $\mathscr{I}(n_k)$ diverges on $n_k^{(0)}$ then this spectrum is not a valid mathematical solution of the kinetic equation, and it cannot be observed experimentally or numerically (unless the divergence is marginal and can be "fixed" by logarithmic corrections of the spectrum). In this case, the interactions of modes in weak WT in real physical situations is likely to be *nonlocal*, e.g. between very long and very short waves bypassing the intermediate scales.

9.2.6.2 Stability

If the spectrum $n_k^{(0)}$ turns out to be local then one should test if it survives small disturbances. Usually in physics, such a test would be called a *stability study* and this is also the case in WT provided that disturbances do not lead to divergence of the collision integral. Disturbances of the stationary solutions can be introduced either in the initial conditions or be generated by introducing an extra small forcing. The latter formulation is particularly useful as it allows us to probe robustness of the solution with respect to the fine properties of the forcing and dissipation (e.g. making forcing slightly anisotropic in an isotropic medium). Following [19–23], an extra small forcing term γ_k can be localized in a range of scales somewhere deeply inside the inertial range,

Fig. 9.7 Evolution of a disturbance $\delta n_k(t)$ of a stationary spectrum $n_k^{(0)} = Ak^\nu$ caused by introducing an extra weak localized forcing γ_k in the inertial interval. Stable case: evolution leads to a steady state with the relative disturbance $\mathscr{A}_k = \delta n/n_k^{(0)}$ decaying toward both small and large k's

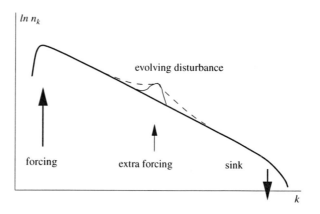

Fig. 9.8 Like in Fig. 9.7, but now for an unstable case: evolution leads to a steady state with the relative disturbance $\mathscr{A}_k = \delta n/n_k^{(0)}$ growing either toward small or toward large k's, or no steady state exists at all

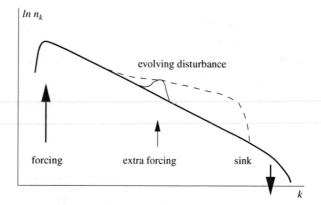

$$\dot{\delta n}_k = \mathscr{I}\left(n_k^{(0)} + \delta n_k\right) + \gamma_k, \tag{9.62}$$

where $\delta n_k \equiv \delta n_k(t) = n_k - n_k^{(0)}$ is the evolving perturbation of the spectrum $n_k^{(0)}$. Then stability can be studied with a linearized version of the above equation, assuming that the relative perturbation of the spectrum is small, $\mathscr{A}_k = \delta n_k/n_k^{(0)} \ll 1$.

First, let us discuss isotropic systems. Figure 9.7 schematically shows the stable case, where a localized additional forcing generates an initially localized disturbance which spreads leading to a steady state in which the relative disturbance $\mathscr{A}_k$ decays toward both the small and the large k's. Figure 9.8 shows instability which can be either *convective* or *absolute*. In the case of convective instability, the evolution leads to a steady state in which the relative disturbance $\mathscr{A}_k$ grows either toward the small or to the large k's. For the absolute instability, $\mathscr{A}_k$ keeps growing at some fixed k's never reaching a steady state.

To formulate the stability criterion in a form applicable to both isotropic and anisotropic systems, let us, following [19], introduce a shorthand tuple notation of the type we have already used, e.g., in (9.43). Namely, for isotropic systems which are scale invariant in $k = |\mathbf{k}|$, we use 1-tuples (i.e. singles,—real numbers) for α, ν, μ, d etc., and d is simply equal to the dimension of the system. For anisotropic systems which are scale invariant simultaneously in all components of $\mathbf{k}$, we write $\boldsymbol{\alpha} = (\alpha_x, \alpha_y)$, $\boldsymbol{\beta} = (\beta_x, \beta_y)$, $\boldsymbol{\nu} = (\nu_x, \nu_y)$, etc. for 2D systems. In this case $\boldsymbol{d} = (1, 1)$, $\boldsymbol{e}_x = (1, 0)$ and

$$k^{\nu} = |k_x|^{\nu_x} |k_y|^{\nu_y}, \quad \boldsymbol{\mu} \cdot \boldsymbol{k} = (\mu_x k_x, \mu_y k_y), \tag{9.63}$$

i.e. like we had before for the drift/Rossby system.

For 3D anisotropic systems which are scale invariant simultaneously in all components of $\mathbf{k}$, we have $\boldsymbol{\alpha} = (\alpha_x, \alpha_y, \alpha_z)$, $\boldsymbol{\beta} = (\beta_x, \beta_y, \beta_z)$, $\boldsymbol{\nu} = (\nu_x, \nu_y, \nu_z)$, $\boldsymbol{d} = (1, 1, 1)$, $\boldsymbol{e}_x = (1, 0, 0)$ and

$$k^{\nu} = |k_x|^{\nu_x} |k_y|^{\nu_y} |k_z|^{\nu_z}, \quad \boldsymbol{\mu} \cdot \boldsymbol{k} = (\mu_x k_x, \mu_y k_y, \mu_z k_z). \tag{9.64}$$

For 3D anisotropic systems with axial symmetry (e.g. waves in magnetized plasmas), around one of the components of $\mathbf{k}$, e.g. k_z and for the modulus of the perpendicular projection, $k_\perp = \sqrt{k_x^2 + k_y^2}$. We have $\boldsymbol{\alpha} = (\alpha_z, \alpha_\perp)$, $\boldsymbol{\beta} = (\beta_z, \beta_\perp)$, $\boldsymbol{\nu} = (\nu_z, \nu_\perp)$, $\boldsymbol{d} = (1, 2)$, $\boldsymbol{e}_x = (1, 0)$ and

$$k^{\nu} = |k_z|^{\nu_z} |k_\perp|^{\nu_\perp}, \quad \boldsymbol{\mu} \cdot \boldsymbol{k} = (\mu_z k_z, \mu_\perp k_\perp). \tag{9.65}$$

Let us expand the perturbation $\mathscr{A}_k$ into harmonics. For isotropic systems we have

$$\mathscr{A}_k = \sum_m \mathscr{A}_k^{(m)}(k, t) \, Y^{(m)}, \tag{9.66}$$

where for 2D systems

$$Y^{(m)} = e^{im\varphi} \quad \text{with} \quad \varphi = \arctan \frac{k_y}{k_x}. \tag{9.67}$$

For 3D systems, there is a similar expansion in terms of the spherical functions.

For anisotropic systems, we will only mention geometry of our main drift/ Rossby wave example (for other geometries see [19]). In this case there are only two harmonics: odd and even with respect to k_y:

$$\mathscr{A}_k = \mathscr{A}_k^{(0)}(k, t) Y^{(0)} + \mathscr{A}_k^{(1)}(k, t) Y^{(1)}, \tag{9.68}$$

where

$$Y^{(0)} = \frac{1}{\sqrt{2}} \quad \text{and} \quad Y^{(1)} = \frac{1}{\sqrt{2}} \, \text{sign} \, k_y. \tag{9.69}$$

In the linear approximation, the harmonics $\mathscr{A}_k^{(m)}$ behave independently of each other, and the character of their evolution, particularly stability or instability, depends on the so-called Mellin function, which we will now define. To be specific, let us deal with three-wave systems.

Definition 9.2 For the m-th harmonic, the Mellin function is defined as

$$
\begin{aligned}
W^{(m)}(s) = \int & \left| V_{12}^k \right|^2 \delta_{12}^k \delta(\omega_{12}^k)(k_1 k_2 k)^\nu \\
\times & \left[\left(k^{-s} Y^{(m)} + k_1^{-s} Y_1^{(m)} + k_2^{-s} Y_2^{(m)} \right) \left(k^{-\nu} - k_1^{-\nu} - k_2^{-\nu} \right) \right. \\
& \left. - \left(k^{-\nu-s} Y^{(m)} - k_1^{-\nu-s} Y_1^{(m)} - k_2^{-\nu-s} Y_2^{(m)} \right) \right] \\
\times & \left(k^{x+s} Y^{(m)*} - k_1^{x+s} Y_1^{(m)*} - k_2^{x+s} Y_2^{(m)*} \right) dk\, dk_1\, dk_2,
\end{aligned}
\tag{9.70}
$$

where the entries of the tuple s are, in general, complex numbers and

$$
x = \alpha - 2\beta - 2d - 2\nu.
\tag{9.71}
$$

Function $W^{(m)}(s)$ has the following important properties.

1. $W^{(0)}(0) \propto \mathscr{I}(k^\nu) = 0$ if ν corresponds to a stationary solution.
2. Function $W^{(m)}(s)$ is analytic in an infinite cylinder in the complex space of variable $s = \sigma + iw$ whose transverse cross-section R_m in given by a set of real $s = \sigma$ at which the integrals in $W^{(m)}(s)$ converge.
3. $W^{(m)}(s)$ and $1/W^{(m)}(s)$ increase utmost algebraically as $\mathfrak{J}(s) = w \to \infty$.
4. Function $W^{(m)}(s)$ becomes asymptotically negative real as $\mathfrak{J}(s) = w \to \infty$, i.e.

$$
\lim_{w \to \infty} \frac{\mathfrak{J}[\mathfrak{W}^{(m)}(s)]}{\mathfrak{R}[W^{(m)}(s)]} = 0.
$$

For formulating the stability criterion, we will need to define *rotation* of the Mellin function.

Definition 9.3 Rotation of the Mellin function $\kappa(s, h)$ is defined as the increment in the argument of $W^{(m)}(s + irh)$ as the parameter r passes the real values from $-\infty$ to $+\infty$ and h is a constant tuple with real entries.

Because of property 4 above the rotation can only take integer values.

Definition 9.4 Zero-rotation region Z_m of the Mellin function $W^{(m)}(s)$ is the region in the space of $\sigma = \mathfrak{R}(s)$ for which $\kappa(s)$ is zero for all $w = \mathfrak{J}(s)$.

Remark 9.2.5 From the definition of the rotation it is obvious that it is enough to calculate rotation for a smaller set of w, namely avoiding w's whose difference is parallel to h.

Remark 9.2.6 The region Z_m is a zero-rotation region if and only if $W^{(m)}(s)$ does not have zeroes in the infinite cylindrical region TZ_m in the s-space with transverse cross-section given by Z_m.

Indeed, rotation at s_1 can only be different from rotation at s_2 only if somewhere in between of these points (on any path in $T Z_m$ connecting s_1 and s_2) function $W^{(m)}(s)$ turns into zero.

Remark 9.2.7 From the previous remark it is clear that Z_m is independent of h. Thus, for practical calculations one can choose h arbitrarily.

Remark 9.2.8 One can show that Z_m is a convex region.

Now we can formulate the stability criterion.

Theorem 9.1 *Stationary spectrum $n_k = A k^\nu$ is stable with respect to the m-thharmonic if and only if the zero rotation region Z_m exists and contains point $\boldsymbol{\sigma = 0}$.*

See illustrations of a stable and an unstable cases in Figs. 9.9 and 9.10 respectively. In the unstable case, the relative perturbations $\mathscr{A}_m$ grow in a cone of unstable directions in space $(\ln k_x, \ln k_y)$ and decays outside of this cone. The boundaries of this cone are perpendicular to the lines tangential to Z_m and passing through the origin, as illustrated in Fig. 9.10.

Remark 9.2.9 Recall that in our formal statement of the stability problem, an extra forcing is introduced deeply in the inertial range, far from the main forcing and the dissipation regions. In such an idealized formulation the cone will always have room for developing in respective directions in space $(\ln k_x, \ln k_y)$. On the other hand, such introduction of extra forcing can be considered as a model for introducing perturbations into the main forcing and the dissipation functions. But to

Fig. 9.9 Zero-rotation region for the m-th harmonic in the stable case, Z_m. Here $0 \in Z_m$

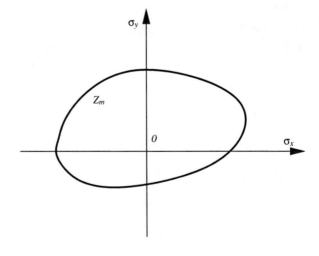

Fig. 9.10 Zero-rotation region for the *m*-th harmonic in the unstable case ($0 \notin Z_m$). The *arrows* show the cone of unstable directions in space ($\ln k_x$, $\ln k_y$). The boundaries of this cone are perpendicular to the lines tangential to Z_m and passing through the origin

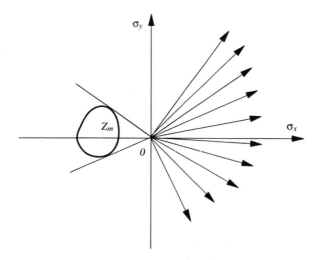

make this picture more realistic, the extra forcing has to be "pushed" closer to the main forcing and the dissipation regions (see e.g. for such regions in Fig. 8.2 in Chap. 8). Thus, one has to place the vertex of the unstable cone at each point of the forcing and the dissipation regions: the instability will be observed in the parts of the respective cascade sector thus covered by such cones; in the parts left uncovered by the cones there will be no instability and the stationary spectrum $n_k = A k^\nu$ will be observed.

From the above stability criterion, one can also deduce a simple necessary condition for stability if

$$W^{(m)}(s) = W^{(m)*}(s^*). \tag{9.72}$$

This property holds either when $Y_m = Y_m^*$, like for the drift/Rossby system with Y_m given by (9.66), or when there is a mirror symmetry: $Y_m \to Y_m^*$ and $|V_{12}^k|^2 \to |V_{12}^k|^2$ when one of the components in the integration variables of $W^{(m)}$ changes sign (e.g. for 2D isotropic systems with Y_m given by (9.67) for which $Y_m^*(k_x, k_y) = Y_m (k_x, -k_y)$).

If condition (9.72) is satisfied and $W^{(m)}(0) > 0$ then (taking into account the property 4 of W above) the rotation $\kappa_m(0)$ is an odd number and the spectrum in question is unstable. Thus we can formulate the following simple necessary condition for stability.

Theorem 9.2 *The stationary spectrum $n_k = A k^\nu$ is stable with respect to the m-thharmonic only if*

$$W^{(m)}(0) \le 0. \tag{9.73}$$

Remark 9.2.10 For the 0-th harmonic this condition is always satisfied since $W^{(0)}(0) = 0$. For this mode, if Z_0 exists, the spectrum can either be unstable, or neutrally stable (i.e. **0** is on the boundary of Z_0 and the relative perturbation $\mathscr{A}_0$ is neither growing nor decaying, see Fig. 9.11).

For reference, we mention that the KZ for the capillary waves (2D isotropic case) is unstable with respect to $m = 1$ perturbations, which means that in *infinite* systems capillary WT should be expected to be anisotropic.

Remark 9.2.11 In finite systems the nonlinear anisotropization can be suppressed by isotropization by wave reflections from the walls. This is an example where the periodic boundary conditions, common for numerics, have a very different effect than the reflecting-wall boundary conditions, which are typical for experiment. Namely, the periodic conditions do not lead to isotropization.

Another important 2D system is WT of deep-water gravity waves. This is an example of a four-wave system which we have not considered here, but for which a similar Mellin-function approach was developed in [20–23]. This analysis showed that both the direct and the inverse cascade KZ's for the deep-water gravity waves are stable.

Now let us consider anisotropic systems, restricting ourselves to the 2D systems which are scale invariant w.r.t. both wavenumber components (e.g. quasi-zonal drift/Rossby waves). In this case, the property (9.72) holds and the necessary stability condition (9.73) is valid. Moreover, in this case it is possible to derive a simple nontrivial necessary condition for stability also for the $m = 0$ mode. As we mentioned before, such stability can be only neutral, in which case the boundary of Z_0 crosses $\sigma = 0$, see Fig. 9.11. For stability it is necessary that somewhere in the small vicinity of $s = 0$:

Fig. 9.11 Zero-rotation region for the 0-th harmonic, Z_0, for a neutrally stable case

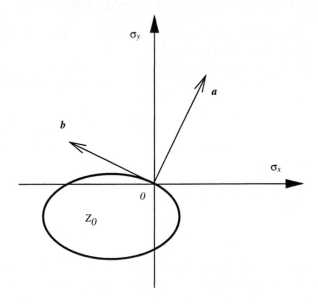

- $\Re[\,W^{(0)}] \le 0$ and $\Im[W^{(0)}] = 0$ (for κ to be even); and
- the points with such a property form a convex region (since the whole Z_0 is convex).

According to the definition (9.70), we can write

$$
\begin{aligned}
W^{(0)}(s) = \frac{1}{\sqrt{2}} \int & |V_{12}^{k}|^2\, \delta_{12}^{k} \delta(\omega_{12}^{k}) \\
& \times \left[-k^{-s}(kk_2)^{\nu} - k^{-s}(kk_1)^{\nu} + k_1^{-s}(k_1k_2)^{\nu} - k_1^{-s}(kk_1)^{\nu} \right. \\
& \left. + k_2^{-s}(k_1k_2)^{\nu} - k_2^{-s}(kk_2)^{\nu} \right] \\
& \times \left(k^{x+s} - k_1^{x+s} - k_2^{x+s} \right) dk\, dk_1\, dk_2 .
\end{aligned}
\tag{9.74}
$$

It is clear that $\Im[W^{(0)}(s)] = 0$ when $w = \Im(s) = 0$. So, let us consider $W^{(0)}(s)$ for real s i.e. for $\Re[s] = \sigma : \Re[W^{(0)}(\sigma)] = W^{(0)}(\sigma)$. Taylor expanding around $\sigma = 0$, we have

$$
\Re[W^{(0)}(\sigma)] = a_x \sigma_x + a_y \sigma_y + \frac{1}{2} L_{xx}\sigma_x^2 + L_{xy}\sigma_x\sigma_y + \frac{1}{2}L_{yy}\sigma_y^2 ,
\tag{9.75}
$$

where

$$
a_x = \left. \frac{\partial W^{(0)}(s)}{\partial s_x} \right|_{s=0} , \qquad
a_y = \left. \frac{\partial W^{(0)}(s)}{\partial s_y} \right|_{s=0} ,
\tag{9.76}
$$

$$
L_{xx} = \left. \frac{\partial^2 W^{(0)}(s)}{2_x} \right|_{s=0} , \quad
L_{xy} = \left. \frac{\partial^2 W^{(0)}(s)}{\partial s_x \partial s_y} \right|_{s=0} , \quad
L_{yy} = \left. \frac{\partial^2 W^{(0)}(s)}{\partial s_y^2} \right|_{s=0} .
\tag{9.77}
$$

Z_0 should be where $W^{(0)}(\sigma) \le 0$, i.e. on the opposite side of vector $a = (a_x, a_y)$ with respect to $\mathbf{0}$, with the boundary of Z_0 being where $\Re[W^{(0)}(\sigma)] = 0$, i.e. perpendicular to a, see Fig. 9.11.

Since Z_0 must be convex, $W^{(0)}(\sigma)$ must be positive in the direction tangential to Z_0, i.e. along the vector $b = (-a_y, a_x)$. This results in the following necessary condition for stability,

$$
L_{xx}a_y^2 + L_{yy}a_x^2 - 2L_{xy}a_x a_y \ge 0 .
\tag{9.78}
$$

This condition has an elegant geometrical interpretation in terms of the solution curves shown in Fig. 9.6. For this, we will need to find relations between the derivatives of $W^{(0)}(s)$ and the derivatives of $f(v)$. We write:

$$
a_x = -\frac{1}{2}\frac{\partial W^{(0)}(0)}{\partial v_x} , \qquad
a_y = -\frac{1}{2}\frac{\partial W^{(0)}(0)}{\partial v_y} ,
\tag{9.79}
$$

$$
L_{xx} = \frac{1}{4}\frac{\partial^2 W^{(0)}(0)}{\partial v_x^2} , \quad
L_{xy} = \frac{1}{4}\frac{\partial^2 W^{(0)}(0)}{\partial v_x \partial v_y} , \quad
L_{yy} = \frac{1}{4}\frac{\partial^2 W^{(0)}(0)}{\partial v_y^2} .
\tag{9.80}
$$

Exercise 9.8 *Prove relations* (9.79) *and* (9.80) *by taking derivatives of* (9.74).

Using relations (9.79) and (9.80) and relation $f(v) = \mathscr{I}(k^v) = C\, W^{(0)}(0)$ (where $C = \mathrm{const} > 0$), we can formulate the following geometrical necessary condition for stability [19].

Theorem 9.3 *The stationary spectrum $n_k = A\, k^{v^*}$ is stable with respect to the harmonic $m = 0$ only if the curve $f(v_x, v_y) = 0$ at point $v = v^*$is convex toward the region where f is positive.*

Exercise 9.9 *Prove this theorem using relations* (9.79) *and* (9.80) *and relation* $f(v) = \mathscr{I}(k^v) = C\, W^{(0)}(0)$.

9.2.6.3 Absolute Instability and Evolutional Nonlocality

If the region of zero rotation Z_m does not exist, then the spectrum in question is either *absolutely unstable* or *evolutionary nonlocal*.

Absolute instability means that the evolution of the perturbations of the spectrum is local (i.e. the collision integral $\mathscr{I}(n_k)$ remains convergent at all finite time) but no steady state is reached within the linear approximation. In this case, the nonlinear effects inevitably become important and subsequent evolution becomes more complicated and harder to study. One can imagine several possibilities for the nonlinear stage: either there will be nonlinear saturation leading to a steady state, or evolution remains non-stationary with regular or irregular oscillations about the stationary solution, or the nonlinear evolution results in nonlocality at large time.

Evolutionary nonlocality means that the collision integral $\mathscr{I}(n_k)$ becomes divergent within the linear approximation either immediately at $t = 0$ when the perturbation is introduced, or at a finite time, or at infinite time.

At present, it is not known how to distinguish conditions for realizability of absolute instability from the ones for the evolutional nonlocality. It is known, however, that a spectrum is evolutionary nonlocal with respect to harmonic m if there is no region of analyticity R_m for the respective Mellin function $W^{(m)}$.

9.2.6.4 Petviashvilli System

Let us now return to our main example, the Petviashvilli wave model, and discuss locality and stability of its power-law solutions in the case of nearly zonal modes, $|k_y| \gg k_x$. Recall that the curve of indices of the infinite family of stationary power-law solutions in this case is shown in Fig. 9.6. This figure also shows the regions of convergence of $W^{(0)}(0)$ and of $W^{(1)}(0)$.

As we already know, convergence of $W^{(0)}$ means stationary locality. Thus we see that all three KZ spectra (describing cascades of E, Ω and Φ respectively) are local in the stationary sense, i.e. they exist as mathematical solutions of the wave kinetic equation. The respective three thermodynamic RJ solutions appear to be on the margin of the stationary locality region. It remains to be understood whether this fact is accidental or if it has nontrivial physical reasons or/and consequences. Note that the RJ spectra always exist as mathematical solutions of the wave kinetic equation, because they turn the integrand of the collision term into zero, but, as we see, small deviations from these solutions lead to divergences.

At the same time, convergence of $W^{(0)}(0)$ indicates that the region of analyticity R_0 of the 0-th Mellin function $W^{(0)}(s)$ exists.

Convergence of $W^{(1)}(0)$ is necessary for existence of the region of analyticity R_1 of the 1-th Mellin function $W^{(1)}(s)$. In Fig. 9.6 we see that R_1 does not exist for any of the three KZ spectra and therefore these spectra are evolutionally nonlocal with respect to the $m = 1$ perturbations.

Remark 9.2.12 Evolutionary nonlocality with respect to the $m = 1$ perturbations means that the evolution of these perturbations will be determined by nonlocal interactions with the ends of the inertial interval. We will see later, however, that this type of nonlocal interaction leads to rapid symmetrization of the spectrum, i.e. suppression of the $m = 1$ harmonics. Thus we learn an important lesson: evolutionary nonlocality of a spectrum does not always imply its non-realizability, and some types of the nonlocal interactions can even enforce stability.

Therefore it makes sense to proceed further and to study behavior of the system with respect to the $m = 0$ perturbations. It turns out that function $f(v)$ is negative in between the two branches of the manifold $f(v) = 0$ in Fig. 9.6. Thus, the geometrical necessary conditions of stability expressed in Theorem 9.3 is satisfied for the RJ branch of solutions in Fig. 9.6 and violated for the KZ branch. Thus, none of the three KZ spectra is stable. For the Ω-cascade spectrum, there appears to be no zero-rotation region Z_0, which means that this spectrum is either evolutionally nonlocal or absolutely unstable. In either of the two possibilities, this spectrum is not realizable and cannot be expected in experiment or numerics.

For the E-cascade and the Φ-cascade spectra, the zero-rotation regions Z_0 exist and are shown in Figs. 9.12 and 9.13 respectively. In these figures, we also show the analyticity regions R_0 for function $W^{(0)}(\sigma)$. The instability in both of these cases is of convective type, i.e. there will be a steady state with $m = 0$ perturbations growing in a sector in space $(\ln k_x, \ln k_y)$: these sectors are shown in Figs. 9.12 and 9.13 with arrows.

Remark 9.2.13 As we mentioned before, to find the regions in the k-space affected by the instability, one has to place the vertex of the instability cone to all possible places where forcing or dissipation are present. Keeping in mind Fig. 8.2 in Chap. 8, we see that the whole of the E-cascade sector is affected by the instability (it is covered by the instability cones with vertices in the dissipation scales of the E-sector).

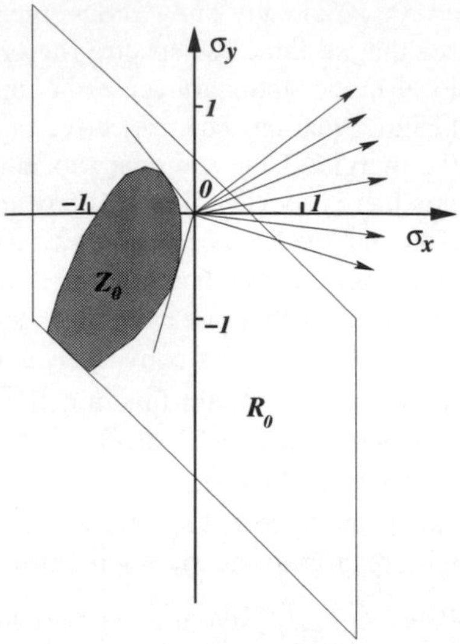

Fig. 9.12 Analyticity (convergence) region R_0 for the Mellin function $W^{(0)}$, and the zero-rotation region Z_0 for the 0-th harmonic of perturbation to the E-cascade spectrum in the Petviashvilli wave system. The cone of unstable directions is shown by *arrows*

Fig. 9.13 Analyticity (convergence) region R_0 for the Mellin function $W^{(0)}$, and the zero-rotation region Z_0 for the 0-th harmonic of perturbation to the Φ-cascade spectrum in the Petviashvilli wave system. The cone of unstable directions is shown by *arrows*

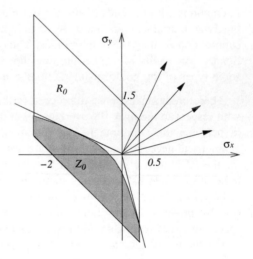

Remark 9.2.14 On the other hand, the instability for the Φ-cascade spectrum is milder that the one for the E-cascade spectrum and it does not totally prevent realizability of this spectrum somewhere in the $\boldsymbol{k}$-space if the inertial range is finite.

Exercise 9.10 *Describe the stability properties of the KZ spectrum corresponding to the cascade of the zonostrophy Φ by placing the vertex of the respective instability cone into the forcing and the dissipation scales of the Φ-cascade spectra.*

References

1. Peierls, R.: Annalen Physik **3**, 1055 (1929)
2. Connaughton, C., Nazarenko, S.: Warm cascades and anomalous scaling in a diffusion model of turbulence. Phys. Rev. Lett. **92**(4), 044501 (2004)
3. L'vov, V.S., Nazarenko, S.V., Rudenko, O.: Bottleneck crossover between classical and quantum superfluid turbulence. Phys. Rev. B **76**, 024520 (2007)
4. Zakharov, V.E.: Weak turbulence in media with decay spectrum. Zh. Priklad. Tech. Fiz. **4**, 35–39 (1965) [J. Appl. Mech. Tech. Phys. **4**, 22–24 (1965)]
5. Zakharov, V.E., Filonenko, N.N.: Energy spectrum for stochastic oscillations of a fluid surface. Dokl. Acad. Nauk SSSR **170**, 1292–1295 (1966) (translation: Sov. Phys. Dokl. **11**, 881–884 (1967))
6. Zakharov, V.E., Filonenko, N.N.: Weak turbulence of capillary waves. Zh. Prikl. Mekh. Tekh. Phys. **4**(5), 62 (1967) (in Russian: J. Appl. Mech. Tech. Phys. **4**, 506)
7. Zakharov, V.E., L'vov, V.S., Falkovich, G.: Kolmogorov Spectra of Turbulence. Series in Nonlinear Dynamics. Springer (1992)
8. Pushkarev, A.N., Zakharov, V.E.: Turbulence of capillary waves—theory and numerical simulation. Phys. D **135**(1–2), 98 (2000)
9. Frisch, U.: Presentation in "International Symposium on Turbulence", Beijing, China, 21–25 Sept 2009
10. Eyink, G., Frisch, U.: Kraichnan. In: Davidson, P.A., Kaneda, Y., Moffatt, H.K., Sreenivasan, K.R. (eds.) Highlights in the History of Turbulence. CUP, Cambridge (2010)
11. Dyachenko, A., Newell, A.C., Pushkarev, A., Zakharov V.E.: Optical turbulence: weak turbulence, condensates and collapsing fragments in the nonlinear Schrodinger equation. Phys. D **57**(1–2), 96 (1992)
12. Nazarenko, S.: Differential approximation for Kelvin-wave turbulence. JETP Lett. **83**(5), 198–200 (2005) (arXiv: cond-mat/0511136)
13. Boffetta, G., Celani, A., Dezzani, D., Laurie, J., Nazarenko, S.: Modeling kelvin wave cascades in superfluid helium. J. Low Temp. Phys. **156**(3–6), 193–214 (2009)
14. Lvov, V., Nazarenko, S.: Differential models for 2D turbulence, JETP Lett. **83**(12), 635–639 (2006) (arXiv: nlin.CD/0605003)
15. Galtier, S., Nazarenko, S.V., Newell, A.C., Pouquet, A.: A weak turbulence theory for incompressible MHD. J. Plasma Phys. **63**(5), 447–488 (2000)
16. Semikoz, D.V., Tkachev, I.I.: Phys. Rev. Lett. **74**, 3093 (1995)
17. Semikoz, D.V., Tkachev, I.I.: Phys. Rev. D **55**, 489 (1997)
18. Sulem, C., Sulem, P.-L., Frisch, H.: Tracing complex singularities with spectral method. J. Comput. Phys. **50**, 138–161 (1983)
19. Balk, A.M., Nazarenko, S.V.: On the physical realisability of anisotropic Kolmogorov spectra of weak turbulence. Sov. Phys. JETP **70**, 1031 (1990)
20. Balk, A.M., Zakharov, V.E.: Stability of weak turbulence: Kolmogorov spectra. In: Plasma Theoryand Nonlinear and Turbulent Processes in Physics, Vol. 1 (World Scientific Publishing Co., Singapore, 1988), pp. 359–376
21. Balk, A.M., Zakharov, V.E.: Stability of weakly turbulent Kolmogorov spectra. Dokl. Acad. Nauk SSSR **299**(5), 1112–1115 (1988) [Sov. Phys. Dokl. **33**(4), 270–272 (1988)]
22. Balk, A.M., Zakharov, V.E.: Stability of Kolmogorov spectra of weak turbulence. In: Integrabilityand Kinetic Equations for Solitons (Naukova Dumka, Kiev, 1990), pp. 417–472.
23. Balk, A.M., Zakharov, V.E.: Stability of weak-turbulent Kolmogorov spectra. Amer. Math. Soc. Transl. Ser. 2 **182**, 31–81 (1998).

Chapter 10
Finite-Size Effects in Wave Turbulence

How does finite size of the bounding volume affect behavior of WT? This question has been asked and addressed recently in a large number of works, including [1–14]. In short, the main effect is that the wave–wave interactions are suppressed because the finite box makes the $\mathbf{k}$-space discrete which makes the exact resonant conditions (e.g. (6.48) and (6.49) in the three-wave case) harder to satisfy. Recall that technically the WT theory based on the kinetic equation is an infinite-box theory because the limit $L \to \infty$ is taken before the weak nonlinearity limit $\varepsilon \to 0$. As we explained in remark Sect. 6.5.1, this means that a large number of quasi-resonant interactions are active and dominant over the exact resonances.

Here we will study the question what happens when, due to the finite size, the number of quasi-resonances is depleted or absent altogether. The finite-size effects in WT can be characterized by considering the nonlinear frequency broadening Γ, which is the inverse correlation time of wave packets. Such a correlation time is roughly equal to the characteristic time of nonlinear evolution. Below we will see that there can be qualitatively different regimes due to the finite box size, which are described by different relationships between Γ and and the frequency spacing in the finite box Δ_ω,

$$\Delta_\omega = \left| \frac{\partial \omega_k}{\partial \mathbf{k}} \right| \frac{2\pi}{L} \sim \frac{\omega_k}{kL}. \tag{10.1}$$

Note that in this estimate and thereafter we do not control numerical factors like 2π, which cumulatively may conspire to give a large number in the final answer. Thus, a more careful approach may be needed if one would like to be more accurate.

The kinetic equation is applicable when $\Gamma \gg \Delta_\omega$, and thus we will call this regime *kinetic*. Naturally, one can guess that a qualitative different behavior will be observed in the opposite limit $\Gamma \ll \Delta_\omega$: this will be called below *discrete wave turbulence*. These two regimes are realized when WT forcing is rather high (but not too high so that the nonlinearity is still weak) and low respectively. However,

S. Nazarenko, *Wave Turbulence*, Lecture Notes in Physics, 825,
DOI: 10.1007/978-3-642-15942-8_10, © Springer-Verlag Berlin Heidelberg 2011

we will also see that there is also a rather wide intermediate range of forcing for which there is a regime with $\Gamma \sim \Delta_\omega$, which we will call *mesoscopic wave turbulence*.

The term *mesoscopic* in the theory of WT was introduced in [15]. It was noted in this paper and, independently in [7], that in existing numerical simulations of the gravity water waves there may be regimes where the statistical properties of the infinite-box systems coexist with effects due to the k-space discreteness associated with a finite computational box. However, the fact that such a mesoscopic regime is active in a *wide intermediate range of wave amplitudes* was first discovered later in the context of MHD turbulence in [11] (see also the Chap. 14 later in this book which is devoted to MHD turbulence). This approach appears to be quite general for WT systems, and below we will describe how it works for any three-wave and four-wave systems. After this, the reader should be able to adapt it easily to the five- and six-wave systems himself. Below we will follow the approach recently suggested in [16]. A more detailed review of the finite-size effects in WT will be published separately in [14].

10.1 Small-Box Regime: Discrete Turbulence

In the discrete WT regime, when $\Gamma \ll \Delta_\omega$, only the terms in the dynamical equations which correspond to *exact* wavenumber and frequency resonances contribute to the nonlinear wave dynamics. All the other terms rapidly oscillate and their net long-term effect is null. The most clear example here is the case when there are no exact resonances, like in the system of the capillary water surface waves. In this case, the averaged (over the fast linear oscillations) nonlinearity is negligible and the turbulent cascade over scales is "frozen" [4]. One can see an analogy with KAM theory which says that trajectories of a perturbed (in our case by nonlinearity) Hamiltonian system remain close to the trajectories of the un-perturbed integrable system (in our case the linear wave system whose trajectories are just harmonic oscillations of the individual modes) if there is no resonances. Of course, this analogy should be taken with caution because even in absence of the lower-order resonances (e.g. triad resonances for the capillary waves) higher-order resonances may be important.

Thus, for the discrete WT regime we have the following reduced dynamical equations

$$i\dot{a}_k = \sum_{1,2} \left(V_{12}^k a_1 a_2 R_{12}^k + 2 V_{1k}^{2*} a_1^* a_2 R_{1k}^2 \right), \tag{10.2}$$

for the three-wave case (Eq. 6.60) with the interaction Hamiltonian (6.63) in which we retain only exact triad resonances) and

$$i\dot{a}_k = \sum_{1,2,3} W_{3k}^{12} a_1 a_2 a_3^* R_{3k}^{12}, \tag{10.3}$$

for the four-wave case (Eq. 6.71 with only exact quadric resonances left).

In Eq. (10.2), factor R_{12}^3 is equal to one when modes $\mathbf{k}_1$, $\mathbf{k}_2$ and $\mathbf{k}_3$ are in exact wavenumber and frequency resonance and it is zero otherwise. Respectively in (10.3), R_{34}^{12} is equal to one when modes $\mathbf{k}_1$, $\mathbf{k}_2$, $\mathbf{k}_3$ and $\mathbf{k}_4$ are in exact wavenumber and frequency resonance, and it is zero otherwise.

Note that some resonant triads/quartets (if at all present) may be isolated, in which case their dynamics is integrable, and the respective nonlinear oscillations can be expressed in terms of the elliptic functions. Some triads/quartets may be linked and form clusters of various sizes, whose dynamics is more complicated and to some extent may be chaotic, especially for larger clusters. Study and classification of such exact resonances and their clusters was initiated by Kartashova and was followed in a large number of works [1–7, 9, 10, 12, 13, 16]. An example of a cluster of eight linked quartets for the gravity water wave system found in [13] is shown in Fig. 10.1.

The nonlinear broadening for the discrete dynamics, which we denote Γ_D, can be estimated from the Eqs. (10.2) and (10.3),

$$\Gamma_D^{(3w)} \simeq |Va_k|\mathcal{M}, \tag{10.4}$$

$$\Gamma_D^{(4w)} \simeq |Wa_k^2|\mathcal{M}, \tag{10.5}$$

where $V = V_{12}^k$ and $W = W_{3k}^{12}$ are the interaction coefficients of the Hamiltonians H_3 and H_4, (6.63) and (6.71) respectively. Superscripts $3w$ and $4w$ stand for "three-wave" and "four-wave" respectively. Here $\mathcal{M}$ is the number of exact resonances which are dynamically important at a fixed $\mathbf{k}$, which is less or equal to the number

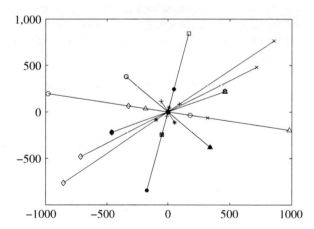

Fig. 10.1 Cluster of eight connected quartets for the system of gravity water waves in a square basin. Each quartet is indicated by its own marker. The coordinates of the eight quartets are given by the following tuples: $\pm[(9, 45), (-49, -245), (55, -115), (-95, -85)]$; $\pm[(-49, -245), (169, 845), (460, 220), (-340, 380)]$; $\pm[(180, -36), (-980, 196), (-460, -220), (-340, 380)]$; $\pm[(320, -64), (855, 765), (460, 220), (715, 481)]$

of modes connected to **k** in the resonant cluster. For simplicity we assumed that all the dynamically important resonances are local, i.e. $k_1 \sim k_2 \sim k_3 \sim k$. Strictly speaking, the estimates (10.4) and (10.5) are only valid if $\mathcal{M}$ is not too large, because when $\mathcal{M} \gg 1$ one should expect statistical cancelations of the effect of different triads/quartets, and our estimates would have to be modified. This is the case for the example of MHD turbulence considered in Chap. 14. Also, our estimates would have to be adjusted for systems with nonlocal in **k** interactions.

Thus, the condition of the discrete turbulence regime, $\Gamma_D \ll \Delta_\omega$, becomes

$$|Va_k| \ll \frac{\omega_k}{kL\mathcal{M}}, \quad \text{for 3-wave systems,} \tag{10.6}$$

$$|Wa_k^2| \ll \frac{\omega_k}{kL\mathcal{M}}, \quad \text{for 4-wave systems.} \tag{10.7}$$

10.2 Infinite-Box Regime: Kinetic Wave Turbulence

In the kinetic regime, the frequency resonance broadening, denoted Γ_K, is determined by the kinetic equation (6.69) for three-wave systems and (6.81) for the four-wave systems. This gives for Γ_K:

$$\Gamma_K^{(3w)} \simeq |V|^2 n_k k^d / \omega_k \simeq |V|^2 |a_k|^2 (kL)^d / \omega_k, \tag{10.8}$$

$$\Gamma_K^{(4w)} \simeq |W|^2 n_k^2 k^{2d} / \omega_k \simeq |W|^2 |a_k|^4 (kL)^{2d} / \omega_k, \tag{10.9}$$

where, for simplicity, we have assumed that the wave spectrum is not too narrow and the range of wavenumbers interacting with **k** is of width $\sim k$.

The upper bound for applicability of the wave kinetic equations follows from the condition of weak nonlinearity $\Gamma_K \ll \omega_k$ which gives

$$|Va_k|(kL)^{d/2} \ll \omega_k, \quad \text{for 3-wave systems,} \tag{10.10}$$

$$|W||a_k|^2 (kL)^d \ll \omega_k, \quad \text{for 4-wave systems.} \tag{10.11}$$

On the other hand, the wave amplitudes should be large enough for the frequency resonance broadening Γ_k to be much greater than the frequency spacing Δ_ω. Taking into account (0), this condition gives

$$|Va_k| \gg \frac{\omega_k}{(kL)^{(d+1)/2}}, \quad \text{for 3-wave systems,} \tag{10.12}$$

$$|W||a_k|^2 \gg \frac{\omega_k}{(kL)^{d+1/2}}, \quad \text{for 4-wave systems.} \tag{10.13}$$

These inequalities can also be interpreted as a condition for the wavepacket mean-free path $\ell = (\partial_k \omega_k)/\Gamma_K$ to be less than the box size L.

10.3   Mesoscopic Turbulence: Sandpile Behavior

Consider first the case when the number of connections of mode $\mathbf{k}$ in its discrete resonant cluster is not large, $\mathcal{M} \gtrsim 1$, as is the case, e.g., for the gravity water waves. Comparing the range of kinetic WT, (10.12), (10.13), and the one of discrete WT, (10.6), (10.7), one can see that there exists a gap,

$$\frac{1}{kL\mathcal{M}} \gg \frac{|Va_k|}{\omega_k} \gg \frac{1}{(kL)^{(d+1)/2}}, \quad \text{for 3-wave systems,} \tag{10.14}$$

$$\frac{1}{kL\mathcal{M}} \gg \frac{|W||a_k|^2}{\omega_k} \gg \frac{1}{(kL)^{d+1/2}}, \quad \text{for 4-wave systems.} \tag{10.15}$$

in which both the conditions for the kinetic WT and for the discrete WT are satisfied. This means that in the region (10.14), (10.15) the wave behavior is neither purely discrete nor purely kinetic WT. Existence of such a gap was first pointed out in [11] in the context of MHD wave turbulence (even though the MHD example is quite different, see Chap. 14). Region (10.14), (10.15) possess the features of both types of turbulent behavior described above. In other words, in this region both types of WT may exist and the system may oscillate in time (or parts of the k-space) between the two regimes giving rise to a qualitatively new type of WT: *mesoscopic wave turbulence*. It was suggested in [12] (in the context of the surface gravity waves) that in forced wave systems the discrete and the kinetic regimes may alternate in time, see Fig. 10.2. Namely, let us consider WT with initially very weak or zero intensity, so that initially WT is in the discrete regime, and let us permanently supply more wave energy via a weak source at small k's. During the discrete phase (with fully or partially "frozen" cascade [4]) the wave energy accumulates until when the resonance broadening Γ_D becomes of order of the frequency spacing Δ_ω. After that the turbulence cascade is released to higher k's in the form of an "avalanche" characterized by predominantly kinetic interactions. At the moment of triggering the avalanche, the broadening Γ jumps up from $\Gamma = \Gamma_D$ to $\Gamma = \Gamma_K \gg \Gamma_D$, see the upper Fig. 10.2. In the process of the avalanche release, the mean wave amplitude lowers so that the value of broadening $\Gamma = \Gamma_K$ becomes of order of the frequency spacing Δ_ω. Remember, for not too large $\mathcal{M}$ in this intermediate range $\Gamma_K \gg \Gamma_D$. Thus, at this point the system returns to the energy accumulation stage in the discrete WT regime, and the cycle repeats, see Fig. 10.2. Because of the obvious analogy, this scenario was called *sandpile behavior* in [12].

Fig. 10.2 "Sandpile" behavior in wave turbulence. Upper graph: the frequency broadening Γ follows the discrete turbulence dependence $\Gamma = \Gamma_D$ until reaching the value $\Gamma = \Delta_\omega$ at time $t = t_1$, at which point it jumps to the kinetic branch $\Gamma = \Gamma_K \gg \Gamma_D$ and rapidly drops in the kinetic regime to the value $\Gamma = \Delta_\omega$ at time $t = t_2$. Then it jumps back to the discrete branch $\Gamma = \Gamma_D \ll \Gamma_K$, after which the cycle repeats. Lower graph: the amplitude gradually grows to $a_k \sim A_1$ for $t < t_1$ and then quickly drops to A_2 for $t_1 < t < t_2$, after which the cycle repeats. For the three-wave systems $A_1 \sim (\omega/V)\,(kL)^{-(d+1)/2}$ and $A_2 \sim \omega/(kLV\mathcal{M})$ and for the four-wave systems $A_1 \sim (\omega/W)^{1/2}\,(kL)^{-(2d+1)/4}$ and $A_2 \sim \sqrt{\omega/(kLW\mathcal{M})}$

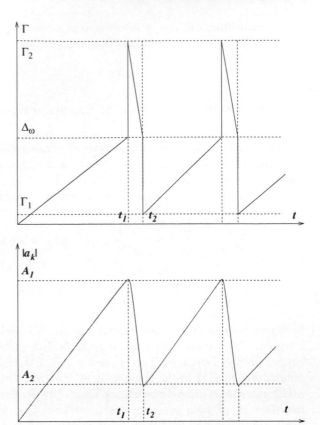

As we see, the sandpile behavior is characterized by a *hysteresis* where in the same range of amplitudes, from A_1 to A_2 in the lower Fig. 10.2, the WT intensity increases in the discrete regime and decreases in the kinetic regime.

For the small-amplitude part of the sandpile cycle, the system will be close to the critical spectrum, where resonance broadening Γ_K is of order of the omega-spacing Δ_ω. This gives the frequency spectrum ω^{-6}, which was predicted in [12] and experimentally confirmed in [8] (c.f. ω^{-4} for the KZ spectrum in this case [17]). Finding a spectrum close to the large-amplitude part of the cycle is not so straightforward because we do not know the dependence of $\mathcal{M}$ on ω.

So far, we only considered the case when $\mathcal{M}$ is not too large. The case $\mathcal{M} \gg 1$ can be very different. Namely, instead of the range where *both* conditions are satisfied simultaneously, the one for the kinetic WT, (10.12), (10.13), and the one for the discrete WT, (10.6), (10.7), one gets a range where *none* of these two conditions are satisfied. This kind of mesoscopic turbulence will be considered using the MHD example in Chap. 14). We will see that in this case the frequency broadening Γ remains of the order of the omega-spacing Δ_ω in a broad (mesoscopic) range of wave amplitudes. Remembering that Γ is a characteristic nonlinear evolution time, we note that constancy of Γ points at a possibility that the energy transfer in such a mesoscopic regime is driven by a hidden effectively linear process, which is yet to be understood.

10.4 Coexistence of Different Regimes in the k-space

The strength of WT typically varies along the turbulent cascade in the **k**-space and therefore one may expect different wave turbulence regimes present in the different parts of the **k**-space at the same instant in time. For example, the nonlinearity increases along the cascade toward high wavenumbers in WT of surface gravity waves and of MHD Alfvén waves. Thus we can expect WT in these systems to be discrete at low k's and kinetic at high k's. Moreover, on the crossover regions one can expect nontrivial gradual transition which involves blending and interaction of different dynamical and statistical mechanisms. This effect is expected to be more pronounced if the interaction of scales is nonlocal, so that some wavenumber(s) from a particular resonant triad (or quartet) could be in the discrete range whereas the other wavenumber(s) from the same triad (or quartet) could be in the kinetic range. As a result, in the cross-over range a continuous spectrum described by the kinetic equation (e.g. KZ) could coexist with selected a few modes belonging to isolated resonant clusters which would evolve coherently at deterministic timescales. Moreover, the same set of modes might randomly alternate in time from being discrete to kinetic and back, as we described above in the sandpile scenario.

10.5 Cascade Tree in the Discrete k-space

Some basic finite-size effects in WT can be seen in an very simple kinematic cascade model which was suggested in [5]. This model builds a "cascade tree" in the following three steps.

- Let us put some energy into a small collection of initial modes. We denote this initial collection of excited modes by S_0 (e.g. in within a circle or a ring at small k's which corresponds to forcing at large scales). One can view set S_0 as the cascade tree's "trunk".
- Next, find the modes which can interact with the initial ones at the given level of nonlinear broadening Γ. Namely, we define a new set of modes S_1 as the union of all k's satisfying the quasi-resonance conditions,

$$|\omega_3 - \omega_2 - \omega_1| < \Gamma, \mathbf{k}_3 - \mathbf{k}_2 - \mathbf{k}_1 = 0, \quad \text{(3-wave case)}, \tag{10.16}$$

$$|\omega_4 + \omega_3 - \omega_2 - \omega_1| < \Gamma, \mathbf{k}_4 + \mathbf{k}_3 - \mathbf{k}_1 - \mathbf{k}_2 = 0, \quad \text{(4-wave case)}. \tag{10.17}$$

with all but one wavenumbers in S_0 and the remaining wavenumber outside of S_0. Provided that Γ is large enough, the set S_1 will be greater than S_0. Set S_1 comprises the cascade tree's "biggest branches".

• Now iterate this procedure to generate a series of cascade generations S_0, S_1, ..., S_N which will mark the sets of active modes as the system evolves. The union of these sets constitutes the whole of the cascade tree with all of its bigger and smaller branches included.

This model is purely kinematic. It does not say anything about how energy might be exchanged dynamically among the active modes, or how rapidly a certain cascade generation is reached. We shall see, however, that the kinematics alone allows one to make some interesting observations.

Let us consider the example of the gravity waves on deep water [7]. Fig. 10.3 shows quasi-resonant generations of modes with generation 0 taken to be in the ring $6 < |k| < 9$. With broadening Γ below a critical value $\Gamma_{crit} = 1.4 \times 10^{-5}$, a finite number of modes outside the initial region get excited (generation 2) but there will be no quasi-resonances to carry energy to outer regions in further generations. If the broadening is larger than Γ_{crit}, the energy cascades infinitely. Note that because of the fractal snowflake structure seen in Fig. 10.3, the active modes are rather sparse in the front of the cascade propagating to higher k, with pronounced anisotropic and intermittent character.

A similar picture of intermittent cascades was also observed for the capillary wave system [5]. However, because there are no exact resonances for this system, the generation 1 and higher appear only if Γ is greater than some minimal value Γ_{crit1}. Further, there exists a second critical value $\Gamma_{crit2} > \Gamma_{crit1}$: the number of generations is finite for $\Gamma_{crit2} > \Gamma > \Gamma_{crit1}$ and the cascade process dies out not reaching infinite k's, whereas for $\Gamma > \Gamma_{crit2}$ the number of generations is infinite and the cascade propagates to arbitrarily high k's. Note that the latter property makes the capillary wave system different from the gravity waves for which the cascade always spread through the wavenumber space infinitely provided $\Gamma > \Gamma_{crit}$.

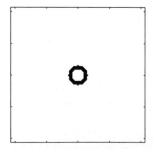 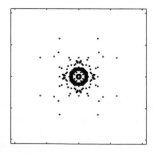 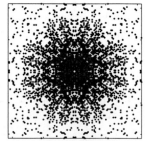

Fig. 10.3 Kinematic cascade of the gravity water waves in the 2D **k**-space. *Left frame*: stage 1 of the cascade with initial modes in the ring $6 \le |k| \le 9$. *Middle* and *right frames*: stages 5 and 9 of the cascade respectively

Another example where the (three-wave) quasi-resonances and the kinematic energy cascades were studied is the system of inertial waves in rotating 3D fluid volumes [18]. This system is anisotropic and the study of the kinematic cascades allows us to find differences between the 2D modes, with wavevectors perpendicular to the rotation axis, and the 3D modes. It appears that the "catalytic" interactions which involve triads including simultaneously 2D and 3D wavevectors dominate over the triads which involve 3D wavevectors only. Note that any catalytic triad contains one 2D vector and two 3D wavevectors whose projections on the rotation axis is the same. Thus, the energy cascade to large k's leaves the parallel wavenumber component unchanged, which at large time leads to a very anisotropic state with x-space structures parallel to the rotation axis (c.f. Taylor columns).

References

1. Kartashova, E.A., Piterbarg, L.I., Reznik, G.M.: Weakly nonlinear interactions between Rossby waves on a sphere. Oceanology **29**, 405 (1990)
2. Kartashova, E.A.: Partitioning of ensembles of weakly interacting dispersing waves in resonators into disjoint classes. Phys. D **46**(1), 43 (1990)
3. Kartashova, E.A.: On properties of weakly nonlinear wave interactions in resonators. Phys. D **54**(1–2), 125 (1991)
4. Pushkarev, A.N.: On the Kolmogorov and frozen turbulence in numerical simulation of capillary waves. Eur. J. Mech. B/Fluids **18**(3), 345 (1999)
5. Connaughton, C., Nazarenko, S.V., Pushkarev, A.: Discreteness and quasi-resonances in capillary wave turbulence. Phys. Rev. E **63**, 046306 (2001)
6. Kartashova, E.A.: Wave resonances in systems with discrete spectra. In: Zakharov, V.E. (ed.) Nonlinear Waves and Weak Turbulence, p. 95. Springer, Berlin (1998)
7. Lvov, Y.V., Nazarenko, S., Pokorni, B.: Discreteness and its effect on water-wave turbulence. Phys. D **218**(1), 24 (2006)
8. Denissenko, P., Lukaschuk, S., Nazarenko, S.: Gravity surface wave turbulence in a laboratory flume. Phys. Rev. Lett. **99**, 014501 (2007)
9. Kartashova, E.A., L'vov, V.S.: A model of intra-seasonal oscillations in the Earth atmosphere. Phys. Rev. Lett. **98**(19), 198501 (2007)
10. Kartashova, E.A., L'vov, V.S.: Triad dynamics of planetary waves. Europhys. Lett. **83**, 50012 (2008)
11. Nazarenko, S.V.: 2D enslaving of MHD turbulence. New. J. Phys. **9**, 307 (2007). doi: 10.1088/1367-2630/9/8/307
12. Nazarenko, S.: Sandpile behaviour in discrete water-wave turulence. J. Stat. Mech. LO2002 (2006)
13. Kartashova, E.A., Nazarenko, S., Rudenko, O.: Resonant interactions of nonlinear water waves in a finite basin. Phys. Rev. E **78**, 016304 (2008)
14. Kartashova, E., Lvov, V., Nazarenko, S., Procacia, I.: Mesoscopic Wave Turbulence: review (2010) (in preparation)
15. Zakharov, V.E., Korotkevich, A.O., Pushkarev, A.N., Dyachenko, A.I.: Mesoscopic wave turbulence. JETP Lett. **82**(8), 487 (2005)
16. Lvov, L., Nazarenko, S.: Discrete and mesoscopic regimes of finite-size wave turbulence. Phys. Rev. E **82**, 056322 (2010)
17. Zakharov, V.E., Filonenko, N.N.: J. Appl. Mech. Tech. Phys. **4**, 506 (1967)
18. Bourouiba, L.: Discreteness and resolution effects in rapidly rotating turbulence. Phys. Rev. E **78**, 056309 (2008)

Chapter 11
Properties of the Higher-Order Statistics. Intermittency and WT Life Cycle

So far we have only considered propertied of WT expressed in terms of the wave spectra, which are second-order moments. For Gaussian wavefields such description would be sufficient, since all statistical properties of the system in this case depend on the spectrum. However, some important features of WT are related with deviations from the Gaussian statistics. To study these features, one has to study higher-order moments and PDFs. As we will see, solutions for the PDFs are not a mere curiosity in WT, but an essential tool that will allow us to incorporate the coherent structures and, therefore, obtain a description for the WT life cycle where random and coherent waves coexist and interact. We will start with the one-mode PDFs.

11.1 Solutions for the One-Mode PDFs and the Moments

First, let us consider the one-mode generating function $\mathscr{Z}_k$ which evolves according to (6.45). If the coefficients η_k and γ_k in this equation were given then it would be a linear first-order PDE which could be easily solved by the method of characteristics. However, these coefficients themselves implicitly (via the spectrum) depend on $\mathscr{Z}_k$ so that the evolution is nonlinear. One can immediately find a steady state solution of (6.45),

$$Z = \frac{1}{1 - \lambda n_k},\qquad(11.1)$$

which corresponds to Gaussian statistics with moments

$$M^{(p)} - p!n_k^p.\qquad(11.2)$$

However, these solutions are invalid at small λ and high p's because large amplitudes $J = |a|^2$, for which nonlinearity is not weak, strongly contribute in

S. Nazarenko, *Wave Turbulence*, Lecture Notes in Physics, 825,
DOI: 10.1007/978-3-642-15942-8_11, © Springer-Verlag Berlin Heidelberg 2011

these cases. Because of the integral nature of definitions of $M^{(p)}$ and Z with respect to the J, the ranges of amplitudes where the weak WT description is applicable are mixed with, and contaminated by, the regions where the weak WT fails. Thus, to clearly separate these regions it is better to work with quantities which are local in J, in particular the one-mode PDF $\mathscr{P}_k$ which evolves according to (6.51). The steady state solution to this equation ($\dot{\mathscr{P}}_k = 0$) corresponds to a probability flux $\mathscr{F}_k$ which is independent of the amplitude variable [2],

$$\mathscr{F}_k = -s_k\left(\gamma_k\mathscr{P}_k + \eta_k\frac{\partial}{\partial s_k}\mathscr{P}_k\right) = \text{const.} \qquad (11.3)$$

This equation has the following general solution,

$$\mathscr{P} = \mathscr{P}_{\text{hom}} + \mathscr{P}_{\text{part}}, \qquad (11.4)$$

where

$$\mathscr{P}_{\text{hom}} = \text{const } e^{-s/n} \qquad (11.5)$$

is the general solution to the homogeneous equation (corresponding to $\mathscr{F} = 0$). This part is called *Rayleigh distribution* and it corresponds to Gaussian statistics of the wavefield a_k. The second term in (11.4), $\mathscr{P}_{\text{part}}$, is a particular solution [2],

$$\mathscr{P}_{\text{part}} = -\frac{\mathscr{F}}{\eta} \, Ei(s/n) \, \exp(-s/n), \qquad (11.6)$$

where $Ei(x)$ is the integral exponential function.

At the tail of the PDF, $s_k \gg n_k$, the solution for $\mathscr{P}_{\text{part}}$ can be represented as series in n/s. In fact, the easiest way to obtain such series is by direct substitution into the (11.3) rather than from (11.6). We have:

$$\mathscr{P}_{\text{part}} = -\frac{\mathscr{F}}{s\gamma} - \frac{\eta\mathscr{F}}{(\gamma s)^2} + \cdots. \qquad (11.7)$$

Thus, the leading order asymptotics of the finite-flux solution is $1/s$ which decays much slower than the exponential part $\mathscr{P}_{\text{hom}}$ and, therefore, it describes an enhanced probability of strong waves if $\mathscr{F} < 0$, see Fig. 11.1. This phenomenon is

Fig. 11.1 One mode PDF in wave turbulence. *Straight dashed line* represents Rayleigh distribution corresponding to Gaussian wave fields. *Solid bold line* shows a typical PDF for $\mathscr{F} < 0$. Shaded area corresponds to enhanced probability of large wave amplitudes,—the WT intermittency phenomenon

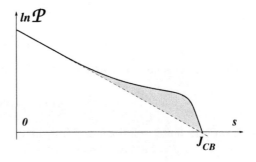

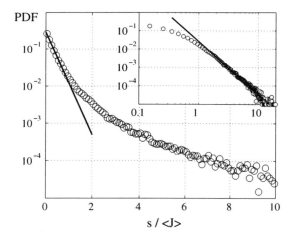

Fig. 11.2 One mode PDF's in wave tank experiments of [2] *Main plot* PDF plot in log-lin coordinates for bandpass filtered signal with frequency window 1 Hz centered at 6 Hz (mean slope of water surface was ~0.16). *Straight line* represents the Rayleigh distribution. The *inset* shows the same plot in log–log coordinates; the *straight line* is a power-law fit $1/s^3$

called *wave turbulence intermittency*, and it loosely corresponds to presence of "freak" or "rogue" waves. PDF's of this kind, with a Rayleigh core and a power-law tail, was indeed observed experimentally for gravity waves in a large wave tank (10 m × 6 m) reported in [4]; see Fig. 11.2 (although the power-law decay was $1/s^3$, not $1/s$, for reasons yet to be understood).

On the other hand, case $\mathscr{F} > 0$ would correspond to a depleted (with respect to Rayleigh) PDF tail, see Fig. 11.3. Both enhanced and depleted PDF tails were observed in numerical simulations of [2, 5]; see Fig. 11.4. We will return to discussing when should we expect each of these cases in next section.

Note that if the weak nonlinearity assumption was valid uniformly to $s = \infty$ then we had to put $\mathscr{F} = 0$ to ensure positivity of $\mathscr{P}$ and the convergence of its normalization, $\int \mathscr{P}\, ds = 1$. In this case, the normalization convergence requirement would lead to the condition $\mathscr{F} = 0$, and as a result we would have $\mathscr{P} = \mathscr{P}_{\text{hom}} = n \exp(-s/n)$ which is a pure Rayleigh distribution corresponding to a Gaussian wave field.

Fig. 11.3 One mode PDF in WT for the case $\mathscr{F} > 0$. *Straight dashed line* represents Rayleigh distribution corresponding to Gaussian wave fields

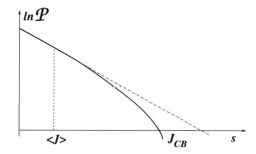

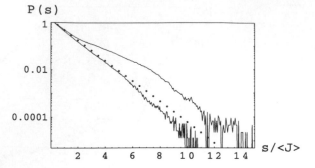

Fig. 11.4 One mode PDF's in DNS of [2] for the gravity water wave turbulence (in log-lin coordinates). *Dashed line* represents Rayleigh distribution. The *upper curve* is the PDF for $k = 15\ k_{min}$. The *lower curve* is the PDF for $k = 35\ k_{min}$

However, the weak WT approach fails for the amplitudes $s \geq J_{CB}$ for which the nonlinear time is of the same order or less than the linear wave period and, therefore, we can expect a *cutoff* of $\mathscr{P}(s)$ at $s = J_{CB}$, see Figs. 11.1 and 11.3. Since J_{CB} is to be found by balancing the linear and the nonlinear time-scales at each k, you may recognize the *critical balance* condition,—hence the subscript *CB* (see Sect. 3.2). Physically, such a PDF cutoff can be interpreted as "wave breaking" or "instability" which make waves with intensities greater that J_{CB} highly improbable, so that one can put $\mathscr{P}(s) = 0$ for $s > J_{CB}$. Obviously, the cutoff enforces convergence of the normalization condition, and makes possible solutions with finite amplitude-space flux $\mathscr{F}$.

An example of a wave breaking process leading to an amplitude cutoff can be found in the system of gravity water waves. In these waves, the nonlinearity cannot get greater than the linear effects, because this would mean that the inertial forces are greater than gravity, in which case water particles would not remain attached to the surface and the waves would break forming droplets and bubbles (sea spray and whitecaps). For the focusing NLS turbulence, "breaking" occurs in the inverse cascade leading to Bose-Einstein condensation. In this case J_{CB} corresponds to the critical intensity at which waves become modulationally unstable, leading to sudden collapses (or filamentation in the nonlinear optics context).

Estimate for the value of J_{CB} can be obtained from either the dynamical equation or, equivalently, by balancing the k-space densities of the Hamiltonians corresponding to the linear and the nonlinear dynamics. For example, for the three-wave systems we balance the densities of Hamiltonians H_2 and H_3 given by expressions (6.62) and (6.63) respectively. In balancing these terms we should take into account that if the wave amplitude is close to the critical value J_{CB} at some k then it will also be of similar value for a range of k's of width k (i.e. the k-modes are strongly correlated when the amplitude is close to critical). This gives:

$$J_{CB} \sim \omega/V^2 k^{2d}. \tag{11.8}$$

For the four-wave systems, we use the expression (6.71). This gives

$$J_{CB} \sim \omega/W k^d. \tag{11.9}$$

Note that in order for our analysis to give the correct description of the PDF tail, the nonlinearity must remain weak for the tail, which means that the breakdown must occur far from the PDF core, i.e. $\langle J \rangle \ll J_{CB}$. On the other hand, when $\langle J \rangle \sim J_{CB}$ one has a strong breakdown predicted in [1, 6] which is hard to describe rigorously due to strong nonlinearity. The scale of strong breakdown k_{nl} was found in [1, 6] by estimating the ration of the linear and the nonlinear timescales using the kinetic equation, in our notations Γ_K / ω_k, and equating this ratio to one (thus finding the scale at which the nonlinearity becomes strong). In [3] it was shown that the same estimate can be obtained from a simple dimensional analysis of the same type as we have used in Chap. 3. For the gravity water waves this gives

$$k_{nl} \sim \epsilon^{-2/3} g. \tag{11.10}$$

Exercise 11.1 *For the gravity water waves k_{nl} could be found from a simple physical argument. For this we note that the energy in the gravity wave is proportional to the mean square of the wave hight, $\langle h^2 \rangle$, and independent of its wavelength. Indeed, in linear and weakly nonlinear waves, according to the virial theorem, the averaged kinetic and potential energies are equal, and thus both are making equal parts in the total wave energy. The potential energy of a sinusoidal wave is proportional to the mean square height and is independent of the wavelength. Therefore, when energy is cascading from long to short waves, the wave amplitude remains equal to its initial value h_0, see Fig. 11.5. Thus the nonlinearity is increasing along the cascade to smaller scales, and it becomes strong when the fluid surface slope becomes ~ 1, i.e. $k_{nl} \sim 1/h_0$.*

Is this argument consistent with the estimate 11.10? If yes, show that it gives the same result. If not, find the flaw.

11.2 Wave Turbulence Life Cycle

Above, we saw that extending the WT description to PDFs leads to revealing new type of solutions with fluxes in the wave amplitude space. Is there a relation between these amplitude fluxes and the k-space fluxes associated with the KZ

Fig. 11.5 Energy cascade in the system of water gravity waves. Conservation of energy means that the wave height remains equal to its initial value h_0

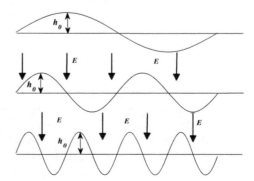

spectra? Can the J-flux and the k-flux be interpreted as projections of the same flux distribution in an extended (k, J)-space? Below, we will argue that a positive answer to this question actually makes sense and it can provide a useful paradigm for the *WT life cycle* comprising coexistence, interaction and mutual transformations of the weak incoherent and the strong coherent wave components. For clarity, we will have in mind the gravity water waves or a similar example with a direct energy cascade in which the nonlinearity increases toward large k's. An example with an inverse cascade in which the nonlinearity increases toward small k's can be found within the NLS model, see Chap. 15.

Logically, the picture of the WT cycle arises from the following two at the first sight paradoxical observations.

- Above, we already related the possibility of an amplitude space flux states with presence of a wavebreaking process. Physically this seems natural, since wavebreaking eliminates waves which reach a certain critical amplitude. Thus the wavebreaking and the WT intermittency phenomenon are related, which to many people would sound natural, and in fact it has been frequently discussed in WT literature, see e.g. [1, 6]. However, since wavebreaking eliminates high-amplitude waves, the respective J-flux $\mathscr{F}$ should be positive, i.e. from small to large J's. But our solutions (11.7) describe intermittency (positive power-law PDF tails) when $\mathscr{F} < 0$, which, at first sight, is inconsistent with what wavebreaking should do.

- As we discussed above, case $\mathscr{F} < 0$ corresponds to an enhanced probability of large wave amplitudes, whereas $\mathscr{F} > 0$ means a depleted probability (with respect to the the Gaussian statistics). Both cases were observed in numerical simulations: Figure 11.4 two examples of wave PDF's obtained in [2, 5] by DNS of surface gravity waves. However, intermittency was observed at lower-k side of the inertial range, whereas for higher k's the PDF tails are depleted. This fact is at odds with the popular view that intermittency must be bigger at higher k's, since this is where the wavebreaking is operating.

These two observations become natural and non-contradictory if we recall that wavebreaking is preceded by creation of strongly nonlinear coherent structures, e.g. Stokes waves on water surface under gravity. When such coherent waves break, their energy is partially dissipated (whitecapping in water gravity waves) and partially it is returned into weak incoherent waves (ripples). Yet some other part of energy often remains for a prolonged time in a singular (broken) coherent structure: recall sharp wave crests with a continuous whitecapping on them. Now, because the coherent structures have a broad spectral content (i.e. they consist of many k-modes which are synchronized with each other), the resulting incoherent component will also have a broad spectrum, including low-k modes.

It is these low-k waves produced by wavebreaking that are responsible of the negative flux $\mathscr{F}$ at low k's. At the same time, at high k's $\mathscr{F}$ is positive and the PDF tail is depleted. In fact, this depletion is natural and easy to understand. Indeed, in gravity WT the nonlinearity gets higher at high k's [1, 6], and the wavebreaking cutoff J_{CB} gets closer to the PDF core. When J_{CB} gets into the core, $J_{CB} \sim \langle J \rangle$, one

Fig. 11.6 Wave turbulence life cycle in the wavenumber-amplitude space. See the text for explanations and discussion

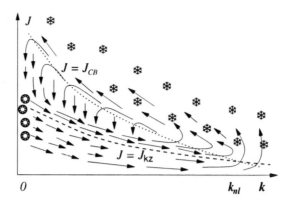

reaches the wavenumber k_{nl} at which the breakdown is strong, i.e. of the kind considered in [1, 6]. Wavenumber k_{nl} corresponds to the highest positive values of $\mathscr{F}$ and, therefore, the most depleted PDF tails with respect to the Gaussian distribution. When one moves away from this region toward lower k's, the value of $\mathscr{F}$ gets smaller and, eventually, changes the sign leading to enhanced PDF tails at low k's.

Putting it all together, we arrive at the picture of the WT cycle in the (k, J)-space sketched on Fig. 11.6. Let us place WT forcing at low k's and J's, as shown by "Sun" symbols in Fig. 11.6. Then, a KZ flux will develop toward large k's along the line $J = J_{KZ}$ corresponding to the KZ spectrum. Since the nonlinearity is increasing along the KZ cascade, it inevitably reaches wavenumber k_{nl} where the nonlinearity value reaches the one of the linear term. Wavebreaking occurs at this scale, marked by snowflakes in Fig. 11.6. Indeed, beyond this scale the nonlinear effects would exceed the linear ones, which, in gravity waves, would mean that the inertial forces are larger than the gravity acceleration, hence the fluid particles cannot stay attached to the water surface. Now, when the energy reaches the scale k_{nl} and the nonlinearity becomes large, different **k**-modes will get correlated. A set of coherent structures will emerge, in which the linear and the nonlinear timescales are critically balanced forming spectrum $J = J_{CB}$. This is Phillips spectrum for the gravity wave example. Thus, the flux will deflect along the curve $J = J_{CB}$, partially reversing toward lower k's on the (k, J)-plane. Breaking of these structures leads to partial dissipation of their energy, moving to the "snowflakes" above the curve $J = J_{CB}$ on Fig. 11.6, and a partial return of the remaining energy into incoherent "ripples" with a broad spectral content. The latter means that the flux once again turns on the (k, J)-plane, this time toward lower J's. At fixed k, this flux toward lower J's corresponds to $\mathscr{F} < 0$ and, therefore, to fat PDF tails corresponding to WT intermittency.

Remark 11.1 As we see, in the above picture fat PDF tails arise because a strongly nonlinear process of wavebreaking injects into the system strong long waves, which further evolve into the lower-amplitude range where the weak WT description is applicable. In the other words, with respect to the system of weakly nonlinear waves the reason for intermittency is external, due forcing by wave-breaking, rather than internal, due to mutual interactions of the weak waves only.

11.3 Solutions for the N-Mode Joint PDF's

So far, we have only considered solutions for the one-mode WT objects: spectra and PDF's. Even at this level we were able to describe important processes: the KZ cascades, the amplitude-space fluxes and the WT intermittency, fluxes in the (k, J)-space and the WT life cycle. On the other hand, the one-point WT closure assumes RPA, implying that the phases ϕ_k and amplitudes J_k are independent random variables. Validity of this assumption can only be examined within the multi-particle description. Moreover, in our description of the WT life cycle the modes J_k become strongly correlated when their amplitudes reach the critical values $J_{CB}(k)$. Obviously, our weak WT description fails for such strongly cor-related and strongly nonlinear modes, but perhaps we can hope that this descrip-tion can still predict a tendency toward development of the amplitude correlations while J_k's remain weaker that J_{CB}. Finally, it would be interesting to know what kind of the multi-point statistics corresponds to the KZ cascade state. All of the above point at the necessity to study the solutions for the N-mode joint PDF's.

Let us consider a general $q \to r$ resonant wave process with $q + r = p$. Then the equation for the N-mode PDF will be as in (6.124) and (6.125), which we will reproduce here for convenience:

$$
\dot{\mathscr{P}} = C\epsilon^{2(p-2)} \int W^{l_1,\dots,l_r}_{j_1,\dots,j_r} \delta\left(\omega^{l_1,\dots,l_r}_{j_1,\dots,j_q}\right) \delta^{l_1,\dots,l_r}_{j_1,\dots,j_q}
$$
$$
\times \left[\frac{\delta}{\delta s}\right]^r_q \left(s_{j_1}\dots s_{j_q} s_{l_1}\dots s_{l_r}\left[\frac{\delta}{\delta s}\right]^r_q \mathscr{P}\right) d\mathbf{k}_{j_1}\dots d\mathbf{k}_{j_q} d\mathbf{k}_{l_1}\dots d\mathbf{k}_{l_r},
$$

(11.11)

where C is a dimensionless constant and

$$
\left[\frac{\delta}{\delta s}\right]^r_q = \frac{\delta}{\delta s_{l_1}} + \dots + \frac{\delta}{\delta s_{l_r}} - \frac{\delta}{\delta s_{j_1}} - \dots - \frac{\delta}{\delta s_{j_q}}.
$$

(11.12)

Easy to see that an arbitrary function of the un-averaged energy $E_i = \int \omega_k s_k \, d\mathbf{k}$ is a steady state solution, $\dot{\mathscr{P}}(E_i) = 0$. (To prove this, one should consider the finite-box version of (11.11) and $E_i = \sum_j^N \omega_j s_j$, and then take the infinite-box limit). Further, if the number of in- and out-waves is the same, $q = r$, then an arbitrary function of the un-averaged waveaction $N_i = \int s_k \, d\mathbf{k}$ is a steady state solution, $\dot{\mathscr{P}}(N_i) = 0$. Same works for any invariant: e.g. for invariant Φ with spectral density ρ_k (see 8.4) and with un-averaged value $\Phi_i = \int \rho_k s_k \, d\mathbf{k}$ we have $\dot{\mathscr{P}}(\Phi_i) = 0$. In general, an arbitrary function of any number (or all) of the un-averaged invariants is a steady solution, e.g.

$$
\dot{\mathscr{P}}(E_i, N_i, \Phi_i) = 0.
$$

(11.13)

   This property is common for all Louisville-type *N*-particle equations, where any distribution function whose level sets are foliated by the constant energy and constant particle surfaces represents a steady state.

   An important special case is given by the exponential function,

$$\mathscr{P} = e^{\int (C_1 \omega_k + C_2 + C_3 \rho_k) s_k \, d\mathbf{k}}, \tag{11.14}$$

where C_1, C_2 and C_3 are arbitrary constants. To understand the meaning of this solutions, let us write its discrete version:

$$\mathscr{P} = \prod_{j}^{N} e^{(C_1 \omega_j + C_2 + C_3 \rho_j) s_j}. \tag{11.15}$$

   We see that this solution describes a thermodynamic equilibrium, corresponding to *N* statistically independent Rayleigh-distributed modes (i.e. Gaussian wave fields) with a mean intensity given by a Rayleigh-Jeans (RJ) spectrum (9.1). It corresponds to the zero flux in the amplitude space in (6.114).

Remark 11.2   These kind of solutions were already discussed in the very first WT paper by Peierls [7]. However, if in (11.15) we replace RJ spectrum with another stationary solution of the kinetic equation, the KZ spectrum, then it is easy to see that the result is not a solution of the PDF (11.11) (not even an approximate solution!) Thus the KZ solution obtained in the one-mode WT closure is valid only in some "coarse-grained" sense. However, it remains to be understood what this coarse-graining is in terms of the multi-mode solutions. It is also possible that for describing the KZ states in the multi-mode space one has to invoke amplitude fluxes (cf. the WT cycle picture in Sect. 11.2).

   Now, we saw in Sect. 11.1 that solutions with finite amplitude fluxes play an important role in describing the WT intermittency phenomenon. For the multi-point statistics, there is a much larger family of solutions with finite amplitude fluxes. Indeed, consider (6.114): $\mathscr{P} = -\int \frac{\delta \mathscr{F}_k}{\delta s_k} \, d\mathbf{k}$. This equation has stationary solutions for all divergence-free fluxes:

$$\mathscr{F}_{k_l} \{s(k)\} = \mathrm{curl}_s A = \int \varepsilon_{lnm} \frac{\delta A_n}{\delta s(k_m)} \, dk_m dk_n, \tag{11.16}$$

where A_n is an arbitrary functional of $s(k)$ and ε_{lnm} is the antisymmetric tensor,

$$\varepsilon_{lnm} = 1 \quad \text{for} \quad k_l > k_m > k_n \text{ and cyclic permutations of } l, m, n,$$
$$\varepsilon_{lnm} = -1 \quad \text{for all other} \quad k_l \neq k_m \neq k_n,$$
$$\varepsilon_{lnm} = 0 \quad \text{if} \quad k_l = k_m, k_l = k_n \text{ or } k_n = k_m.$$

In the other words, probability flux $\mathscr{F}$ can be an arbitrary solenoidal field in the functional space of $s(k)$. From Eqs. (6.115) and (6.112) one can see that (11.16) is a first order equation with respect to the *s*-derivative. Special cases of the steady solutions are the zero-flux and the constant-flux solutions which, as we know, correspond to the cases of Gaussian and intermittent WT respectively.

On the other hand, condition (11.16) has plenty of "vortex" solutions. For example, by taking $A_k = f(\mathbf{k})s(\mathbf{k}_1)s(\mathbf{k}_2)$, with an arbitrary integrable function $f(\mathbf{k})$, we get a "solid body" rotation in the (s_1, s_2)-plane:

$$\mathscr{F}_{k_1} = as_2, \quad \mathscr{F}_{k_2} = -as_1, \quad \mathscr{F}_{k_j} = 0 \text{ for } \quad j \neq 1, 2,$$

where $a = \int f_k \, d\mathbf{k}$. Obviously, in such states the modes $\mathbf{k}_1$ and $\mathbf{k}_2$ are correlated. For further study of such mode correlations and vortex-like solutions in the two-mode amplitude space see paper [8].

11.4 Validity of RPA

One of the most important uses of (11.11) for the N-mode PDF is for examining validity of an important assumption in the WT closure, namely that the RPA property survives over the nonlinear evolution time. First of all, since (11.11) does not depend on the angular variables, we conclude that the phase independence does survive over the nonlinear time.

Situation with the amplitude variables is more subtle, because variables s_j do not separate in the above equation for the PDF. Let us consider a product factorization of the N-mode PDF:

$$\mathscr{P}^{(N)} = \mathscr{P}_{j_1}\mathscr{P}_{j_2}\ldots\mathscr{P}_{j_N} \tag{11.17}$$

into the discrete version of (11.11). We already saw a stationary solution of this kind: the thermodynamic solution (11.15). However, it is not a solution if the one-mode PDF's $\mathscr{P}_j$ correspond to the cascade-type KZ spectrum n_j^{KZ}, i.e. $\mathscr{P}_j = \left(1/n_j^{KZ}\right)\exp\left(-s_j/n_j^{KZ}\right)$, nor it is a solution for any other PDF of form (11.17). This means that the amplitudes, even initially independent, will correlate with each other at the nonlinear time. Does this mean that the existing WT theory, and in particular the kinetic equation, is invalid?

To answer to this question let us consider the three-wave example using the discrete version of the equation for the N-mode PDF (6.114) with flux (6.115). Integrating out all the amplitude variables but one, we get an expression for $\dot{\mathscr{P}}^{(1)}$ in terms of $\mathscr{P}^{(2)}$ and $\mathscr{P}^{(3)}$. Similarly, integrating out all the amplitude variables but two, we get an expression for $\dot{\mathscr{P}}^{(2)}$ in terms of $\mathscr{P}^{(3)}$ and $\mathscr{P}^{(4)}$.

Exercise 11.2 *Derive evolution equations for $\mathscr{P}^{(1)} \equiv \mathscr{P}$ and for $\mathscr{P}^{(2)}$. Show that*

$$\partial_t\left(\mathscr{P}^{(2)}_{j_1,j_2}(s_{j_1}, s_{j_2}) - \mathscr{P}_{j_1}(s_{j_1})\mathscr{P}_{j_2}(s_{j_2})\right) = O(\epsilon^4) \quad (j_1, j_2 \in \mathscr{B}_N) \tag{11.18}$$

if $\mathscr{P}^{(4)}_{j_1,j_2,j_3,j_4}(s_{j_1}, s_{j_2}, s_{j_3}, s_{j_4}) = \mathscr{P}_{j_1}(s_{j_1})\mathscr{P}_{j_2}(s_{j_2})\mathscr{P}_{j_3}(s_{j_3})\mathscr{P}_{j_4}(s_{j_4}).$

One can see that, with an ϵ^2-accuracy, the Fourier modes will remain independent of each other in any pair over the nonlinear time if they were independent in every triad at $t = 0$.

Similarly, one can show that the modes will remain independent over the nonlinear time in any subset of $M < N$ modes with accuracy M/N (and ϵ^2) if they were initially independent in every subset of size $M + 2$. Namely

$$\mathscr{P}^{(M)}_{j_1, j_2, \ldots, j_M}\left(s_{j_1}, s_{j_2}, s_{j_M}\right) - \mathscr{P}_{j_1}\left(s_{j_1}\right)\mathscr{P}_{j_2}\left(s_{j_2}\right)\ldots\mathscr{P}_{j_M}\left(s_{j_M}\right)$$
$$= O(M/N) + O\left(\epsilon^2\right) \quad (j_1, j_2, \ldots, j_M \in \mathscr{B}_N) \tag{11.19}$$

if $\mathscr{P}^{(M+2)}_{j_1, j_2, \ldots, j_{M+1}} = \mathscr{P}_{j_1}\mathscr{P}_{j_2}\ldots\mathscr{P}_{j_{M+1}}$.

Mismatch $O(M/N)$ arises from some terms in the PDF equation with coinciding indices j. For $M = 2$ there is only one such term in the N-sum and, therefore, the corresponding error is $O(1/N)$ which is much less than $O(\epsilon^2)$ (due to the order of the limits in N and ϵ). However, the number of such terms grows as M and the error accumulates to $O(M/N)$ which can greatly exceed $O(\epsilon^2)$ for sufficiently large M.

We see that the accuracy with which the modes remain independent in a subset is worse for larger subsets and that the independence property is completely lost for subsets approaching in size the entire set, $M \sim N$. One should not worry too much about this loss because N is the biggest parameter in the problem (size of the box) and the modes will be independent in all M-subsets no matter how large. Thus, the statistical objects involving any *finite* number of modes are factorizable as products of the one-mode objects and, therefore, the WT theory reduces to considering the one-mode objects.

Thus, strictly speaking we should re-define RPA in a more relaxed form, and postulate the amplitude independence property as a product factorization of the M-mode PDF's for any $M \ll N$. Indeed, in this form RPA is sufficient for the WT closure and, on the other hand, it remains valid over the nonlinear time. In particular, only property (11.18) is needed, as far as the amplitude statistics is concerned, for deriving the 3-wave kinetic equation, and this fact validates this equation and all of its solutions, including the KZ spectrum.

The situation were modes can be considered independent when taken in relatively small sets but should be treated as dependent in the context of much larger sets is not so unusual in physics. Consider for example a distribution of electrons and ions in plasma. The full N-particle distribution function in this case satisfies the Louisville equation which is, in general, not a separable equation. In other words, the N-particle distribution function cannot be written as a product of N one-particle distribution functions. However, an M-particle distribution can indeed be represented as a product of M one-particle distributions if $M \ll N_D$ where N_D is the number of particles in the Debye sphere. We see an interesting transition from a an individual to collective behavior when the number of particles approaches N_D. In the special case of the one-particle function we have here the famous mean-field Vlasov equation which is valid up to $O(1/N_D)$ corrections (representing particle collisions).

References

1. Biven, L., Nazarenko, S.V., Newell, A.C.: Breakdown of wave turbulence and the onset of intermittency. Phys. Lett. A Vol **280**/1–2, 28–32. (2001)
2. Choi, Y., Lvov, Y., Nazarenko, S.V., Pokorni, B.: Anomalous probability of large amplitudes in wave turbulence. Phys. Lett. A **339**(3–5), 361 (2004)
3. Connaughton, C., Nazarenko, S., Newell, A.C.: Dimensional analysis and weak turbulence. Phys. D **184**, 86–97 (2003)
4. Denissenko, P., Lukaschuk, S., Nazarenko, S.: Gravity surface wave turbulence in a laboratory flume. Phys. Rev. Lett. **99**, 014501 (2007)
5. Lvov, Y.V., Nazarenko, S., Pokorni, B.: Discreteness and its effect on water-wave turbulence. Phys. D **218**(1), 24 (2006)
6. Newell, A.C., Nazarenko, S.V., Biven, L.: Wave turbulence and intermittency. Phys. D **152–153**, 520–550 (2001)
7. Peierls, P.: Ann. Phys. **3**, 1055 (1929)
8. Choi, Y., Jo, S.G., Kim, H.I., Nazarenko, S.V.:Aspects of two-mode probability density function in weak wave turbulence, J. Phys. Soc. of Jpn, **78**(8), Article Number 084403, (2009).

Chapter 12
Solutions to Exercises

12.1 Zonostrophy Invariant: Exercise 8.1

Let us introduce auxiliary variables,

$$p = k_y + k_x \sqrt{3}, \tag{12.1}$$

$$q = k_y - k_x \sqrt{3}. \tag{12.2}$$

Substituting these expressions into ω_k we have

$$\omega_k = k_x k^2 = \frac{1}{6\sqrt{3}} (p^3 - q^3). \tag{12.3}$$

Using this expression in the frequency resonance condition (8.6), we get for this resonant condition in terms of the auxiliary variables,

$$p_1 p_2 p = q_1 q_2 q, \tag{12.4}$$

where

$$p_{1,2} = k_{1,2y} + k_{1,2x} \sqrt{3}, \tag{12.5}$$

$$q_{1,2} = k_{1,2y} - k_{1,2x} \sqrt{3}. \tag{12.6}$$

Thus we have

$$\frac{p^2}{q} - \frac{p_1^2}{q_1} - \frac{p_2^2}{q_2} = -\frac{(p_1 q_2 - p_2 q_1)^2}{q_1 q_2 q} = -\frac{(p_1 q_2 - p_2 q_1)^2}{p_1 p_2 p} = \frac{q^2}{p} - \frac{q_1^2}{p_1} - \frac{q_2^2}{p_2}. \tag{12.7}$$

S. Nazarenko, *Wave Turbulence*, Lecture Notes in Physics, 825,
DOI: 10.1007/978-3-642-15942-8_12, © Springer-Verlag Berlin Heidelberg 2011

This means that the quantity

$$\frac{p^2}{q} - \frac{q^2}{p} \tag{12.8}$$

satisfies the resonant condition and is, therefore, a density of a conserved quantity. Substituting back expressions (12.1) and (12.2), we finally have

$$\rho_k = \frac{1}{6\sqrt{3}} \left(\frac{p^2}{q} - \frac{q^2}{p} \right) = \frac{k_x k^2}{k_y^2 - 3k_x^2}. \tag{12.9}$$

12.2 Waveaction Conservation for the Four-Wave Systems: Exercise 8.3

Four-wave Hamiltonian (6.71) yields Eq. (6.72), which, after multiplying by $-i\, a_k^*$ (and leaving the diagonal terms in the sum), becomes:

$$\dot{a}_k a_k^* = -i\omega_k a_k a_k^* - i \sum_{\mathbf{k}_1,\mathbf{k}_2,\mathbf{k}_3} W_{3k}^{12} \delta_{3k}^{12} a_1 a_2 a_3^* a_k^*. \tag{12.10}$$

Adding to this equation its complex conjugate and summing the result over $\mathbf{k}$, we get $\dot{\mathcal{N}}$ on the LHS, whereas the RHS becomes zero. Thus, $\mathcal{N} = $ const.

For the water gravity waves, conservation of $\mathcal{N}$ is approximate, because Hamiltonian (6.71) represents only the leading order term in an infinite series (in small nonlinearity) of the exact interaction Hamiltonian.

12.3 Rayleigh-Jeans Solutions: Exercise 9.1

Let us introduce notation for the density of the invariant involved in the RJ spectrum (9.1): $\sigma_k = \sum_j C_j\, \rho_k^{(j)}$. The integrand of the kinetic equation for a general $q \to r$ resonant wave process contains factor

$$\delta\left(\omega_{j_1,\dots,j_q}^{l_1,\dots,l_r}\right) \delta_{j_1,\dots,j_q}^{l_1,\dots,l_r} \left[\frac{1}{n}\right]_{j_1,\dots,j_q}^{l_1,\dots,l_r}. \tag{12.11}$$

Substituting here the RJ spectrum (9.1), we get

$$\delta\left(\omega_{j_1,\dots,j_q}^{l_1,\dots,l_r}\right) \delta_{j_1,\dots,j_q}^{l_1,\dots,l_r} \sigma_{j_1,\dots,j_q}^{l_1,\dots,l_r}. \tag{12.12}$$

But this expression is zero because, the invariant's density σ satisfies $\sigma^{l_1,\dots,l_r}_{j_1,\dots,j_q} = 0$ on the resonant manifold. Therefore, the RHS of the kinetic equation is zero, which means that the RJ spectrum (9.1) is a steady state solution.

12.4 Energy Flux Direction in Systems with a Single Relevant Dimensional Parameter: Exercise 9.2

For wave systems with a single dimensional parameter, the 1D energy spectrum has an exponent found in (3.10). For the waveaction spectrum exponent, for $N = 3$, this gives[1] $v = \alpha - 5 + \frac{5-d-3\alpha}{2}$. From this expression we see that $v < -\alpha$ if (and only if) $\alpha < 5 + d$, which is the condition for the energy cascade to be direct.

12.5 Zakharov Transform for the Four-Wave Systems: Exercise 9.3

Like in the three-wave case, the ω δ-function contributes into index x with α and the interaction coefficient with -2β. The part containing n's contributes with $-3v$, the ZT Jacobean gives -4, the wavenumber δ-function gives d, factor $(k_1\,k_2\,k_3)^{d-1}$ gives $4(d-1)$. Adding all these contributions together we get (9.35).

12.6 Geometrical Condition of Stability: Exercise 9.9

This solution is illustrated on Fig. 12.1. Let us consider a stationary solution $n_k = Ak^{v^*}$ and let us expand function $f(v_x, v_y)$ in small neighborhood of $v = v^*$. Using relations (9.79) and (9.80) and relation $f(v) = \mathscr{I}(k^v) = CW^{(0)}(0)$, we have

$$f(v^* + \delta) = \nabla_v \cdot \delta + \frac{1}{2}\frac{\partial^2 f}{\partial v_x^2}\delta_x^2 + \frac{1}{2}\frac{\partial^2 f}{\partial v_y^2}\delta_y^2 + \frac{\partial^2 f}{\partial v_x \partial v_y}\delta_x \delta_y$$
$$= -2C\boldsymbol{a}\cdot\delta + 2C(L_{xx}\delta_x^2 + L_{yy}\delta_y^2 + 2L_{xy}\delta_x\delta_y). \qquad (12.13)$$

Let us consider deviations δ along the tangential vector to the curve $f(v_x, v_y) = 0$, i.e. $\delta = \lambda b$ with small λ and $b = (-a_y, a_x)$. This gives

$$f(v^* + \lambda b) = 2C\lambda^2(L_{xx}a_y^2 + L_{yy}a_x^2 - 2L_{xy}a_x a_y). \qquad (12.14)$$

---

[1] For simplicity, we consider only the cases where the wave energy is distributed over the physical-space manifold of the same dimension as the dimension of the k-space. For a more general treatment see [1].

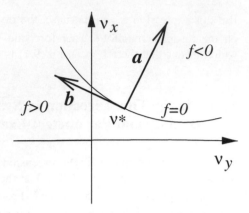

Fig. 12.1 Geometrical condition of stability

According to the condition (9.78), for stability of the spectrum it is necessary for this expression to be positive. Thus, the function f must be positive for the deviations from $v = v^*$ in the tangential direction to the curve $f(v_x, v_y) = 0$. This means that the curve $f(v_x, v_y) = 0$ must be convex toward the region where f takes positive values.

Reference

1. Connaughton, C., Nazarenko, S., Newell, A.C.: Dimensional analysis and weak turbulence, Physica D **184**, 86–97, (2003)

Part IV
Selected Applications

In this part, we will present a detailed description of the three major applications of Wave Turbulence: Drift/Rossby wave turbulence, MHD turbulence and WT in Bose–Einstein condensates. These applications were chosen because in these examples WT forms a major unremovable part of the general turbulence picture, rather than a particular additional process which is not essential for explaining the main turbulence mechanisms. By necessity, to describe these applications we will develop further theoretical approaches to deal with situations when WT is nonlocal, when it is strongly affected by presence of boundaries, condensation into large scales or/and zonal flows, generation of coherent structures and vortices, interaction with strongly turbulent component. We will conclude this part by giving a list and a brief description of further topics which could be use in setting up graduate projects or essays.

Chapter 13
Nonlocal Drift/Rossby Wave Turbulence

13.1 When is Turbulence Nonlocal?

KZ solutions are important because they highlight similarity between WT and classical Kolmogorov picture of turbulence, which brings out an understanding that such turbulent systems are strongly non-equilibrium and the notion of the energy flux through scales is much more relevant to their description than "temperature" or "chemical potential" used for systems close to thermodynamic equilibrium.

However, as we saw in Sect. 9.2.6, not all KZ spectra are relevant and realizable because of instability and nonlocality. This is especially true for anisotropic systems. Besides, the KZ solutions can only be expected in a rather artificial setup of sources and sinks, which often does not correspond to any real physical mechanisms of forcing and dissipation. For example, the inverse cascade setup assumes presence of dissipation at large scales which would efficiently absorb the turbulent flux. Often in physical systems, there is no such large-scale dissipation, and the turbulence "piles up" at the largest available scale (e.g. size of the physical system). In anisotropic systems, the source and sink distributions in the KZ setup are even more artificial, as e.g. in the drift wave (Petviashvili) example in Fig. 8.2, where the dissipation is assumed near the k_x- and k_y-axes which is totally unrealistic. Thus in this example the energy will accumulate near anisotropic large scales with dominant *zonal component*, as it follows from the Fjørtoft's argument illustrated in Fig. 8.2.

What can we expect when such "condensation" of turbulent spectrum occurs at one or several edges of the inertial interval? It is natural to expect that for some large enough condensate level the locality property will break down and further evolution will be determined by nonlocal interaction with the condensate scales.

S. Nazarenko, *Wave Turbulence*, Lecture Notes in Physics, 825,
DOI: 10.1007/978-3-642-15942-8_13, © Springer-Verlag Berlin Heidelberg 2011

13.2 Nonlocal Weak Drift/Rossby Turbulence

Let us consider our favorite example of drift turbulence, and this time we will keep it more general than just the Petviashvilli's model, without specifying the inter- action coefficient [thus including possibility of the vector nonlinearity as in (6.13)] and including the finite deformation radius effects [also as in (6.13)]. Besides, we will consider all scales and not only $|k_y| \gg k_x$ as in the KZ setup. In what follows, we will describe a turbulence scenario suggested and developed in papers [1–4]. In this scenario, small-scale Drift/Rossby waves and zonal flows form two essential coexisting and interacting components in turbulence. Small-scale Drift/Rossby waves are produced by an external linear instability. Their mutual nonlinear interactions lead to generation of zonal flows. In turn, zonal flows feed back onto the small-scale waves and suppress them. In the magnetic fusion community, this mechanism is presently considered [6] as the main explanation of the turbulent transport suppression[1] which was observed in tokamaks and which is known as Low-to-High (LH) transition in confinement [5]. Transport blocking by zonal flows is also important in geophysical flows [7]. Below, we will also highlight recent numerical evidence for the suggested drift-wave/zonal-flow scenario [8].

13.2.1 Nonlocal Interaction with Large Scales

Thus let us start with the kinetic equation (6.57) in which the dispersion relation is (7.7) and the interaction coefficient is (6.12) or (7.8). Let us suppose that the main interaction of mode $\mathbf{k}$ is with large scales, i.e. with wavenumbers $\mathbf{p}$ such that $p \ll k$ and $p \ll 1/\rho$. Then the kinetic equation (6.57) can be re-written as

$$\frac{\partial n_k}{\partial t} = 8\pi \int\limits_{p \ll k, 1/\rho} |V_{\mathbf{p},\mathbf{k}-\mathbf{p},\mathbf{k}}|^2 \delta(\overline{\omega}(\mathbf{k}) - \overline{\omega}(\mathbf{k} - \mathbf{p}))[n(\mathbf{k} - \mathbf{p}) - n(\mathbf{k})]n(\mathbf{p}) \, d\mathbf{p}.$$

(13.1)

where

$$\overline{\omega}(\mathbf{k}) = \rho^2 \beta k_x + \omega_k = \frac{\beta k_x \rho^4 k^2}{1 + \rho^2 k^2}.$$

(13.2)

In (13.1), we neglected term $n_k \, n_2$ assuming that $[\, n(\mathbf{k} - \mathbf{p}) - n(\mathbf{k})]\, n(\mathbf{p}) \approx (\nabla_k n_k \cdot \mathbf{p})\, n_p \gg n_k \, n_2$ i.e. the spectrum is so large at scales $p \ll k$ that $n_p/n_k \gg k/p$. We have also neglected $\overline{\omega}(\mathbf{p})$ in the frequency δ-function, because it is $\sim p^2$ and the remaining part in this δ-function is $\sim p$.

---

[1] Suppression of small-scale wave turbulence obviously results in suppressing of the turbulence-produced anomalous transport of the energy and particles across the confined plasma.

Let us define function $F(\mathbf{k}, \mathbf{p})$ as

$$F(\mathbf{k}, \mathbf{p}) = 4\pi |V_{\mathbf{p},\mathbf{k}-\mathbf{p},\mathbf{k}}|^2 \delta(\overline{\omega}(\mathbf{k}) - \overline{\omega}(\mathbf{k} - \mathbf{p}))[n(\mathbf{k} - \mathbf{p}) - n(\mathbf{k})]n(\mathbf{p}). \quad (13.3)$$

Taking into account that $F(\mathbf{k}, \mathbf{p}) = -F(\mathbf{k} - \mathbf{p}, -\mathbf{p})$ and Taylor expanding in small p, we have:

$$\int\limits_{p \ll k, 1/\rho} [F(\mathbf{k}, \mathbf{p}) - F(\mathbf{k} + \mathbf{p}, \mathbf{p})]\, d\mathbf{p} = -\int\limits_{p \ll k, 1/\rho} \left(\frac{\partial F}{\partial \mathbf{k}} \cdot \mathbf{p}\right) d\mathbf{p} = \frac{\partial}{\partial k_i} \hat{D}_{ij} \frac{\partial n_k}{\partial k_j},$$

$$\quad (13.4)$$

where

$$\hat{D}_{ij} = 4\pi \int |V_{\mathbf{p},\mathbf{k}-\mathbf{p},\mathbf{k}}|^2 \delta(\overline{\omega}(\mathbf{k}) - \overline{\omega}(\mathbf{k} - \mathbf{p})) n_{\mathbf{p}}\, p_i p_j\, d\mathbf{p}. \quad (13.5)$$

From the resonance condition $\overline{\omega}(\mathbf{k} + \mathbf{p}) = \overline{\omega}(\mathbf{k})$, one can see that for small p:

$$p_x = -\theta p_y, \quad (13.6)$$

$$\theta = \frac{\partial \overline{\omega}_k}{\partial k_y} \bigg/ \frac{\partial \overline{\omega}_k}{\partial k_x} = \frac{k_x k_y}{k_y^2 + 3k_x^2 + \rho^2 k^4}. \quad (13.7)$$

Performing integration in (13.5) using the frequency δ-function, we get

$$\hat{D} = A \frac{\partial \overline{\omega}_k}{\partial k_x} \begin{pmatrix} \theta^2 & -\theta \\ -\theta & 1 \end{pmatrix}. \quad (13.8)$$

with

$$A = 4\pi \left(\frac{\partial \overline{\omega}_k}{\partial k_x}\right)^{-2} \int\limits_{-\infty}^{+\infty} \left[|V_{\mathbf{p},\mathbf{k}-\mathbf{p},\mathbf{k}}|^2 n_{\mathbf{p}}\right]_{p_x = -\theta p_y} p_y^2\, dp_y. \quad (13.9)$$

Matrix $\hat{D}$ is degenerate. Changing variables $(k_x, k_y) \rightarrow (k_y, v = \overline{\omega}_k)$ in (13.1), we bring the evolution equation to the following form,

$$\frac{\partial n_k}{\partial t} = \frac{\partial \overline{\omega}_k}{\partial k_x} \left(\frac{\partial}{\partial k_y} A \left(\frac{\partial n_k}{\partial k_y}\right)_v\right)_v, \quad (13.10)$$

where $\left(\frac{\partial}{\partial k_y}\right)_v$ means the derivative with respect to k_y with v kept constant, i.e.

$$\left(\frac{\partial}{\partial k_y}\right)_v = -\theta \frac{\partial}{\partial k_x} + \frac{\partial}{\partial k_y}. \quad (13.11)$$

Equation (13.11) describes a one-dimensional diffusion along the curves

$$\overline{\omega}_k = \text{const}, \quad (13.12)$$

Fig. 13.1 Curves (13.12)

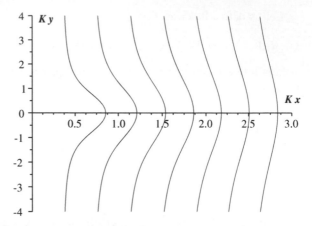

which are shown in Fig. 13.1. The diffusion coefficient of this process, A, is proportional to the WT intensity at the largest scales.

This diffusion process has a simple interpretation as a (1D) random walk of small-scale drift-wave packets in the **k**-space due to random shearing produced by the large scales. If, on the other hand, the large scales were not random (e.g. a laminar shear flow) then we would expect a deterministic shearing which would be described by a first-order equation in the **k**-space. Describing this process would be beyond the WT approach, but it could be done using a ray tracing (WKB) description based on the scale separation between the small and the large scales [4].

13.2.2 Evolution of Nonlocal Rossby/Drift Turbulence: a Feedback Loop

Equation (13.11) allows us to find some important qualitative features of evolving nonlocal turbulence which have far-reaching consequences for both the nuclear fusion and the GFD applications.

First of all we note that (13.11) conserves the total waveaction of the small-scale turbulence,

$$\mathcal{N} = \int n_k \, d\mathbf{k} = \text{const}, \tag{13.13}$$

whereas the total energy of the small-scale turbulence is not conserved,

$$E = \int |\omega_k| n_k \, d\mathbf{k} \neq \text{const}. \tag{13.14}$$

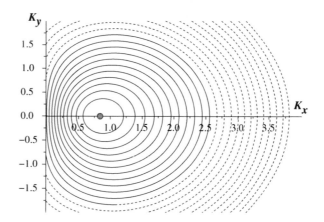

Fig. 13.2 Typical instability growth rate for Rossby/drift waves

However, the total energy of the system, including both small and large scales, must be conserved. When moving along the curves (13.12) (shown in Fig. 13.1) away from the k_x-axis toward large k's, the small-scale wavepackets loose their energy because ω_k is decreasing along these curves. Thus, the energy lost at the small scales is transferred to the the large scales.

If there is no efficient mechanism of dissipation at the large scales then the energy transfer from the small scales will lead to accumulation of the turbulent spectrum at the large scales. In turn, according to (13.9), this will lead to growth of the diffusion coefficient A and, therefore, acceleration of the transfer of the small-scale wavepackets toward large k's and the rate of the energy loss by these wavepackets.

To find qualitative consequences of such coupled dynamics of the small and the large scales, let us specify turbulence forcing and dissipation mechanisms. Namely, we will assume that both the turbulence sources and the sinks are linear, i.e. produced by a relevant physical instability and viscosity, which can be modeled by adding a term $\gamma_k n_k$ to the RHS of (13.11),

$$\frac{\partial n_k}{\partial t} = \frac{\partial \overline{\omega}_k}{\partial k_x} \left(\frac{\partial}{\partial k_y} A \left(\frac{\partial n_k}{\partial k_y} \right)_v \right)_v + \gamma_k \, n_k, \tag{13.15}$$

where the forcing region corresponds to wavevectors **k** at which $\gamma_k > 0$ and the dissipation is where $\gamma_k < 0$.

Typical shape of the instability/dissipation function γ_k (e.g. ITG instability in plasma [9] or baroclinic instability in GFD [10]) is shown in Fig. 13.2. Maximum of the instability, shown by a grey blob, is located on the k_x-axis at $k_x \sim 1/\rho$. The neutrally stable modes (where $\gamma_k = 0$) are shown by the thick solid curve which passes through $k = 0$. The instability region is located inside this curve (shown by thin solid level lines $\gamma_k = \text{const} > 0$) and the dissipation scales are outside (shown by dashed level lines $\gamma_k = \text{const} < 0$).

Evolution of the spectrum n_k described by the (13.15) can be found by considering independent eigenvalue problems for each curve v. Asymptotically, on curve ν the evolution will be determined by the maximum eigenvalue on this curve. If such an eigenvalue is positive, the spectrum n_k will grow exponentially at each point on this curve. Correspondingly, if the eigenvalue is negative then the spectrum will exponentially die out on this curve. It is clear that the biggest positive eigenvalues will be on the curves which pass close to the maximum of γ_k, and the negative eigenvalues will be for the curves away from this maximum, where the negative values of γ dominate over the positive ones, see Fig. 13.3.

Growth of n_k on the curves with positive eigenvalues will lead to accelerated energy transfer to the large scales, see Figs. 13.3 and 13.4. As we said above, this will lead to an increase of the diffusion coefficient A, faster motion of the wavepackets from the instability to the dissipation regions and, therefore, decreasing of the eigenvalues on all the curves. Thus, the evolution will lead to shrinking of the region where n_k grows, until eventually it shrinks to a jet-like spectrum concentrated on a single k-space curve, and a steady state will be reached, see Figs. 13.3, 13.4 and 13.5.

Fig. 13.3 Initial evolution of nonlocal Rossby/drift turbulence. Turbulence grows in the region of the k-space (shaded *grey*) which correspond to those curves ν in the family (13.12) for which the eigenvalues of the problem (13.15) are positive. Outside of these regions turbulence decays. Energy is transferred to the zonal flow scales as shown by the *arrows*

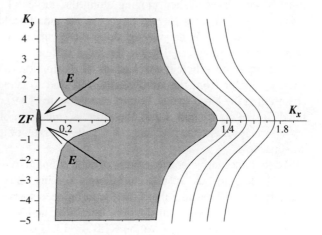

Fig. 13.4 Same as in Fig. 13.3, but now for intermediate times. The region in which turbulence grows (*grey*) is shrinking, while the energy continues to flow from the small scales to the large zonal scales

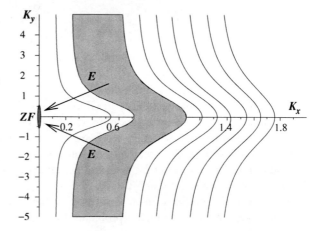

Fig. 13.5 Same as in Fig. 13.3, but now for long times. The turbulence growth region has shrunk to a single curve, and the energy of the zonal flows has saturated at the level estimated by (13.18)

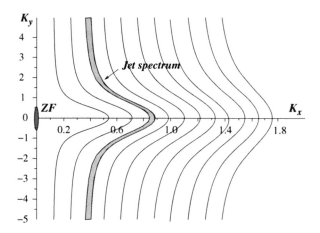

If the dissipation at large scales is small then the jet spectrum will have a small amplitude,—just enough to maintain the energy flux to the large scales to balance their dissipation. In turn, the large scales will saturate at a finite intensity determined by the condition that the maximum eigenvalue in the system (including all the curves) is equal to zero. This allows us to estimate the saturation intensity of the large-scale turbulence from balancing the two terms on the right-hand side of (13.15), which gives

$$n_p \sim \frac{\gamma_{max}\beta}{4\pi p^3 |V_{p,k-p,k}|^2}. \tag{13.16}$$

From this relation one can get an estimate of the saturation large-scale velocity U using relation

$$n_p \sim \frac{U^2}{p^5 \beta \rho^4}. \tag{13.17}$$

Exercise 13.1 *Derive estimate* (13.17).

For the CHM model (6.13), using the expression for the interaction coefficient (6.13), we have from (13.16) and (13.17):

$$U \sim \sqrt{v_* \gamma_{max} L / \pi}, \tag{13.18}$$

where $v_* = \beta \rho^2$ (so called drift velocity) and $L \sim 1/p$ is the typical scale of the large-scale flow, which could be of the order of the physical size of the system.

13.2.3 Nonlocal Interaction with Small-Scale Zonal Flows

Besides the scales $p \ll k$, there is another interesting range of scales with which there may be nonlocal interaction, namely $\mathbf{p} \to (0, 2k_y)$. These scales correspond to small-scale zonal flows. This follows from the studies of nonlocality of Rossby/drift wave spectra, which showed that most spectra of interest, including all the KZ spectra, have undefined Mellin functions $W^{(1)}$ because the defining these functions integral diverges for $\mathbf{p} \to (0, 2k_y)$, see [1].

Thus, let us now, following [2], consider contributions to the evolution of the small-scale Rossby/drift turbulence arising from nonlocal interactions with scales $\mathbf{p}$ in vicinity of $(0, 2k_y)$ by Taylor expanding about this point in the collision integral of the kinetic equation (6.57). This gives:

$$\frac{\partial n_k}{\partial t} = S[n(k_x, -k_y) - n(k_x, k_y)] + \frac{\partial \eta_k}{\partial k_x} \left(\frac{\partial}{\partial k_y} B \left(\frac{\partial n_k}{\partial k_y} \right)_w \right)_w . \tag{13.19}$$

where we changed variables $(k_x, k_y) \to (k_y, w = \eta_k)$ with η_k given by

$$\eta_k = \arctan \frac{k_y + k_x \sqrt{3}}{\rho k^2} - \arctan \frac{k_y - k_x \sqrt{3}}{\rho k^2} - 2\sqrt{3} \omega_k / (\beta \rho) = \text{const.} \tag{13.20}$$

In (13.19), notation $\left(\frac{\partial}{\partial k_y} \right)_w$ means the derivative with respect to k_y with w kept constant. Functions $S \equiv S(\mathbf{k})$ and $B \equiv B(\mathbf{k})$ are defined as

$$S = 8\pi \left| \frac{\partial \overline{\omega}_k}{\partial k_y} \right|^{-1} \int\limits_{|q_x| \ll k_x} \left[|V_{\mathbf{q}, \mathbf{k}-\mathbf{q}, \mathbf{k}}|^2 n_q \right]_{\mathbf{q}=(q_x, -2k_y)} dq_x, \tag{13.21}$$

$$B = 4\pi \left| \frac{\partial \overline{\omega}_k}{\partial k_y} \right|^{-1} \left(\frac{\partial \eta_k}{\partial k_x} \right)^{-1} \frac{4k_x^2 k_y^2 (1 + 4\rho^2 k_y^2)^2}{\left[3k_y^2 - 3k_x^2 + \rho^2 (3k_y^4 - 6k_x^2 k_y^2 - k_x^4) \right]^2}$$

$$\times \int\limits_{|q_x| \ll k_x} \left[|V_{\mathbf{q}, \mathbf{k}-\mathbf{q}, \mathbf{k}}|^2 n_q \right]_{\mathbf{q}=(q_x, -2k_y)} q_x^2 \, dq_x. \tag{13.22}$$

The first and the second terms on the RHS of (13.19) arise from the leading and the first sub-leading terms of the Taylor expansion around point $(0, 2k_y)$ in the collision integral of the kinetic equation (6.57). The first term describes a rapid relaxation of the spectrum to a symmetric equilibrium $n(k_x, -k_y) = n(k_x, k_y)$. Thus, the leading effect of the nonlocal interaction with the small-scale zonal flows is the elimination of the asymmetric perturbation $Y^{(1)}$. This is why, when studying locality and stability of the Rossby/drift wave KZ spectra in Sect. 9.2.6, we said that it makes sense to study evolution of the even perturbations $Y^{(0)}$, even though the odd ones, $Y^{(1)}$, were shown to lead to nonlocality.

Remark 13.2.1 One can easily recognize in (13.20) the density of the zonostrophy invariant, c.f. (8.16) and (8.17). Thus, in this case the small-scale wavepackets move in the $\mathbf{k}$-space in such a way that their zonostrophy is conserved, which once again highlights importance of the zonostrophy in the Rossby/drift turbulence.

Fig. 13.6 "Sea shell" curves (13.20) along which the wavepackets move due to their interactions with small-scale zonal flows (i.e. with the scales near the k_y-axis, but not too close to 0)

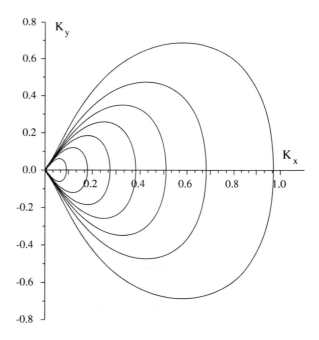

Respective curves along which the wavepackets move, $\eta(k_x, k_y) = $ const are shown in Fig. 13.6. As we see, these "sea-shell" shaped curves are connected with the point $\mathbf{k} = 0$ and, therefore, the nonlocal interactions with the small-scale zonal flows provide an additional route to supplying energy to large scales. However, not all of the initial small-scale energy end up at the large scales because, when moving along these curves toward zero, the waves loose energy. Indeed, during this motion the wave action n_k is conserved and the frequency ω_k is decreasing and, since the total energy is conserved, the energy is transferred to the zonal flows. Thus we see similar effects as we have already observed before for the energy transfer from the small-scale turbulence to the large scales due to the nonlocal interaction with the large scales. The only essential difference in that the intermediate scales do not die out anymore. Another difference that could possibly have observable consequences is that all of the sea-shell curves have the same asymptote near zero with $k_y = \pm\sqrt{3}\,k_x$, which mean focusing of the wave energy on these directions.

13.3 Beyond Weak Turbulence: Two Regimes of Zonal-Flow Growth

Here, we will use the insights obtained from considering the nonlocal weak Rossby/drift wave turbulence, reinterpret them in a qualitative physical way, and extend such a qualitative picture to stronger wave turbulence. This will lead us to describing two different possible regimes of excitation of zonal flows by drift-wave turbulence, saturation intensities for the zonal flows corresponding to these regimes, and the condition of transition from one regime to another.

Let us consider a zonal flow (ZF) with velocity $\mathbf{v} = (U(y), 0)$ which varies randomly in y with a certain correlation length L. Roughly speaking, L corresponds to a typical wavelength in profile $U(y)$. Let us consider a drift-wave packet (DWP) propagating on background of this ZF as shown in Figs. 13.7 and 13.8.

In our model, drift turbulence is nonlocal. Namely, we ignore interactions of drift waves among themselves and only take into account their interaction with the ZF. For $kL \gg 1$, this process is described by a WKB approach, as introduced in [4]. In this approach, the waveaction density in the wavenumber-coordinate space is described by the following (linear with respect to the small scales) equation,

$$D_t n = 0, \tag{13.23}$$

where D_t denotes time derivative taken along wave rays,

$$\dot{\mathbf{x}} = \frac{\partial \omega}{\partial \mathbf{k}}, \tag{13.24}$$

$$\dot{\mathbf{k}} = -\frac{\partial \omega}{\partial \mathbf{x}}, \tag{13.25}$$

Fig. 13.7 Drift-wave packet propagating on background of a zonal flow

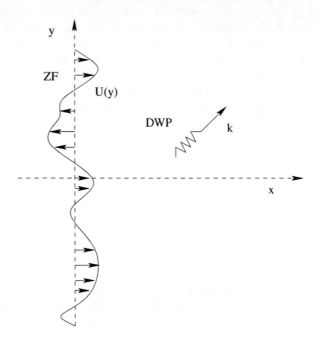

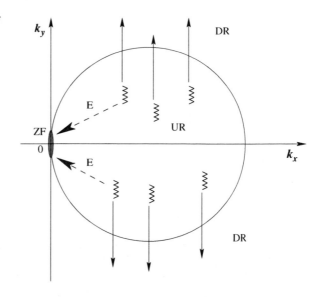

Fig. 13.8 K-space motion of drift-wave packets from UR to DR during which they transfer energy to the zonal flow

where $\mathbf{k} = (k_x, k_y)$ is the 2D wavevector, and the frequency is given by

$$\omega = \frac{k_x v_* + k_x U \rho^2 k^2}{1 + \rho^2 k^2}. \tag{13.26}$$

Here $v_* = \beta \rho^2$, so called drift velocity.

Note that the energy density E is related to waveaction n as $E = n/k^2$, i.e. n happens to be the potential enstrophy density [4].

Remark 13.3.1 If the motion of DWP in Fourier space is restricted by the curves (13.12), then expression $E = n/k^2$ coincides with the usual WT expression $E = \omega n$. However, expression $E = n/k^2$ is more general because it does not restrict the large scales to be weakly nonlinear.

Substituting (13.26) into (13.24) and (13.25) we have

$$k_x = \text{const}, \tag{13.27}$$

$$\dot{k}_y = -\frac{k_x U'(y) \rho^2 k^2}{1 + \rho^2 k^2}, \tag{13.28}$$

$$\dot{x} = \text{not important}, \tag{13.29}$$

$$\dot{y} = \frac{2 k_x k_y \rho^2 (U - v_*)}{(1 + \rho^2 k^2)^2}. \tag{13.30}$$

13.3.1 Weak ZF: Diffusive Regime

Here, we will re-consider the weak Drift/Rossby wave turbulence and obtain the same results as in Sect. 13.2 in a more qualitative way.

Suppose that ZF is weak so that DWP can travel through many correlation lengths L is direction y. Then DWP will experience a random walk in wavenumber component k_y and, therefore, evolution of mean waveaction n will be described by a diffusion equation

$$\dot{n} = \frac{\partial}{\partial k_y}\left(D\frac{\partial n}{\partial k_y}\right) + \gamma_k n, \tag{13.31}$$

where we added a term describing instability in the unstable region (UR where growth rate $\gamma_k > 0$) and dissipation in the dissipative region (DR where $\gamma_k < 0$); see Figs. 13.2 and 13.8. Diffusion coefficient D can be estimated by a standard random walk argument,

$$D = (\dot{k}_y)^2\tau, \tag{13.32}$$

where $\dot{k}_y$ is given by (13.28) and τ is the correlation time

$$\tau = \frac{L}{|\dot{y}|} = \frac{L(1 + \rho^2 k^2)^2}{2\rho^2 k_x k_y |U - v_*|}. \tag{13.33}$$

Thus,

$$D = \frac{\rho^2 k^4 k_x (U')^2 L}{2 k_y |U - v_*|} \sim \frac{\rho^2 k^4 U^2}{2|U - v_*|L}, \tag{13.34}$$

where we we used estimates $k_x \sim k_y$ ($\sim k/\sqrt{2}$) and $U' \sim U/L$. Then, the characteristic time associated with the diffusion process (and therefore the characteristic time of the ZF growth) is

$$\tau_{\text{dif}} = k_y^2/D = \frac{|U - v_*|L}{\rho^2 k^2 U^2} \tag{13.35}$$

Now, let us follow the argument of Sect. 13.2.2 that ZF will be excited by DW turbulence via an inverse cascade until a saturated value determined from the condition of balance of the diffusion and instability terms in model (13.31), i.e.

$$D/k_y^2 = \gamma_{\text{max}}, \tag{13.36}$$

where γ_{max} is the maximum value of the growth rate of the underlying instability. Easy to see that this condition can be also be written as

$$\gamma_{\text{max}}\tau_{\text{dif}} \sim 1. \tag{13.37}$$

Assuming $U \ll v_*$ (case $U \sim v_*$ will be considered separately) and substituting (13.34) in (13.36) we have

$$U_{\text{dif}} \sim \frac{\sqrt{v_* L \gamma_{\max}}}{\rho k}. \tag{13.38}$$

This estimate coincides with the previously obtained estimate (13.18) for the case $k\rho \sim 1$.

13.3.2 Strong ZF: Rapid Distortion Regime

Now we will consider the case when turbulence is not weak in the ZF scales, and therefore the theory developed in Sect. 13.2.2 is inapplicable.

A regime of rapid evolution occurs when the zonal flow is so strong that DWP wavenumber experiences order-one changes (and therefore gets swept from UR to DR) in a time shorter than time τ needed for DWP to cross one ZF oscillation. The characteristic time of such rapid distortions (RD) follows from (13.28):

$$\tau_{rd} \sim k_y / \dot{k}_y \sim \frac{(1 + \rho^2 k^2)L}{\rho^2 k^2 U}. \tag{13.39}$$

Thus, the RD regime kicks in when

$$\tau_{rd} < \tau, \tag{13.40}$$

or

$$U > \frac{v_*}{(1 + \rho^2 k^2)} \quad \text{[Critical ZF amplitude when RD occurs]}. \tag{13.41}$$

Thus, RD regime will at some point replace the diffusive regime if the diffusive estimate for the saturated ZF is greater than the above estimate, i.e. when

$$U_{\text{dif}} = \frac{\sqrt{v_* L \gamma_{\max}}}{\rho k} > \frac{v_*}{(1 + \rho^2 k^2)}, \tag{13.42}$$

or finally

$$L\gamma_{\max} > \frac{\rho^2 k^2 v_*}{(1 + \rho^2 k^2)^2} \quad \text{[Condition when RD is expected]}, \tag{13.43}$$

In case $\rho k \lesssim 1$, according to (13.41), RD only occurs if U is of order of v_* or greater. So, we conclude that for ZF with $U \ll v_*$ the growth is always diffusive, which justifies this assumption made when obtaining (13.38).

Thus, RD occurs when U is of order of v_* or greater and now we are going to consider this case. In critical points where the profile $U(y)$ crosses $U = v_*$ there will be stagnation of wavepackets because according to (13.30) $\dot{y} = 0$ at these points. It is not yet clear what special role is played by these regions in forming the profile $U(y)$. Saturation of the RD growth can be estimated as $U = U_{rd}$ such that

$$\tau_{rd}\gamma_{\max} \sim 1, \tag{13.44}$$

which gives

$$U_{rd} = (1 + \frac{1}{\rho^2 k^2})L\gamma_{\max}. \tag{13.45}$$

For $\rho^2 k^2 \sim 0.2$ we have $U_{rd} \sim 6 L \gamma_{\max}$. In principle U_{rd} can reach values exceeding v_* if $L \gamma_{\max} > v_*/6$. We know that for DW to be proper waves we need $\gamma < \omega \sim v_* k$, but on the other hand $kL \gg 1$ so $L \gamma_{\max}/v_*$ can in principle be large.

13.3.3 Transition Between the Two Regimes of the Zonal Flow Generation

The results for the two regimes of the zonal flow generation are summarized on Fig. 13.9. Transition occurs when the strength of the instability reaches $\gamma_c = \frac{\rho^2 k^2 v_*}{(1+\rho^2 k^2)^2 L}$, and the transitional value of the zonal flow velocity is $U = \frac{v_*}{(1+\rho^2 k^2)}$.

13.4 Numerical Modeling of the Forced-Dissipated CHM Equation

Numerical results of the forced dissipated CHM equation were recently performed in order to test the theoretical predictions described in the previous sections [8]. Here we will briefly highlight some of the key findings.

Fig. 13.9 Velocity of the zonal flow U as a function of the instability growth rate γ. Transition from the slow diffusive regime to the rapid distortion regime occurs at $\gamma = \gamma_c = \frac{\rho^2 k^2 v_*}{(1+\rho^2 k^2)^2 L}$

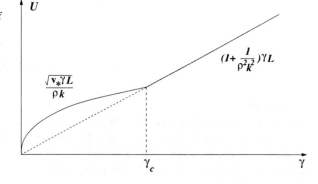

As we described above, the forcing and the dissipation are introduced via adding an instability/viscosity term $\gamma_k \, \psi_k$ to the CHM equation in Fourier space. The particular function chosen in [8] is

$$\gamma_k = \text{const} \; k_x \sqrt{\frac{1 - \rho^2 k^2}{1 + \rho^2 k^2}}. \tag{13.46}$$

This function satisfies the basic properties described above and illustrated in Fig. 13.2.

To characterize the strength of turbulence, the following dimensionless parameter was chosen,

$$\chi = \frac{\gamma}{\rho \beta}. \tag{13.47}$$

It measures the relative size of the linear and nonlinear terms. The initial wave amplitudes where chosen to be very small within the **k**-space region corresponding to positive γ, and zero outside of this region.

Figure 13.10 shows successive (in time) frames of the WT intensity in the 2D **k**-space in a run with $\chi = 0.027$. Evolutions of the zonal energy (defined as the total energy of the modes with $|k_y| > k_x$), the off-zonal energy (defined as the total energy of the modes with $|k_y| \leq k_x$) and the energy of the forced modes (defined as the total energy of the modes with $|\mathbf{k} - \mathbf{k}^*| < 20$) are shown in Fig. 13.11.

We see that initially the energy of the forced modes grows exponentially, as prescribed by the linear instability. The first sign of the nonlinear instability is seen in propagation of the 2D spectrum along the curve which corresponds to the wavevectors which are in three-wave resonances with the most unstable mode, $\mathbf{k}^*$.

Fig. 13.10 Successive (in time) frames of the WT intensity in the 2D **k**-space. Curves (13.12) (corresponding to nonlocal evolution due to interaction with large-scale ZF) are shown by *grey dashes*. "Sea-shell" curves (13.20) (corresponding to nonlocal evolution due to interaction with small-scale ZF) are shown by *black dashes*. *Solid black curve* marks the set of modes which are in triad resonances with the mode **k** = 50.0) corresponding to the maximum of γ_k (courtesy of Brenda Quinn)

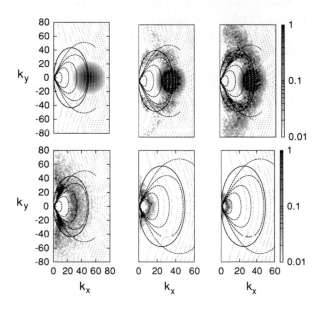

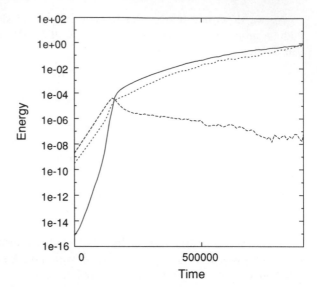

Fig. 13.11 Evolution of the zonal energy (*solid line*), the off-zonal energy (*dotted line*) and the energy of the forced modes (*dashed line*). See the main text for the definitions of the respective energies. By courtesy of Brenda Quinn

After some zonal modes are excited, at $|k_y| > k_x$, the nonlocal interactions with both the large scale and the small-scale zonal flows kick in. This is seen by a characteristic propagation of the wave spectrum along the curves shown in Figs. 13.1 and 13.6 (corresponding to nonlocal evolution due to interaction with large-scale and small-scale ZF's respectively). In fact, we can see that propagation along these two sets of curves occurs simultaneously, which can bee seen in two separate branches along which WT moves in the **k**-space.

Further, we can see that the zonal spectrum tend to concentrate at the line $k_y = \pm\sqrt{3}\,k_x$, as was theoretically predicted in the previous section.

We see a clear gap between the zonal scales and the small-scale turbulence (although with a more complicated 2D structure than in the simple theoretical picture above).

Finally, we see in Fig. 13.11 the predicted suppression of the small-scale modes in the instability range, and thus we confirm the main prediction of our theoretical scenario. Namely we predict that the zonal flows, generated by the nonlinear interaction of the short waves, feed back on the short scales and suppress them, thereby switching off the external instability and, as a result, suppress the anomalous turbulent transport. The zonal energy is visibly dominant over the off-zonal one for most of the run except for the latest time. The zonal-dominated anisotropy is more pronounced in runs with higher βs where it remains for arbitrarily large times. The zonal flow energy E_{zonal} saturates at a value which is in a good agreement with the estimate (13.18), which for this run gives $E_{zonal} = U^2/(\rho p)^2 \sim 1$. This estimate is also confirmed in the runs with different values of β. In particular, with all other parameters fixed, $E_{zonal} \propto \beta$, i.e. the systems with weaker nonlinearities (larger βs) produce stronger zonal flows (which is somewhat counterintuitive).

We shall finish this chapter with two final remarks.

Remark 13.4.1 In plasmas, the CHM model was derived assuming Boltzmann distribution of electrons along the magnetic field. It was noted in [11] that for the wave mode with the zero parallel (to the magnetic field) wavevector component such an approach is invalid and must be modified. Importantly, such a modification leads to a more pronounced zonation effect. It is natural to expect that the described above feedback loop exists (perhaps in an even stronger form) in such a modified CHM system too, although this remains to be demonstrated in future.

Remark 13.4.2 Above, we have used the CHM model in which the instability was introduced empirically, based on the linear analysis of the higher-level models. However, the higher-level models differ from the CHM not only in the linear but also in the nonlinear dynamics. The latter difference becomes essential if the nonlinearity is not weak. Thus, in future it would be interesting to study the zonal-flow/drift-wave feedback loop directly in the higher-level nonlinear models, such as e.g. the two-layer model in GFD [10] or Hasegawa–Wakatani model in plasmas [12].

13.5 Solution to Exercise

13.5.1 Relation Between the Spectrum and the Velocity of the Large Scales: Exercise 13.1

We use the relation $n_p = E_p/\omega_p \sim E_p/(\beta\rho^2 p)$. Energy spectrum E_p is the Fourier transform of the second correlator: $E_p = \rho^{-2} \int \langle \psi(\mathbf{x}) \, \psi(0) \rangle e^{-i\mathbf{p} \cdot \mathbf{x}} \, dx \sim \rho^{-2} p^{-2} \psi^2$ (this is the large-scale limit of the energy expression in CHM). Velocity and the stream function are related as $U \sim p\,\psi$. Putting it all together we get the required relation (13.17).

References

1. Balk, A.M., Nazarenko, S.V., Zakharov, V.E.: On the nonlocal turbulence of drift type waves. Phys. Lett. A **146**, 217–221 (1990)
2. Balk, A.M., Zakharov, V.E., Nazarenko, S.V.: Nonlocal drift wave turbulence. Sov. Phys. JETP **71**, 249–260 (1990)
3. Nazarenko, S.V.: On the nonlocal interaction with zonal flows in turbulence of drift and Rossby waves. Sov. Phys. JETP Lett. **25**, 604–607 (1991)
4. Dyachenko, A.I., Nazarenko, S.V., Zakharov, V.E.: Wave-vortex dynamics in drift and beta-plane turbulence. Phys. Lett. A **165**, 330–334 (1992)
5. Wagner, F., et al.: Phys. Rev. Lett. **49**, 1408 (1982)
6. Diamond, P.H., Itoh, S.-I., Itoh, K., Hahm, T.S.: Zonal flows in plasma—a review. Plasma Phys. Control. Fusion **47**(5), R35–R161 (2005)
7. James, I.N.: Suppression of baroclinic instability in horizontally sheared flows. J. Atmos. Sci **44**(24), 3710 (1987)

8. Connaughton, C., Nazarenko, S., Quinn, B.: The life-cyle of dift-wave turbulence driven by small scale instability. arXiv; 1008.3338 (Submitted to Phys. Rev. Lett.)
9. Rudakov, L.I., Sagdeev, R.Z.: Dokl. Akad. Nauk. SSR **138**, 581 (1961) [Sov. Phys. Dokl. **6**, 415 (1961)]
10. McWilliams, J.C.: Fundamentals of Geophysical Fluid Dynamics. Cambridge University Press, Cambridge (2006)
11. Dorland, W., Hammett, G.: Gyrofluid turbulence models with kinetic effects. Phys. Fluids B **5**, 812 (1993)
12. Hasegawa, A., Wakatani, M.: Plasma edge turbulence. Phys. Rev. Lett. **50**, 682–686 (1983)

Chapter 14
Magneto-Hydrodynamic Turbulence

14.1 Introduction

WT theory for weak Alfvén waves, i.e. for weak deviations from a strong uniform external magnetic field, was developed in [1–3] within the incompressible MHD model. They found a spectrum describing an energy cascade from large to small scales, see expression (3.15) in Chap. 3. This spectrum was found to be consistent with observational data [4]. This theory also provided a conceptual framework for further extensions to other MHD systems, i.e. compressible MHD [5], Electron MHD [6] and Hall MHD [7]. As usual, the WT description was first developed for the wave *energy spectra* assuming that wave turbulence is statistically uniform in an *unbounded* physical space. Later, it was extended to describe the higher-order statistical objects (including wave PDF's and higher amplitude moments) and the finite-box effects in [8]. In particular, for the first time three different regimes regimes of WT in MHD were demonstrated: *discrete WT* in which the 3D waves are slaved to the purely 2D modes, *kinetic WT* for which the wave spectrum and PDF are described by kinetic equations in the infinite-box limit, and a *mesoscopic* regime, in which the discreteness effects are crucially important in a *wide* range of amplitudes. In other contexts, e.g. WT of the water gravity waves, mesoscopic effects were previously noted in papers [9, 10]. However, the fact that the mesoscopic regime arises in a wide range of wave amplitudes was first noted in [8].

In Chap. 10 we already described the finite-box effects in WT, and how in general they lead to the three different WT regimes: discrete, kinetic and mesoscopic. Detailed consideration was devoted to the systems where the discrete resonant clusters which have sparse connections between the wave modes, $M \gtrsim 1$. In this case, the mesoscopic regime occurs in the range of wave amplitudes given by the inequalities (10.14), (10.15). In this range, both the conditions for the kinetic WT and for the discrete WT are satisfied. We argued that in this case (in presence of forcing) one expects a repetitive switching between the discrete and

S. Nazarenko, *Wave Turbulence*, Lecture Notes in Physics, 825,
DOI: 10.1007/978-3-642-15942-8_14, © Springer-Verlag Berlin Heidelberg 2011

the kinetic regimes, such that the wave energy is accumulated in the discrete regime and is released to the small scales in a sandpile-like fashion.

We also mentioned in Chap. 10 that case $\mathcal{M} \gg 1$ is different in that neither the condition for the discrete WT nor the condition for the kinetic WT can be satisfied. In case of the MHD turbulence we do have $\mathcal{M} \gg 1$, and here we will use the MHD example to study the mesoscopic WT regime when $\mathcal{M} \gg 1$. We will see that in this case the mesoscopic regime exhibits a plateau behavior for the frequency broadening Γ in which this quantity is insensitive to the turbulence intensity and it remains of the order of the inter-mode frequency spacing Δ_ω.

In what follows in this section, we will follow the formalism developed in this book in Chap. 6 (and earlier in papers [11–13] and [8]), in which the WT theory (which is traditionally dealing only with the wave spectrum) is extended to the wave PDF's. This will allow us to obtain not only solutions corresponding to Gaussian statistics, but also solutions with power-law PDF tails which correspond to an anomalously high probability of strong waves ("freak" MHD waves).

Because of the mentioned degeneracy of the dispersion law and because of presence of two different wave polarizations, the MHD system is somewhat different from the systems we considered before in Chap. 6. Thus we will re-derive WT for MHD "from scratch" rather than simply apply the results of Chap. 6.

14.2 Reduced MHD Model

Treatment of MHD wave turbulence in the case $k_\perp \approx k_z$ is very lengthy, and even the final kinetic equations take two journal pages to write [1]. This situation is not ideal for those who would like to have an easy introduction into the WT approach, as well as to understand the main theoretical assumptions and their range of validity. On the other hand, MHD turbulence in presence of a strong external magnetic field, $\mathbf{B} = B_0 \mathbf{e}_z$, has a tendency to evolve to states with $k_\perp \gg k_z$ and, therefore, this limit turns out to be the most important one.

Thus, let us following papers [2, 8] consider very anisotropic MHD wave turbulence with $k_\perp \gg k_z$. Specifically, let us start with so-called reduced MHD model (RMHD) obtained by Kadomtsev and Pogutse [14] and, independently, by Strauss [15],

$$\partial_t \theta + \{\psi, \theta\} = \partial_z \psi, \tag{14.1}$$

$$\partial_t \nabla_\perp^2 \psi + \{\psi, \nabla_\perp^2 \psi\} = \partial_z \nabla_\perp^2 \theta + \{\theta, \nabla_\perp^2 \theta\}, \tag{14.2}$$

where ψ and θ are the velocity and the magnetic stream-functions respectively,

$$\mathbf{u}_\perp = \mathbf{e}_z \times \nabla_\perp \psi, \tag{14.3}$$

$$\tilde{\mathbf{b}}_\perp = \mathbf{e}_z \times \nabla_\perp \theta, \tag{14.4}$$

and the curly bracket means the 2D Jacobian, i.e. $\{\psi, \theta\} = \mathbf{e}_z \cdot (\nabla_\perp \psi \times \nabla_\perp \theta)$. For simplicity, we put $B_0 = 1$.

The conditions of applicability of RMHD are (see e.g. [8]):

$$k_z \ll k_\perp$$

and

$$b_\perp, u_\perp \gg \tilde{b}_z \frac{k_z}{k_\perp}, u_z \frac{k_z}{k_\perp}.$$

Note that there is no requirement $\tilde{b}_\perp, u_\perp \gg B_0 k_z / k_\perp$, i.e. the linear time is allowed to be smaller than the nonlinear time. This means that WT approach can be used within RMHD model for sufficiently small amplitudes.

Let us introduce the Elsässer stream-functions as

$$\zeta^\pm = \phi \mp \theta. \tag{14.5}$$

Then (14.1) and (14.2) can be rewritten as

$$\nabla_\perp^2(\partial_t \zeta^\pm \pm \zeta^\pm) = -\frac{1}{2}[\{\zeta^-, \nabla_\perp^2 \zeta^+\} + \{\zeta^+, \nabla_\perp^2 \zeta^-\} \pm \nabla_\perp^2 \{\zeta^+, \zeta^-\}]. \tag{14.6}$$

In the linear approximation, these equations have the following wave solutions,

$$\zeta^\pm \propto e^{i\mathbf{k}\cdot\mathbf{x} - i\omega^\pm t},$$

with frequencies $\omega^\pm$ related to the wavevector $\mathbf{k} = (\mathbf{k}_\perp, k_z)$ via the dispersion relations

$$\omega^\pm = \pm k_z. \tag{14.7}$$

From these relations we see that super-scripts "+" and "−" correspond to the waves propagating along and against the external field respectively, with group velocities $u_g^\pm = \pm B_0$ (remember that we took $B_0 = 1$), as shown in Fig. 14.1.

Let us place the system in a 3D periodic box with dimensions $L_\perp \times L_\perp \times L_z$ and denote the Fourier transform of $\zeta^\pm$ by $\hat{\zeta}_k^\pm$,

$$\hat{\zeta}_k^\pm = L_\perp^{-2} L_z^{-1} \int_0^{L_\perp} \int_0^{L_\perp} \int_0^{L_z} \zeta^\pm(\mathbf{x}) \, e^{-i\mathbf{k}\cdot\mathbf{x}} \, dx \, dy \, dz.$$

with wavevector $\mathbf{k}$ taking values on a 3D lattice,

$$\mathbf{k}_l = \left(\frac{2\pi l_x}{L_\perp}, \frac{2\pi l_y}{L_\perp}, \frac{2\pi l_z}{L_z}\right), \quad \mathbf{l} \equiv (l_x, l_y, l_z) \in \mathbb{Z}^3,$$

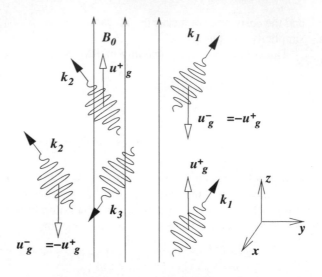

Fig. 14.1 Alfvén wave packets with "+" and "−" polarizations. The "+" and the "−" wavepackets have their group velocities directed strictly along and strictly against the external magnetic field B_0 respectively, $u_g^\pm = \pm B_0$ (independent of the wavenumber **k**)

(cf. (5.3)). Here, we chose a box with different parallel and perpendicular dimensions because, as we will see later, these dimensions play drastically different roles. In particular, only $L_z \to \infty$ limit is needed for the kinetic equation to work, whereas the perpendicular wavevector components may remain discrete.

Let us now introduce the interaction representation action variables:

$$c_k^\pm = ik_\perp \hat{\zeta}_k^\pm \, e^{i\omega^\pm t}/\epsilon, \qquad (14.8)$$

where ϵ is a positive real number which will help us later to keep track of the nonlinearity order (cf. 6.14). As before in Chap. 6 (Sect. 6.3), we will consider a weak nonlinearity limit by accepting ordering $c^\pm \sim 1$ and $\epsilon \ll 1$ corresponding to disturbances which are weak compared to the external field. Later, we will perform expansions in small ϵ. In such a weakly nonlinear system the variables $c^\pm$ will be useful because they change in time much slower than the linear oscillations, which will allow us to separate the linear and the nonlinear time-scales. Thus, we let us re-write our equations in terms of these variables,

$$\dot{c}_k^\pm = \epsilon \sum_{1,2} V_{k12} \, e^{\pm 2ik_{1z}t} c_1^\mp c_2^\pm \delta_{12}^k, \qquad (14.9)$$

where for brevity $c_{1,2}^\pm = c^\pm(\mathbf{k}_{1,2})$, and

$$V_{k12} = V(\mathbf{k}_\perp, \mathbf{k}_{1\perp}, \mathbf{k}_{2\perp}) = \frac{(\mathbf{k}_\perp \cdot \mathbf{k}_{2\perp})(\mathbf{k}_{1\perp} \times \mathbf{k}_{2\perp})_z}{k_\perp k_{1\perp} k_{2\perp}} \qquad (14.10)$$

is the interaction coefficient.

So far, we have not made any additional assumptions, i.e. (14.9) is equivalent to the original RMHD (14.1) and (14.2).

14.3 Very Weak WT: Discrete Regime and 2D Enslaving

In this section, we will consider the limit of very small amplitudes, when the finite-box effects are crucial. We will use the same approach as in Sect. 10.1, but we will see very distinct effects specific for the MHD system, namely enslaving of the 3D waves to the 2D modes. These effects appear due to a degenerate form of the dispersion relations (14.7).

Thus, let us consider the case where the nonlinearity is very weak so that the characteristic evolution time of $c^{\pm}$ is long enough to satisfy the condition

$$\Gamma \sim 1/\tau_{nl} \ll \Delta_\omega = \Delta k_z = 2\pi/L_z, \qquad (14.11)$$

i.e. the nonlinear frequency broadening Γ is much less than the frequency spacing between adjacent modes Δ_ω. Let us introduce an intermediate time T such that

$$\frac{2\pi}{\omega} \ll T \ll \tau_{nl}$$

and average (14.9) over this time. The amplitudes $c^{\pm}$ are slow and are not changed by such an averaging, and the only factor to be averaged is $e^{\pm 2ik_{1z}t}$. We have:

$$\dot{c}^{\pm}(\mathbf{k}_\perp, k_z) = \epsilon \sum_{\mathbf{k}_{1\perp}, \mathbf{k}_{2\perp}} V_{k12}\, c^{\mp}(\mathbf{k}_{1\perp}, 0) c^{\pm}(\mathbf{k}_{2\perp}, k_z) \delta(\mathbf{k}_\perp - \mathbf{k}_{1\perp} - \mathbf{k}_{2\perp}). \qquad (14.12)$$

This is the MHD example of the discrete regime considered in Chap. 10; cf. (10.2) and (10.3).

As we see, according to (14.12) the nonlinear transfer occurs via the pure 2D ($k_z = 0$) modes and, i.e., the $k_z \neq 0$ modes are *slaved* to the 2D component. Indeed, the solution to (14.12) has the form

$$c^{\pm}(\mathbf{k}_\perp, k_z) = c_\perp^{\pm}(\mathbf{k}_\perp) c_z^{\pm}(k_z), \qquad (14.13)$$

where $c_z^{\pm}(k_z) = \text{const}$ and for $c_\perp^{\pm}(\mathbf{k}_\perp)$ we have the following equation,

$$\dot{c}_\perp^{\pm}(\mathbf{k}_\perp) = \epsilon \sum_{\mathbf{k}_{1\perp}, \mathbf{k}_{2\perp}} V_{k12}\, c_\perp^{\mp}(\mathbf{k}_{1\perp}) c_\perp^{\pm}(\mathbf{k}_{2\perp}) \delta(\mathbf{k}_\perp - \mathbf{k}_{1\perp} - \mathbf{k}_{2\perp}). \qquad (14.14)$$

These equations are identical to the pure 2D version of RMHD equations, and they are completely decoupled from the 3D wave component. Obviously, these equations are strongly nonlinear, and the resulting motion represents strong 2D MHD turbulence. In particular, in case when there is no 2D component of the magnetic field present, such 2D MHD turbulence becomes 2D Navier-Stokes turbulence characterized by the dual cascade behavior (see Sect. 2.1.3 of Chap. 2).

Interaction of the 3D component with the 2D component can be viewed as a *linear* process of distortion of Alfvén wave packets by shear and strain produced by 2D turbulence. This interaction does not change the parallel wavenumbers of these wave packets, as reflected in the time independence of $c_z^{\pm}(k_z)$ in (14.13).

Fig. 14.2 Anisotropization
of MHD spectrum due to
suppressed energy transfer in
the k_z-direction

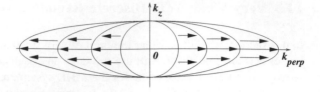

On the other hand, according to expression (14.13) the perpendicular component $c_\perp^\pm(\mathbf{k}_\perp)$ is the same as in the pure 2D modes, $c_\perp^\pm(\mathbf{k}_\perp, t) = c^\pm(\mathbf{k}, t)|_{k_z=0}$, and therefore we can expect the dual cascade behavior for this component. Namely, for the integrated over k_z energy spectrum we have the following. If there is no magnetic field present at the $k_z = 0$ modes then the k_z-integrated spectrum can coincide with either the direct enstrophy or inverse enstrophy cascade spectra of 2D Navier-Stokes turbulence (described in Sect. 2.1.3).

Independence of the k_z-part of the 3D spectrum of time means that there are no energy transfers in the k_z-direction, and all cascades occur in the perpendicular directions only. Such transverse cascade lead to increase of WT anisotropy, $k_z/k_\perp \to 0$, as illustrated in Fig. 14.2. This justifies our use of the RMHD model (i.e. an anisotropic limit of the MHD model).

Now, let us estimate the frequency broadening Γ, which, in accordance with the notations introduced in Chap. 10, we will call Γ_D (discrete regime). As we mentioned in Chap. 10 and in the beginning of this chapter, in the MHD case the number of dynamically important waves $\mathcal{M}$, which are in resonance with mode $\mathbf{k}$, is large. Indeed, due to the degenerate (non-dispersive) dispersion relation $\omega^\pm = \pm k_z$, the frequency resonance is satisfied by every three modes two of which have the same k_z's and the third one has $k_z = 0$, and which are in a triad $\mathbf{k}_\perp$-resonance. There are infinite number of these resonances for each $\mathbf{k}$, but, assuming locality, the number of *dynamically important* interactions is finite, $\mathcal{M} \sim (k_\perp L_\perp)^2$. Using this estimate in (14.12) (i.e. replacing the sum with $\mathcal{M}$), and taking $V \sim k_\perp$, we get

$$\Gamma_D \sim (k_\perp L_\perp)^2 k_\perp c_\perp \sim (k_\perp L_\perp)^2 k_\perp^2 \hat{\zeta}_k \sim k_\perp^2 \zeta \sim k_\perp \tilde{b} \sim k_\perp u. \qquad (14.15)$$

This estimate is typical for hydrodynamic turbulence, which is natural considering our previous observation that the 3D modes are slaved to the purely 2D (strong) turbulence.

Thus, the discrete regime condition (14.11) becomes

$$\frac{\tilde{b}}{B_0} \ll \frac{2\pi}{k_\perp L_z} \equiv \frac{\Delta k_z}{k_\perp}, \qquad (14.16)$$

where we put back B_0 ($B_0 = 1$ above for simplicity).

This very simple inequality has a profound meaning: in weakly nonlinear systems with $\tilde{b}/B_0 < 1$ the largest scales are likely to be in the discrete regime, i.e. they are enslaved to the $k_z = 0$ mode. For example, in numerical simulation of

MHD turbulence in work [16], for the strongest value of the external field they had $\tilde{b}/B_0 \sim 10$, and the parallel resolution was 128. Thus, the first decade in the k_z-direction was in the discrete regime, which is quite large considering that the whole inertial range was about two decades wide (corresponding to their perpendicular resolution of 512). This example teaches us that the discrete regime must be very typical for all the current numerical simulations of MHD turbulence and the currently available resolution levels.

14.4 Large-Box Limit: Kinetic Regime

Now let us consider the case when the disturbances are strong enough for the nonlinear broadening Γ to be much greater than the mode spacing Δ_ω. Note that the estimate for Γ in this regime will be different from what we obtained in (14.15) and it will be estimated later. On the other hand, we want the nonlinear broadening Γ to remain smaller than the linear frequency ω_k, i.e. we want the nonlinearity to be small, $\epsilon \ll 1$.

14.4.1 Weak Nonlinearity Expansion

Following the approach we described in Chap. 6, let us introduce time T which is intermediate between the fast linear period and the slow nonlinear time,

$$\frac{2\pi}{\omega_k} \ll T \ll \tau_{nl} = \frac{1}{\Gamma},$$

and let us seek for solution at $t = T$ in terms of series in small ϵ,

$$b_k^\pm(T) = c_k^{\pm(0)} + \epsilon c_k^{\pm(1)} + \epsilon^2 c_k^{\pm(2)} + \cdots \qquad (14.17)$$

The leading order corresponds to $\epsilon = 0$, i.e. to the linear approximation. In this order, the interaction representation amplitude is time-independent and equal to its initial value at $t = 0$, i.e. $c_k^{\pm(0)} = c_k^\pm(0)$. The next order is obtained by substituting $c_k^{\pm(0)}$ into the RHS of (14.19) and integrating it from zero to T,

$$c_k^{\pm(1)} = \sum_{1,2} V_{k12} \Delta_{1T}^\pm c_1^{\mp(0)} c_2^{\pm(0)} \delta_{12}^k, \qquad (14.18)$$

where

$$\Delta_{1T}^\pm = \frac{e^{\pm 2ik_{1z}T} - 1}{\pm 2ik_{1z}}.$$

In the next order we get

$$\dot{c}_k^{\pm(2)} = \sum_{1,2} V_{k12} \, e^{\pm 2ik_{1z}t} \, (c_1^{\mp(0)} c_2^{\pm(1)} + c_1^{\mp(1)} c_2^{\pm(0)}) \delta_{12}^k. \qquad (14.19)$$

Substituting here from (14.18) and integrating, we get

$$c_k^{\pm(2)} = \sum_{1,2,3,4} V_{k12} \delta_{12}^k \left(c_1^{\mp(0)} c_3^{\mp(0)} c_4^{\pm(0)} V_{234} \delta_{34}^2 E_{31}^+ + c_2^{\pm(0)} c_3^{\pm(0)} c_4^{\mp(0)} V_{134} \delta_{34}^2 E_{31}^- \right),$$

$$(14.20)$$

where

$$E_{31}^{\pm} = \int_0^T \Delta_{1t}^{\pm} \, e^{\pm 2ik_{1z}t} \, dt.$$

14.4.2 Statistical Averaging

Statistical averaging will be done as in Chap. 6, i.e. assuming that the initial wave field is RPA, as defined in Chap. 5. First, let us write the initial values of the Fourier coefficients in the amplitude-phase representation $c_k^{\pm(0)} = \sqrt{J_k^{\pm}} \phi_k^{\pm}$ where $J_k^{\pm} \in \mathbb{R}^+$ and $\phi_k^{\pm} \in \mathbb{S}^1$ and let us require these quantities to to be independent random variables for all **k**-modes. Additionally, let us take the phase factors ϕ_k to be uniformly distributed on the unit circle in the complex plane $\mathbb{S}^1$. However, recall that RPA does not specify the shape of PDF's for the intensities J_k because this is not necessary for the WT closure. In the other words, the statistics is not fixed to be Gaussian (or close to Gaussian), and we will derive an evolution equation for the amplitude PDF which describes evolution of non-Gaussian WT states. For this, let us introduce the generating function

$$\mathcal{Z}_k^{\pm}(\lambda, t) = \langle\langle e^{\lambda \mathcal{V} J_k^{\pm}} \rangle_\phi \rangle_J, \qquad (14.21)$$

where the $\mathcal{V} = L_\perp^2 L_z$ and the angle brackets with subscripts ϕ and J mean the RPA averaging over the statistics of $\phi_k^{\pm}$ and of $J_k^{\pm}$ respectively (done separately since they are statistically independent). In terms of the generating function, the wave spectrum and the higher moments of the wave intensity are,

$$n_k^{\pm} = \mathcal{V} \langle |c_k^{\pm}|^2 \rangle = \partial_\lambda \mathcal{Z}^{\pm}|_{\lambda=0}, \qquad (14.22)$$

$$M_k^{(p)} = \mathcal{V}^p \langle |c_k^{\pm}|^{2p} \rangle = \partial_\lambda^p \mathcal{Z}^{\pm}|_{\lambda=0}, \quad p \in \mathbb{N}, \qquad (14.23)$$

and the intensity PDF is

$$\mathcal{P}_k^{\pm}(s) = \langle \delta(s - J_k^{\pm}) \rangle = \mathcal{L}^{-1} \mathcal{Z}_k^{\pm}(\lambda), \tag{14.24}$$

where $\mathcal{L}^{-1}$ denotes the inverse Laplace transform operator.

Remark 14.4.1 Strictly speaking, spectrum $n^{\pm}$ is not a waveaction spectrum, because c_k is not a Hamiltonian waveaction variable. Instead, the physical meaning of $n^{\pm}$ is a (3D) energy spectrum of the "+" and "−" waves. The energy spectrum is related to the waveaction spectrum via factor $|\omega_k^{\pm}|$ (cf. Chap. 6). The absolute value sign arises when the MHD system is reformulated in a Hamiltonian way (see [17–19]): something we have avoided here because it does not simplify the WT derivations in this example. However, since $|\omega_k^{\pm}| = |k_z|$ and since, as we will see later, there is no evolution in k_z, we would not make a big mistake by calling $n_k^{\pm}$ a waveaction spectrum.

Remark 14.4.2 Let us briefly remind about the relation about the k-space moments $M_k^{(p)}$ considered here and the structure functions $\mathcal{S}^{(p)}$ introduced in (5.24) as a class of two-point x-space correlators. Recall, e.g. relation for the forth-order objects (5.32). The structure functions $\mathcal{S}^{(p)}$ are more common in the turbulence literature. In turn, the k-space moments are more similar to the two-point rather than one-point x-space correlators. Thus, similarly to the turbulence structure functions, the k-space moments are a sensitive measure of intermittency. In addition, the k-space moments are more natural objects in the system of weakly nonlinear waves than the x-space moments.

To derive an equation for $\mathcal{Z}$ let us find its value at $t = T$ by substituting (14.17) into (14.21) and expanding in small ϵ,

$$\mathcal{Z}_k^{\pm}(T) = \left\langle e^{\lambda \mathcal{V}|c_k^{\pm(0)} + \epsilon c_k^{\pm(1)} + \epsilon^2 c_k^{\pm(2)}|^2} \right\rangle = \mathcal{Z}_k^{\pm}(0) + \left\langle e^{\lambda \mathcal{V} J_k^{\pm}} \left[\epsilon \lambda \mathcal{V} \langle c_k^{\pm(1)} c_k^{\pm(0)*} + \text{cc} \rangle_{\phi} \right. \right.$$
$$+ \epsilon^2 \left\langle (\lambda \mathcal{V} + \lambda^2 \mathcal{V}^2 J_k^{\pm})|c_k^{\pm(1)}|^2 + \lambda \mathcal{V} \left(c_k^{\pm(2)} c_k^{\pm(0)*} + \text{cc} \right) \right.$$
$$\left. \left. \left. + \frac{\lambda^2 \mathcal{V}}{2} \left(c_k^{\pm(1)2} (c_k^{\pm(0)})^2 + \text{cc} \right) \right\rangle_{\phi} \right] \right\rangle_J + O(\epsilon^3). \tag{14.25}$$

Here, we have to substitute $c_k^{\pm(1)}$ and $c_k^{\pm(2)}$ from (14.18) and (14.20) respectively and perform averaging, first over ϕ and second over J. Linear in ϵ terms turn into zero upon ϕ-averaging because they contain products of an odd number (three) of ϕ's. Further, the term containing $c_k^{\pm(1)2}(c_k^{\pm(0)*})^2$ also turns into zero upon ϕ-averaging because they contain non-equal number of ϕ^+'s and ϕ^-'s.

The ϕ-averaging of the remaining terms gives

$$\langle |c_k^{\pm(1)}|^2 \rangle_{\psi} = \sum_{1,2,3,4} V_{k12} V_{k34} \, \Delta_{1T}^{\pm} \Delta_{3T}^{\pm*} \sqrt{J_1^{\mp} J_2^{\pm} J_3^{\mp} J_4^{\pm}} \langle \phi_1^{\mp} \phi_2^{\pm} \phi_3^{\mp*} \phi_4^{\pm*} \rangle_{\phi} \delta_{12}^k \delta_{34}^k$$
$$= \sum_{1,2} V_{k12}^2 J_1^{\mp} J_2^{\pm} |\Delta_{1T}^{\pm}|^2 \delta_{12}^k, \tag{14.26}$$

and

$$\langle c_k^{\pm(0)*} c_k^{\pm(2)}\rangle_\phi = \sum_{1,2,3,4} V_{k12} V_{234} \delta_{12}^k \delta_{34}^2 \sqrt{J_k^\pm J_1^\mp J_3^\mp J_4^\pm} \langle \phi_k^{\pm*} \phi_1^\mp \phi_3^\mp \phi_4^\pm \rangle_\phi E_{31}^+$$

$$= -\sum_{1,2} V_{k12}^2 \delta_{12}^k J_k^\pm J_1^\mp E_{-11}^+, \tag{14.27}$$

where we took into account that $c_k^* = c_{-k}$ because c_k^* arises from Fourier transform of a real function. Note that the second term in (14.20) did not contribute to (14.27) because it leads to a product of non-equal number of ϕ^+'s and ϕ^-'s and, therefore, it has a zero average. Substituting (14.26) and (14.27) into (13.7), we have:

$$\mathcal{Z}_k^\pm(T) - \mathcal{Z}_k^\pm(0) = \epsilon^2 \langle e^{\lambda \mathcal{V} J_k^\pm} [(\lambda \mathcal{V} + \lambda^2 \mathcal{V}^2 J_k^\pm) \sum_{1,2} V_{k12}^2 J_1^\mp J_2^\pm |\Delta_{1T}^\pm|^2 \delta_{12}^k$$

$$- 2\lambda \mathcal{V} \sum_{1,2} V_{k12}^2 \delta_{12}^k J_k^\pm J_1^\mp \, \mathcal{R} E_{-11}^+] \rangle_J, \tag{14.28}$$

where, as before, $\mathcal{R}$ denotes the real part. Usually in the WT theory, the next step would be averaging over the independent amplitudes J_k, followed by the large-box limit, followed by $\epsilon \to 0$ ($T \to \infty$) limit. This sequence of taking limits is essential because the resulting frequency resonance should be broad enough to cover many wave modes. However, the Alfvén wave frequency, and therefore the frequency resonance, depend on the parallel wavenumber only. Thus, for the resonance function $\Delta_T^\pm$ to cover many modes, it is enough to take limit $L_z \to \infty$ leaving $L_\perp$ finite, which allows one to generalize description to systems bounded in the transverse direction. Thus, let us perform averaging of (14.28) over amplitudes J_k and take limit $L_z \to \infty$, which gives:

$$\mathcal{Z}_k^\pm(T) - \mathcal{Z}_k^\pm(0) = \epsilon^2 [(\lambda \mathcal{Z}_k^\pm + \lambda^2 \partial_\lambda \mathcal{Z}_k^\pm) L_\perp^{-2} \sum_{1_\perp, 2_\perp} \int V_{k12}^2 n_1^\mp n_2^\pm |\Delta_{1T}^\pm|^2 \delta_{12}^{\perp k} \, \delta(k_z - k_{1z} - k_{2z}) \, dk_z$$

$$- 2\lambda \partial_\lambda \mathcal{Z}_k^\pm L_\perp^{-2} \sum_{1_\perp, 2_\perp} \int V_{k12}^2 n_1^\mp \, \mathcal{R} E_{-11}^+ \delta_{12}^{\perp k} \, \delta(k_z - k_{1z} - k_{2z}) \, dk_z], \tag{14.29}$$

where the summation is performed over the transverse wavenumber components and $\delta_{12}^{\perp k}$ is the Kronecker symbol with respect to the transverse wavenumber coordinates. Now let us take limit $\epsilon \to 0$ ($T \to \infty$) taking into account that

$$\lim_{T \to \infty} |\Delta_{1T}^\pm|^2 = \pi T \, \delta(k_{1z}), \quad \text{and} \quad \lim_{T \to \infty} (\Re E_{-11}^+) = \frac{\pi T}{2} \, \delta(k_{1z}).$$

Replacing $(\mathcal{Z}_k^\pm(T) - \mathcal{Z}_k^\pm(0))/T$ with $\dot{\mathcal{Z}}_k^\pm$ we finally have

$$\dot{\mathcal{Z}}_k^\pm = \lambda [\eta_k^\pm (\mathcal{Z}_k^\pm + \lambda \partial_\lambda \mathcal{Z}_k^\pm) - \gamma_k^\pm \partial_\lambda \mathcal{Z}_k^\pm], \tag{14.30}$$

where

$$\eta_k^\pm = \pi\epsilon^2 L_\perp^{-2} \sum_{1_\perp,2_\perp} V_{k12}^2 n^\mp(k_{1\perp},0)n^\pm(k_{2\perp},k_z)\delta_{12}^{\perp k}, \tag{14.31}$$

$$\gamma_k^\pm = \pi\epsilon^2 L_\perp^{-2} \sum_{1_\perp,2_\perp} V_{k12}^2 \, n^\mp(k_{1\perp},0)\, \delta_{12}^{\perp k}, \tag{14.32}$$

cf. (6.45), (6.46) and (6.47). Equation (14.30) contains the complete information about the evolution of the one-mode statistics. In particular, taking the inverse Laplace we obtain equation for the PDF,

$$\dot{\mathcal{P}}^\pm(s) + \partial_s \mathcal{F}^\pm = 0, \tag{14.33}$$

where

$$\mathcal{F}^\pm = -s(\gamma_k^\pm \mathcal{P}^\pm + \eta_k^\pm \partial_s \mathcal{P}^\pm), \tag{14.34}$$

cf. (6.51) and (6.52). Taking the first moment of (14.33), we obtain evolution equations for the wave spectra,

$$\dot{n}_k^\pm = \int s\dot{\mathcal{P}}^\pm \, ds = -\int s\partial_s \mathcal{F}^\pm \, ds = \int \mathcal{F}^\pm \, ds = -\gamma_k^\pm n_k^\pm + \eta_k^\pm. \tag{14.35}$$

Substituting from (14.32) and (14.31), we rewrite this as:

$$\dot{n}^\pm(k_\perp,k_z) = \pi\epsilon^2 L_\perp^{-2} \sum_{1_\perp,2_\perp} V_{k12}^2 \, n_1^\mp(k_\perp,0)[n^\pm(k_{2\perp},k_z) - n^\pm(k_\perp,k_z)] \, \delta_{12}^{\perp k}, \tag{14.36}$$

cf. (6.55), (6.57), (6.58), (6.59), (6.69) and (6.70). One can see that evolution in (14.30), (14.33) and (14.36) is always mediated by interaction with $k_z = 0$ modes and, as a result, k_z enters as an external parameter into these equations. In the other words, there is no energy transfer between modes with different finite k_z's. For the energy spectra, this property was found in [1] (compare also with the similar property of the inertial waves in rotating fluids [20, 21]) and for PDF it was found in [8]. Thus, in the kinetic regime we can refer to an anisotropization due to the purely perpendicular character of the energy transfer illustrated in Fig. 14.2, like we did before for the discrete regime. Also like for the discrete regime, we note that such an anisotropization justifies using the RMHD model.

Remark 14.4.3 Despite the fact that the turbulent transfers in k_z are absent in both the discrete and the kinetic regimes, with consequences for the respective shared properties in these two regimes, one must realize that these two regimes are still very different in their transverse evolutions. Particularly, the characteristic evolution times (and the respective frequency broadenings Γ) are different.

Absence of interactions across k_z allows us to separate the non-evolving k_z-dependence and an evolving $k_\perp$-part in these equations,

$$n^\pm(k_\perp,k_z,t) = n_z^\pm(k_z)n_\perp^\pm(k_\perp,t) \tag{14.37}$$

with $n_z^\pm|_{k_z=0} = 1$, and

$$\mathcal{P}^\pm(k_\perp, k_z, s, t) = \mathcal{P}_z^\pm(k_z)\mathcal{P}_\perp^\pm(k_\perp, I, t), \tag{14.38}$$

where $I = s/n_z^\pm(k_z)$. Then, for the evolving perpendicular part of the PDF we have the following equation,

$$\dot{\mathcal{P}}_\perp^\pm(I) + \partial_I \mathcal{F}_\perp^\pm = 0, \tag{14.39}$$

where

$$\mathcal{F}_\perp^\pm = -I(\gamma_k^\pm \mathcal{P}_\perp^\pm + \eta_{\perp k}^\pm \partial_I \mathcal{P}_\perp^\pm), \tag{14.40}$$

with

$$\eta_{\perp k}^\pm = \pi \epsilon^2 L_\perp^{-2} \sum_{1_\perp, 2_\perp} V_{k12}^2 n_{1\perp}^\mp n_{2\perp}^\pm \delta_{12}^{\perp k}, \tag{14.41}$$

and for the spectrum we have

$$\dot{n}_{\perp k}^\pm = \pi \epsilon^2 L_\perp^{-2} \sum_{1_\perp, 2_\perp} V_{k12}^2 \, n_{1\perp}^\mp [n_{2\perp}^\pm - n_{k\perp}^\pm] \delta_{12}^{\perp k}. \tag{14.42}$$

We emphasize again that in the case of Alfvén waves, taking the $L_\perp \to \infty$ limit is technically un-necessary and, therefore, WT description in this case works for bounded in the transverse direction systems. Here, one could think of such naturally occurring MHD waveguides as coronal loops on the Sun [22] and solar-wind "spaghetti" structures [23]. Of course, considering these applications would require inclusion of the fluid compressibility effects and a specific geometry of the bounding volume.

Naturally, (14.42) can also be used for unbounded systems after taking the limit $L_\perp \to \infty$. This yields the continuous version of the above equation [1],

$$\dot{n}_{\perp k}^\pm = \pi \epsilon^2 \int V_{k12}^2 \, n_{1\perp}^\mp [n_{2\perp}^\pm - n_{k\perp}^\pm] \, \delta(\mathbf{k}_\perp - \mathbf{k}_{1\perp} - \mathbf{k}_{2\perp}) \, d\mathbf{k}_{1\perp} d\mathbf{k}_{2\perp}. \tag{14.43}$$

14.4.3 Conditions of Realizability of the Kinetic Regime

Let us estimate the characteristic nonlinear frequency broadening for the kinetic regime which, following the conventions introduced in Sect. 6.10 we denote as Γ_K. Using (14.42) or (14.43), we get

$$\Gamma_K \sim \frac{k_\perp^2 \tilde{b}^2}{k_z}. \tag{14.44}$$

Easy to see that

$$\Gamma_K \sim \Gamma_D^2 / \omega_k. \tag{14.45}$$

For realizability of the kinetic regime, Γ_K must be greater than the ω-spacing, $\Delta_\omega = \Delta k_z = 2\pi/L_z$, and at the same time to remain less than the wave frequency ω_k. This results in the following realizability condition,

$$\frac{k_z}{k_\perp} \gg \frac{\tilde{b}}{B_0} \gg \frac{2\pi k_z^{1/2}}{k_\perp L_z^{1/2}}. \tag{14.46}$$

Remarkably, there is a factor $\sqrt{k_z L_z}$ difference in the RHS of this inequality and RHS of inequality (14.16), which means that there is a gap between the limits of applicability of the discrete and the kinetic regimes. Thus, like in Chap. 10 we conclude that there exists an intermediate range of *mesoscopic* wave turbulence. However, this regime is different from what we had in Chap. 10, because there we had *both* conditions for the discrete and for the kinetic regimes satisfied, and now we have *none* of these two conditions satisfied. We will discuss such mesoscopic MHD turbulence a little later, in Sect.14.5. For now we mention that for the characteristic parameters of the numerical simulations of [16] the kinetic WT conditions (14.16) are not satisfied. Therefore, the wave modes in these simulations were in the discrete regime (at low k's) and in the mesoscopic regime (at high k's).

Now we will discuss some important solutions in the kinetic regime.

14.4.4 Spectra in the Kinetic Regime: Energy Cascades—Balanced and Imbalanced Turbulence

Let us now find solutions of the equation for the spectrum, (14.36). Since the k_z-part of these solutions is arbitrary and time-independent, we will only need to solve the equation for the perpendicular part, (14.42) or (14.43). First of all, let us consider steady-state power-law spectra,

$$n_{\perp k}^\pm \propto k_\perp^{\nu^\pm}. \tag{14.47}$$

A trivial solution of this kind with $\nu^\pm = 0$ describes a thermodynamics equipartition of energy and it is valid for both discrete and continuous systems because is corresponds to expressions under the sum of (14.42) and in the integrand of (14.43) which are zero point-wise.

A more interesting solution is the one that corresponds to a KZ cascade of energy from low to high $k_\perp$'s. Unfortunately, the discrete system is much harder to examine analytically in this case, and we will have to restrict ourselves to analysis of the continuous (infinite-box) system (14.42).

Remark 14.4.4 We have already obtained the KZ spectrum for the MHD turbulence for the symmetric case $n_k^+ = n_k^-$ using the dimensional analysis in Chap. 3, Sect. 3.1.3.3. This exercise taught us that anisotropy is crucial for the MHD system: ignoring the anisotropy the dimensional approach leads to the erroneous Iroshnikov-Kraichnan spectrum. Further, the dimensional analysis implies WT *locality*, which cannot be checked within the dimensional approach. Fortunately, this can be done by finding the KZ spectra as *exact* solutions of the kinetic equation (14.42).

Let us split the integral in (14.42) in two equal parts and in the integrand of one of these parts change variables via a Zakharov-type transformation [1],

$$k'_{1\perp} = \frac{k_\perp k_{1\perp}}{k_{2\perp}}, \quad k'_{2\perp} = \frac{k_\perp^2}{k_{2\perp}}, \tag{14.48}$$

leaving the directions of $\mathbf{k}_{\perp 1}$ and $\mathbf{k}_{\perp 2}$ unchanged. This results is the following condition for the power-law steady state spectra,

$$\int V_{k12}^2 \, k_{\perp 1}^{v^\mp} (k_{\perp 2}^{v^\pm} - k_\perp^{v^\pm}) \left[1 - \left(\frac{k_\perp}{k_{\perp 2}} \right)^{v^\mp + v^\pm + 6} \right]$$
$$\times \, \delta(\mathbf{k}_\perp - \mathbf{k}_{\perp 1} - \mathbf{k}_{\perp 2}) \, d\mathbf{k}_{\perp 1} d\mathbf{k}_{\perp 2} = 0. \tag{14.49}$$

Exercise 14.1 *Derive equation* (14.49).

Requiring the square bracket in the integrand of (14.49) to be identical zero, we get a one-parametric family of KZ solutions with exponents satisfying [1]

$$v^+ + v^- = -6. \tag{14.50}$$

Writing explicitly the prefactors, we have for the KZ solutions:

$$n_{\perp k}^\pm = C(v^\pm)(\epsilon_\perp^\pm B_0)^{1/2} \, k_\perp^{v^\pm}, \tag{14.51}$$

where $\epsilon_\perp^\pm$ are the values of the fluxes through scales of the "perpendicular" energies $n^\pm$ corresponding to the waves with "+" and "−" polarizations and $C(v^\pm)$ is a dimensionless function. Again, we have explicitly written the external field B_0 for the dimensional correctness.

Remark 14.4.5 According to the expression (14.37) and because (14.36) is linear with respect to the parallel spectrum, the fluxes of the total energies and the ones of the perpendicular energies are related as $\epsilon^\pm = n_z \epsilon_\perp^\pm$.

Remark 14.4.6 Note that $n_{\perp k}^\pm$ is a 2D energy spectrum (in the $\mathbf{k}_\perp$-plane). Often in literature one deals with 1D energy spectra (see e.g. our Chap. 3). For the Alfvén waves, the 1D energy spectra are:

$$E_{\perp k}^\pm = 2\pi k_\perp n_{\perp k}^\pm = 2\pi C(v^\pm)(\epsilon_\perp^\pm B_0)^{1/2} \, k_\perp^{v^\pm + 1}. \tag{14.52}$$

In particular, for the balanced WT, when $v^+ = v^-$, we recover the result of Sect. 3.1.3.3 (3.15):

$$E_\perp^\pm \propto k_\perp^{-2}. \tag{14.53}$$

Locality of the scale interactions is checked by finding the range of convergence of the integral in the RHS of the kinetic equation (14.42), i.e. confirming that the dimensionless constants $C(v^\pm)$ (corresponding to these values of the spectral indices) are finite.

Exercise 14.2 *Substitute the spectra (14.47) with indices satisfying the (14.50) into the integral on the RHS of the kinetic equation (14.42) and find the condition of convergence of this integral. Thereby obtain the following locality condition* [1]:

$$-4 < v^+, v^- < -2. \tag{14.54}$$

Remark 14.4.7 Different exponents in the one-parametric family (14.50) correspond to different values of the function C, i.e. to different degrees of imbalance between the fluxes of the forward and and backward propagating Alfvén waves, $\epsilon_\perp^\pm$. The limiting values of the spectral indices -4 and -2 correspond to zero and infinite ratios of the wave forcings. The dependence relating the fluxes and the spectral indices was found numerically in [1], and the interested reader is referred to this paper for details.

Remark 14.4.8 Fluxes $\epsilon_\perp^+$ and $\epsilon_\perp^-$, and therefore their ratio, may depend on k_z. This is determined by the forcing distribution for the "+" and "−" waves over different k_z's. Thus, the spectral indices $v^\pm$ may depend on k_z (while still satisfying condition (14.50)). This interesting property of imbalanced MHD turbulence is often overlooked in literature.

Remark 14.4.9 For some reason (e.g. by an initial condition) "+" or "−" waves may have a spectrum steeper than $k_\perp^{-4}$. In this case, WT will be nonlocal and such a case was studied in [3] within an approach similar to the one described in Chap. 13 for the nonlocal Rossby/drift turbulence. The main result: if the "−" waves have a spectrum steeper than $k_\perp^{-4}$ then the "+" waves develop a steady state spectrum with exponent -2, and vice versa.

14.4.5 Cross-Helicity

Consider the cross-helicity invariant, whose perpendicular spectrum is $H_c(k_\perp) = n^+ - n^-$. Solutions corresponding to imbalanced WT with $v^+ \neq v^-$ may display the effect of a sign change in $H_c(k_\perp)$, as shown in Fig. 14.3. This occurs if the point k^*, where n^+ and n^- intersect, falls inside the inertial range. This effect is rather

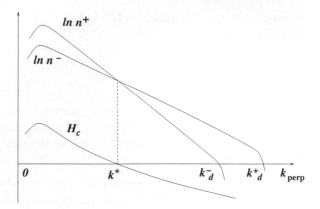

Fig. 14.3 Case when "+" and "−" Alfvén waves have different spectra. Cross-helicity spectrum H_c changes sign at wavenumber $k_\perp^*$ where spectra n^+ and n^- intersect

general, and it can occur also in strong MHD turbulence, e.g. it was obtained numerically within EDQNM model in paper [24]. On the other hand, a claim was made specifically for the weak MHD turbulence in [25] that the dissipation scale always adjusts itself to the point k^*, so that the cross-helicity spectrum can never change sign. We believe that this claim is unjustified, because it implies that the dissipation scale is independent of the kinematic viscosity and the resistivity coefficients.

14.4.6 Transient Evolution Leading to Formation of the KZ Spectrum

Let consider balanced WT, $n^+ = n^-$, forced or unforced, with an initial spectrum in a finite range of wavenumbers. Because the KZ spectrum (14.53) has finite capacity (i.e. it is integrable at infinity), we expect an explosive (finite-time) formation of this spectrum in the system. The character of the evolution of the finite capacity systems was described in Sect. 9.2.3. Namely, a self-similar front will reach infinite $k_\perp$ in finite time t^*. For $t < t^*$, the propagating front leaves behind a power-law spectrum with an anomalous slope which is steeper than the one of the KZ spectrum, and for $t > t^*$ the KZ spectrum forms as a backscatter wave propagating from the dissipative scales toward lower $k_\perp$'s, see Fig. 9.5. This picture of transient evolution was obtained by numerical simulation of the kinetic equation (14.43) in [1].

14.4.7 PDF's in the Kinetic Regime: Turbulence Intermittency

Let us now find solutions of the PDF equation, (14.33), (14.34). Since the k_z-part of these solutions is arbitrary and t-independent, like for the spectra, we will only need to solve the respective equation for the perpendicular part, (14.39).

Fig. 14.4 Plateu behavior in mesoscopic Alfvén wave turbulence. Frequency broadening is constant, $\Gamma \sim \Delta_\omega$, in a wide range of amplitudes, from $\tilde{b}_1 \sim B_0/(k_\perp L_z)$ to $\tilde{b}_2 \sim 2\pi B_0 k_z^{1/2}/(k_\perp L_z^{1/2})$

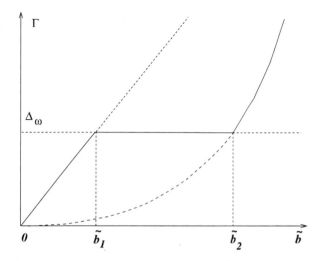

Up to the specifics of the MHD system (i.e. absence of evolution in k_z), in what follows we will obtain solutions similar to the ones found in Sect. 11.1. One of such solutions corresponds to $\mathcal{F}_\perp^\pm = 0$ in (14.39). This gives

$$\mathcal{P}_\perp^\pm \propto \exp\left(-\frac{\gamma_k^\pm I}{\eta_{\perp k}^\pm}\right) = e^{\frac{-I}{n_\perp^\pm}} = e^{\frac{-s}{n_k^\pm}}, \qquad (14.55)$$

which is the Rayleigh distribution of intensity J corresponding to the Gaussian statistics of the wavefield.

However, there is also a steady solution which corresponds to non-zero values of the amplitude-space flux $\mathcal{F}_\perp^\pm$. This solution can be obtained in terms of the integral exponential functions, and its large s asymptotics is given by

$$\mathcal{P}_\perp^\pm \approx \frac{-\mathcal{F}_\perp^\pm n_z}{s}, \qquad s \gg n_\perp^\pm, \qquad (14.56)$$

which corresponds to strong *turbulence intermittency* with anomalously high probability of strong (much stronger than mean) waves. Obviously, this result cannot be extended to the arbitrarily large intensities s because the WT approach based on small nonlinearity would break down. It is important however that for weak wavefields the WT description can be valid for intensities which are in the PDF tail, $s > \langle s \rangle = n_k$, if these s remain within the weakly nonlinear range. For larger s which correspond to strong nonlinearity the behavior is much more complicated and hard to be treated rigorously. However, on the phenomenological level one could argue that there has to be a PDF cutoff at some limiting large s because waves of greater intensities do not exist. Such a cutoff amplitude could be related, e.g., with reaching the *critical balance state* (see Sect. 3.2). Because the power-law solution for the PDF tail implies non-zero amplitude-space fluxes $F^\pm$, the physical mechanisms responsible for the amplitude cutoff must at the same time produce some amplitude-dependent dissipation. In some systems, like e.g. the water gravity waves, this mechanism is wavebreaking. Indeed, wavebreaking at the same time limits achievable wave amplitudes, and it redistributes the wave energy over different modes (a breaking wave produces ripples with a broad spectral content). It is presently less clear what mechanisms could play a similar role in MHD.

14.5   Mesoscopic MHD Wave Turbulence

Comparing condition of applicability of the discrete regime, (14.46), and the one for the kinetic regime, (14.16), we see that both of these conditions fail in the following range of amplitudes,

$$\frac{1}{k_\perp L_z} < \frac{\tilde{b}}{B_0} < \frac{2\pi k_z^{1/2}}{k_\perp L_z^{1/2}} . \tag{14.57}$$

This is a rather large range (with max/min ratio of amplitudes of $\sqrt{k_z L_z}$) where WT is neither purely discrete nor purely kinetic, and we will call the respective state as *mesoscopic* WT. We have already mentioned in Sect. 14.4.3 that the mesoscopic regime is most typical to the numerical simulation of MHD turbulence at the current level of resolution, i.e. the infinite-box behavior is not achieved.

In Sect. 10.3 we already discussed the mesoscopic WT. However, there we concentrated on the case with sparse discrete resonances, when the number of actively important modes $\mathcal{M}$ resonantly connected to mode $\mathbf{k}$ is small, which is not the case for the MHD waves. We saw that for small $\mathcal{M}$ the mesoscopic regime was associated with a range of amplitudes where *both* the discrete and the kinetic turbulence may exist, which implies possibility of a hysteresis effect, when these two regimes would alternate in time.

What is different for the mesoscopic regime in MHD, when *neither the discrete nor the kinetic* regimes can exist? The only way to reconcile that WT is neither discrete, $\Gamma \ll \Delta_\omega$, nor kinetic, $\Gamma \gg \Delta_\omega$, is to accept that in the range of amplitudes (14.57) the frequency broadening remains of the same order as the frequency spacing, $\Gamma \sim \Delta_\omega$.

Mesoscopic regime of WT is rather complicated, mostly unstudied, and only little is known about it. However, some basic things can be said. Since condition $\Gamma \sim 2\pi/L_z$ is independent of the amplitude, this suggests a linear dynamics. Indeed, in linear systems the characteristic evolution time is amplitude independent. This could be realized, e.g., if the MHD turbulence in this regime would contain two components,—a strong 2D component (condensate) and a weak 3D wave component whose evolution would be mainly due to the (linear) process of shearing by the 2D vortices. Testing this prediction (e.g. numerically) and developing an effective linear theory for such a mesoscopic Alfvén wave turbulence would be an interesting subject for future study.

For now, we could apply a simple dimensional argument to obtain the energy spectrum. In linear systems, the energy spectrum $E_\perp^\pm$ must be proportional to the energy flux $\epsilon_\perp^\pm$. The rest of the dimensions will be uniquely reproduced using Γ and $k_\perp$, and the result is

$$E_\perp^\pm \sim \frac{\epsilon_\perp^\pm}{\Gamma k_\perp}. \tag{14.58}$$

In fact, this is an anisotropic version of the passive scalar cascade spectrum in Batchelor's regime, which is not surprising, because such a system is also linear and also characterized by an amplitude- and scale-independent evolution time (inverse strain of the advecting velocity).

A word of caution should be said about applying scaling (14.58). It assumes that the mesoscale regime is realized in a certain range of k's. On the other hand, several regimes may coexist simultaneously, e.g. the discrete regime at low k's and and mesoscopic regime at high k's, as it seems to be the case in the numerical simulations of paper [16].

14.6 Summary

Different regimes of MHD turbulence are summarized in table. Question marks indicate unanswered questions.

MHD regime	Equations	Properties	Spectra
Discrete WT	(14.12)	3D waves enslaved to 2D turbulence	In 3D modes, E_z is arbitrary, and perpendicular spectrum mimics the one of the 2D modes
Mesoscopic WT	Two component?	Frequency broadening $\Gamma \sim$ frequency spacing Δ_ω	$E_\perp \propto k_\perp^{-1}$
Kinetic WT	(14.33), (14.34), (14.36), (14.39), (14.40), (14.42), (14.43)	KZ cascades, balanced and imbalanced. Gaussian PDF's and power-law tails	$E_\perp^\pm \propto k_\perp^{\nu^\pm}$; $\nu^- + \nu^+ = -4$; E_z – arbitrary
Strong WT	?	Critical balance, see Sect. 3.2	$E_\perp \propto k_\perp^{-5/3}$; $k_z \propto k_\perp^{2/3}$

14.7 Further Reading

Our description of MHD turbulence was with an emphasis on the WT theory. This is an important but obviously not the only approach, and a vast literature exists on different theoretical, observational and numerical issues in MHD turbulence in various astrophysical and laboratory applications. Here, we will suggest an incomplete list of further reading. This list is in addition to the papers we have already cited in this chapter and in the respective discussions of the MHD turbulence in Chap. 3. Paper [19] developed WT for low-beta MHD, i.e. when gas pressure is much less than the magnetic pressure (opposite to our incompressible MHD, which corresponds to high beta). This paper has a great emphasis on the Hamiltonian setup and is of an independent interest for people interested to learn about the Hamiltonian formulation in MHD.

A compact and easy to read review of MHD turbulence, including aspects of the dynamo theory and astrophysical applications, with some historical perspective, can be found in [26]. A book on MHD turbulence [27] contains many useful facts, from basic theory to numerical aspects, transition to turbulence, discussion of reconnections, compressible MHD, astrophysical applications (but no discussion of the WT theory).

An interesting modification of the critical balance (CB) approach for strong MHD turbulence (see Sect. 3.2) was suggested in [28] based on the idea that nonlinearity is weakened in scale-dependent due to a certain alignment of the fluctuating fields. Another interesting variation on the CB approach was developed in [29] based on an idea that not only critically balanced modes contribute to the spectrum, but also all the modes which are in resonance with the CB modes.

Rapidly increasing power of modern computers have lead to new interesting numerical results. Recent simulations of weak MHD wave turbulence reported in [30] appear to confirm the WT theory predictions for the wave spectra, including the prediction for the imbalanced WT expressed by (14.5). This paper also emphasized an observed pinning of the "+" and "−" spectra at the dissipation scale, but we are convinced that in their simulation this was an artefact of insufficient inertial range. (In some sense, pinning at the dissipation scale is trivial because all spectra get eventually "pinned" to zero by dissipation, but our point is that the WT theory allows the cross-helicity to change sign within the inertial range, see Sect. 14.4.5).

Recent numerical studies of strong MHD turbulence in an external magnetic field reported in [31] seem to confirm the CB theory, with adjustments to alignment as suggested in [28]. Strong MHD turbulence without an external magnetic field was recently studied by high-resolution DNS in [32] (in this book we have not covered such MHD turbulence at all). Finally, let us mention a couple of recent applications of the WT approach to particular physical processes. In [33], WT theory was used to describe solar coronal heating. In [34], an inverse transfer of MHD energy (from small to large scales) was studied as a scenario for regenerating the large-scale magnetic fields in galaxies.

The above list is far from being exhaustive, but it should hopefully be enough as an initial guidance into the MHD turbulence subject, thereby facilitating further independent learning.

References

1. Galtier, S., Nazarenko, S.V., Newell, A.C., Pouquet, A.: A weak turbulence theory for incompressible MHD. J. Plasma Phys. **63**(5), 447–488 (2000)
2. Galtier, S., Nazarenko, S.V., Newell, A.C., Pouquet, A.: Anisotropic turbulence of Shear-Alfv. n waves. Astrophys. J. **564**, L49–L52 (2002)
3. Nazarenko, S.V., Newell, A.C., Galtier, S.: Non-local MHD turbulence. Physica D **152–153**, 646–652 (2001)
4. Saur, J., Politano, H., Pouquet, A., Matthaeus, W.H.: Evidence for weak MHD turbulence in the middle magnetosphere of Jupiter. Astron. Astrophys. **386**, 699–708 (2002)
5. Galtier, S., Nazarenko, S.V., Newell, A.C.: On wave turbulence in MHD. Nonl. Proc. Geophys. **8**(3), 141–150 (2001)
6. Galtier, S., Bhattacharjee, A.: Anisotropic weak whistler wave turbulence in electron MHD. Phys. Plasma. **10**(8), 3065–3076 (2003)
7. Galtier, S.: Wave turbulence in incompressible hall magnetohydrodynamics. J. Plasma. Phys. **72**, 721–769 (2006)
8. Nazarenko S.V.: 2D enslaving of MHD turbulence. New J. Phys. **9**, 307 (2007). doi: 10.1088/1367-2630/9/8/307
9. Zakharov, V.E., Korotkevich, A.O., Pushkarev, A.N., Dyachenko, A.I.: Mesoscopic wave turbulence. JETP Lett. **82**(8), 487 (2005)
10. Lvov, Y.V., Nazarenko, S., Pokorni, B: Discreteness and its effect on water-wave turbulence. Physica D **218**(1), 24 (2006). arxiv: math-ph/0507054 (July 2005)
11. Choi, Y., Lvov, Y., Nazarenko, S.V.: Probability densities and preservation of randomness in wave turbulence. Phys. Lett. A **332**(3–4), 230–238 (2004)
12. Choi, Y., Lvov, Y., Nazarenko, S.V.: Joint statistics of amplitudes and phases in wave turbulence. Physica D **201**, 121–149 (2005)
13. Choi Y., Lvov Y., Nazarenko S.V., Pokorni B.: Anomalous probability of large amplitudes in wave turbulence. Phys. Lett. A **339**(3–5), 361–369
14. Kadomtsev, B.B., Pogutse, O.P.: Sov. Phys. JETP **38**, 283 (1974)
15. Strauss, H.R.: Phys. Fluids **19**, 134 (1976)
16. Muller, W.-C., Biskamp, D., Grappin, R.: Statistical anisotropy of magneto-hydrodynamic turbulence. Phys. Rev. E **67**, 066302 (2003)
17. Zakharov, V.E., Kuznetsov, E.A.: Variational principle and canonical variables in magnetohydrodynamics. Sov. Phys. Dokl. **15**, 913–914 (1971)
18. Zakharov, V.E., Kuznetsov, E.A.: Hamiltonian formalism for nonlinear waves. Usp. Fiz. Nauk. **167**(11), 1137–1168 (1997)
19. Kuznetsov, E.A.: Weak magnetohydrodynamic turbulence of a magnetized plasma. J. Exp. Theor. Phys. **93**(5), 1052–1064 (2001)
20. Waleffe, F.: Phys. Fluids A **4**, 350 (1992)
21. Galtier, S.: Phys. Rev. E **68**, 015301 (R) (2003)
22. Nakariakov, V.M., Roberts, B.: Solar Phys. **168**, 273 (1996)
23. Nakariakov, V.M., Roberts, B., Mann, G.: Astron. Astrophys. **311**, 311 (1996)
24. Grappin, R., Leorat, J., Pouquet, A.: Dependence of MHD turbulence spectra on the velocity field-magnetic field correlation. Astron. Astrophys. **126**, 51–58. ADS (1983)
25. Lithwick, Y., Goldreich, P.: Imbalanced weak magnetohydrodynamic turbulence. Astrophys. J. **582**, 1220–1240 (2003)
26. Schekochihin, A.A., Cowley, S.C.: Turbulence and magnetic fields in astrophysical plasmas. In: Molokov, S., Moreau, R., Moffatt, H.K. (eds.) Magnetohydrodynamics: Historical Evolution and Trends, p. 85. Springer, Berlin (2007)
27. Dieter Biskamp, MHD Turbulence. Cambridge University Press, United Kingdom (2003). ISBN: 9780521810111, 312p
28. Boldyrev, S.: Spectrum of magnetohydrodynamic turbulence. Phys. Rev. Lett. **96**(1–4), 115002 (2006)

29. Alexakis, A.: Nonlocal phenomenology for anisotropic magnetohydrodynamic turbulence. Astrophys. J. **667**, L93–L96 (2007)
30. Boldyrev S., Perez J.C.: Spectrum of weak magnetohydrodynamic turbulence. Phys. Rev. Lett. **103**, 225001 (2009)
31. Mason J., Cattaneo F., Boldyrev S.: Numerical measurements of the spectrum in magnetohydrodynamic turbulence. Phys. Rev. E **77**, 036403 (2008)
32. Mininni, P.D., Pouquet, A.G., Montgomery, D.C.: Small-scale structures in three-dimensional magnetohydrodynamic turbulence. Phys. Rev. Lett. **97**, 244503 (2006)
33. Bigot, B., Galtier, S., Politano, H.: An anisotropic turbulent model for solar coronal heating. Astron. Astrophys. **490**(1), 325–337 October IV (2008)
34. Galtier, S., Nazarenko, S.: Large-scale magnetic field re-generation by resonant MHD wave interactions. J. Turbulence **9**(40), 1–10 (2008)

Chapter 15
Bose-Einstein Condensation

15.1 Introduction

In this chapter we will describe general properties of Bose-Einstein condensation (BEC) phenomenon common for cold atom and optical systems, based on the fact that both of these systems can be studied within the nonlinear Schrödinger (NLS) model given by (5.1). An extension of this model to systems bounded by a potential $U(\mathbf{x})$ (modeling BEC confinement in magnetic traps or a light confinement in optical cavities due to variable refractive index or mirrors) is:

$$i\,\partial_t \psi + \nabla^2 \psi \pm |\psi|^2 \psi - U\psi = 0. \tag{15.1}$$

Here signs "+" and "−" correspond to cases with atoms with attractive and repelling potentials respectively, or, for optical systems, focusing and de-focusing Kerr media. De-focusing case (e.g. Rubidium or Sodium gas) has been studied more frequently because it corresponds to stable BEC whereas focusing case (e.g. Lithium) is unstable in dimensions two and three which leads to sudden bursts/collapses. Study of such collapses, particularly within the NLS model, is an interesting subject in itself, and interested readers can learn about this subject in review [1] and book [2]. In most that follows we will study the de-focusing model, except our discussion of the 1D optical turbulence in the end of this chapter. In BEC theory, the de-focusing NLS is usually called Gross-Pitaevskii equation [3, 4].

15.2 Kinetic Equation for the Wave Spectrum

Let us ignore for now the trapping potential, i.e. put $U \equiv 0$ in (15.1). We begin with an exercise:

Exercise 15.1 *Consider the NLS system in a periodic box and apply Fourier transform to* (15.1) *with* $U \equiv 0$. *Show that this gives an equation of form* (6.72) *with frequency* $\omega_k = k^2$ *and interaction coefficient* $W_{34}^{12} \equiv 1$.

Thus, this is a special case of the general case of four-wave systems we considered in Sect. 6.9.1. In particular, we derived a four-wave kinetic equation (6.81) which for the NLS case becomes [5]:

$$
\dot{n}_k = 4\pi \int n_1 n_2 n_3 n_k \left[\frac{1}{n_k} + \frac{1}{n_3} - \frac{1}{n_1} - \frac{1}{n_2} \right]
$$
$$
\times \delta(\mathbf{k} + \mathbf{k_3} - \mathbf{k_1} - \mathbf{k_2}) \delta(k^2 + k_3^2 - k_1^2 - k_2^2) \, d\mathbf{k_1} d\mathbf{k_2} d\mathbf{k_3}. \quad (15.2)
$$

We will now discuss stationary solutions of this equation.

15.3 Role of Thermodynamic Solutions

First of all, let us discuss the role of the thermodynamic Rayleigh-Jeans (RJ) states (9.1). This discussion is especially important for the NLS model, since thermodynamics framework is most used by the BEC research community, and we need to reconcile it with the WT approach whose main objects are KZ cascading states. Considering that we have two NLS invariants, the energy and the number of particles, the RJ spectra (9.1) become

$$
n_k = \frac{T}{k^2 + \mu}, \quad (15.3)
$$

where $T > 0$ and $\mu > 0$ are constants having a meaning of temperature and chemical potential (to be precise, the negative chemical potential) respectively.

As we already mentioned in Sect. (9.1), the RJ spectra are of limited relevance in WT because the energy integral diverges on them for any finite initial energy and particle physical-space densities. This manifests the *"Ultraviolet (UV) Catastrophe"* when the wave energy is transferred to higher and higher frequencies so that the system cools down to the absolute zero of temperature, $T = 0$. Indeed, T is the energy per degree of freedom, and the number of degrees of freedom (modes), over which the finite initial energy is to be distributed, experiences an unbounded growth.

To make the RJ spectrum (15.3) a legitimate solution, a cutoff wavenumber k_c is sometimes introduced in BEC literature; see e.g. [6]. The authors of paper [6] reasoned that such a cutoff is natural in numerical simulations and it may correspond to a dissipative scale in real physics. This approach is rather dangerous, because, as well-known from the general turbulence, numerical cutoffs lead to backscatter of energy toward the lower-k modes with resulting un-physical energy accumulation near the highest k's,—so called *bottleneck effect*. In turbulence this numerical artefact is usually solved by introducing a high-k dissipation (e.g. by using "hyperviscosity"). This problem has not yet been well appreciated by people doing numerical simulations of BEC using the NLS model, and often no high-k dissipation is used to prevent the bottleneck [6, 7].[1] On the other hand, if a physical dissipation is present at high k's in a real system, its effect will be different from a cutoff, because it will lead to an energy dissipation rather than a backscatter at high k's. Thus we conclude that *introducing a fixed cutoff wave-number k_c is not physically relevant* within the BEC context.

On the other hand, the thermodynamic equilibrium states do make sense in BEC. To understand this let us recall how the Plank's law has provided a solution to the UV Catastrophe in the thermodynamic theory of electromagnetic waves. Similarly, a resolution of the UV divergence of energy on the RJ spectrum can be obtained by replacing the latter by the Bose-Einstein (BE) distribution,

$$n_k = \frac{1}{e^{(k^2+\mu)/T} - 1}. \qquad (15.4)$$

Easy to see, this distribution becomes RJ spectrum when $k^2 + \mu \ll T$, whereas in the opposite limit $k^2 + \mu \gg T$ it becomes Maxwell-Boltzmann (MB) distribution:

$$n_k = \text{const}\, e^{-\frac{k^2}{T}}. \qquad (15.5)$$

The exponential (Gaussian) decay of the MB distribution (15.5) can be qualitatively viewed as an effective spectrum cutoff k_c determined by the condition $n(k_c) \sim 1$, i.e. $k_c = \sqrt{T - \mu}$. Importantly, this cutoff is not fixed but depends on T and μ (and it can even move in time for non-stationary spectra).

Remark 15.3.1 Introduction of the variable cutoff k_c where $n(k_c) \sim 1$ may be seen as phenomenological way to take into account the quantum effects in the "classical" wave kinetic equation (15.2) and, therefore, in the NLS model. However, it remains to be understood if such a variable cutoff could be implemented in numerical simulations of the NLS equation.

The BE distribution (15.4) is not a solution of the kinetic equation (15.2), so introduction of the variable cutoff k_c where $n(k_c) \sim 1$ may be seen as phenomenological way to take into account the quantum effects in the "classical" model. On the other hand, BE distribution (15.4) is a solution of more a general quantum kinetic equation:

$$\dot{n}_k = 4\pi \int n_1 n_2 n_3 n_k \left[\left(\frac{1}{n_k}+1\right)\left(\frac{1}{n_3}+1\right) - \left(\frac{1}{n_1}+1\right)\left(\frac{1}{n_2}+1\right) \right]$$
$$\times \, \delta(\mathbf{k}+\mathbf{k_3}-\mathbf{k_1}-\mathbf{k_2})\delta(k^2+k_3^2-k_1^2-k_2^2)\, d\mathbf{k_1}d\mathbf{k_2}d\mathbf{k_3}. \qquad (15.6)$$

This equation follows from a standard *second quantization* technique for bosons and it was first obtained by Nordheim [8] (as well as it's fermionic version with

---

[1] Paper [7] filtered out high k's at the data processing, but not at the computational stage. Obviously, this could not prevent the backscatter/bottleneck effect discussed here.

minus signs instead of the pluses in the integrand). The Nordheim equation (15.6) becomes the kinetic equation (15.2) in limit $n_k \to \infty$ and it becomes a classical Boltzmann equation[2] for $n_k \to 0$:

$$\dot{n}_k = 4\pi \int (n_1 n_2 - n_3 n_k)\delta(\mathbf{k} + \mathbf{k_3} - \mathbf{k_1} - \mathbf{k_2})\delta(k^2 + k_3^2 - k_1^2 - k_2^2)d\mathbf{k_1}d\mathbf{k_2}d\mathbf{k_3}.$$

(15.7)

Let us, following paper [6], integrate the RJ spectrum (15.3) over the **k**-space assuming a cutoff at $k = k_c$. This gives for 3D systems:

$$\mathcal{N} = 4\pi T k_c \left[1 - \frac{\sqrt{\mu}}{k_c} \arctan \frac{k_c}{\sqrt{\mu}} \right],$$

(15.8)

$$E = \frac{4\pi T k_c^3}{3} \left[1 - 3\frac{\mu}{k_c^2} + 3\left(\frac{\mu}{k_c^2}\right)^{3/2} \arctan \frac{k_c}{\sqrt{\mu}} \right].$$

(15.9)

Also, following [6], we introduce the particle containing scale λ_N which in the RJ state is $\lambda_N \sim 1/\sqrt{\mu}$, and note that the BE condensation occurs when $\lambda_N \to \infty$. For a fixed $\mathcal{N}$, this limit is achieved at a finite critical temperature, which can be obtained by putting $\mu = 0$ and $k_c = \sqrt{T - \mu} = \sqrt{T}$ in (15.8):

$$T_{cr} = (\mathcal{N}/4\pi)^{2/3}.$$

(15.10)

Up to an order-one constant, this is a familiar expression for the critical temperature of an ideal (non-interacting) Bose gas [9]. Corresponding critical energy E_{cr} is related to T_{cr} according to (15.9) (with $\mu = 0$ and $k_c = \sqrt{T}$):

$$E_{cr} = \frac{4\pi T_{cr}^{5/2}}{3}.$$

(15.11)

Below the critical temperature, for $T < T_{cr}$, a standard condensation argument says that a finite fraction for the particles is in the ground state with $k = 0$ and the rest of the particles are RJ-distributed with $\mu = 0$:

$$n_k = \mathcal{N}_0 \delta(\mathbf{k}) + T/k^2.$$

(15.12)

Integrating over **k** with cutoff at k_c we have

$$\mathcal{N} - \mathcal{N}_0 = 4\pi T k_c = 4\pi T^{3/2},$$

(15.13)

or

$$\frac{\mathcal{N}}{\mathcal{N}_0} = 1 - \left(\frac{T}{T_{cr}}\right)^{3/2} = 1 - \left(\frac{E}{E_{cr}}\right)^{3/5}.$$

(15.14)

---

[2] To be precise, this is a special case of the Boltzmann equation for rigid-sphere particles.

This is a familiar result for the non-interacting Bose gas, see e.g. Fig. 3.2 in book [10].

Remark 15.3.2 Thus, we have obtained the correct dependence for the condensate fraction using the classical wave model (NLS equation (5.1), kinetic equation (15.2), RJ spectrum (15.3)) by introducing a variable cutoff $k_c = \sqrt{T - \mu}$. In paper [6], the cutoff k_c was assumes to be fixed, which led them to a different dependence: $\mathcal{N}/\mathcal{N}_0 = 1 - T/T_0$. This dependence will also change when the gas is placed in a trapping potential U (see e.g. Chap. 10 in book [10] for BEC in a harmonic trap).

Let us now consider the 2D case. Instead of (15.8) and (15.8) we now have:

$$\mathcal{N} = \pi T \, \ln(1 + k_c^2/\mu), \tag{15.15}$$

$$E = \pi T \left[k_c^2 + \mu \ln(1 + k_c^2/\mu) \right]. \tag{15.16}$$

By considering limit $\mu \to 0$ we see that there is no finite critical temperature in this case. Thus, one could conclude that there is no BE condensation in 2D [11]. Importantly, this conclusion is valid only if the gas is *non-interacting* and *homogeneous* (no finite-size trap). In finite traps, BE condensation of non-interacting 2D gases was shown to exist in [12] (see also book [10]). Another condensation-promoting effect appears to be the mutual interaction of atoms, which in the NLS model corresponds to the nonlinearity.

One has to be cautious with using the RJ spectrum (15.3) for predicting BE condensation because the weak interaction (nonlinearity) assumption breaks down for small μ. Indeed, kinetic equation (15.2) leads to the following estimate for the nonlinear frequency Γ_K (inverse interaction time):

$$\Gamma_K \sim n_k^2 k^{2d}/\omega_k. \tag{15.17}$$

According to the RJ spectrum (15.3), for $k \sim \sqrt{\mu}$ we have $n_k \sim T/\mu$, so

$$\frac{\Gamma_K}{\omega_k} \sim T^2 \mu^{d-4}. \tag{15.18}$$

Thus, for both 3D and 2D systems the nonlinear to linear ratio becomes large as $\mu \to 0$, and both the kinetic equation (15.2) and its RJ solution (15.3) become inapplicable. This is why the above condensation argument is rather shaky for the NLS model. In particular, it fails to predict BE condensation in 2D, which is clearly observed numerically in [13–15] (see later in this chapter). Sometimes, nonlinear interaction is taken into account by correcting the RJ-state via Bogoliubov transformation (also to be discussed later). However, this approach is applicable only when the condensate is much stronger than the un-condensed component and, thus, cannot be used for predicting T_{cr}.

Remark 15.3.3 The above condensation argument is more solid in 3D than in 2D. Indeed, the numerics of the truncated 3D NLS [6] show that the condensation argument based on the RJ spectra works reasonably well provided the initial nonlinearity is small. Namely, if $T_{cr} L \ll 1$, the gravest modes remain weakly nonlinear (according to (15.18)) and T_{cr} is L-independent.

In 2D, no L-independent T_{cr} is possible unless we take into account the effect of finite nonlinearity. Interestingly, in the 2D NLS the condition for breakdown of the weakly nonlinear assumption $\Gamma_K/\omega_k \sim 1$ appears to provide the condensation criterion, as it was demonstrated numerically for truncated 2D NLS in [15].

15.4 Non-Equilibrium Condensation and KZ Spectra

From the previous section we saw that the thermodynamic equilibrium states can only form if the (classical wave) NLS model is "fixed" by introducing quantum effects via a phenomenological (T- and μ-dependent) cutoff k_c. This cutoff represents the scale where $n_k \sim 1$, at which the BE distribution (15.3), (15.4) changes from Lorentzian RJ dependence to exponential/Gaussian dependence of the MB distribution (15.5). Thermodynamic states are also possible for a simple fixed cutoff k_c, although they are un-physical. What matters is that the cutoff, variable or fixed, makes the effective number of degrees of freedom finite, and this prevents the *UV catastrophe*.

Suppose we do not want to "fix" the NLS model, and we want to know what happens to the system of waves which always remain purely classical (in the other words, $n_k \gg 1$ in all the relevant scales). In this case, the UV catastrophe becomes a real physical effect rather than a paradox to be resolved, and its meaning is that no finite-temperature thermodynamic equilibrium is possible in such systems. Instead, the statistics will remain strongly non-equilibrium, and should be described in terms of the WT cascades rather than the equilibrium distributions. As we discussed in Sect. 8.2.1, NLS is a dual cascade system: the energy E cascades toward larger wavenumbers (direct cascade) and the particles $\mathcal{N}$ cascade toward smaller wavenumbers (inverse cascade); see Fig. 8.1. In BEC, dual behavior has a nice physical interpretation [16]. Consider a system in a trap, as shown in Fig. 15.1. In this setup, the forward cascade will correspond to an energy transfer toward larger energy levels. When such an energy cascade reaches highes available levels in the trap, it will "spill" over the potential barrier. This corresponds to *evaporative cooling*, a technique used experimentally in BEC experiments.

Remark 15.4.1 Evaporative cooling is also active when the occupation numbers of the highest levels is small $n_k \lesssim 1$. (This is probably the most typical regime in BEC experiments). In this case, particles in the higher levels could be close to thermodynamical state with MB distributions. When the gaussian tail of this distribution reaches the energies higher than the potential barrier, the respective particles evaporate. This is similar to evaporation and cooling of tea, where most energetic water molecules leave the liquid surface into ambient air.

On the other hand, the inverse cascade corresponds to a non-equilibrium transfer of particles toward lower energy levels, see Fig. 15.1.

Fig. 15.1 Non-equilibrium BE condensation in a trap as a dual cascade turbulent system. The direct cascade takes the energy to higher energy levels in the trap and eventually spills it over the barrier,—this is an "evaporative cooling" process. The inverse cascade takes the particles toward the ground state,—this is the condensation

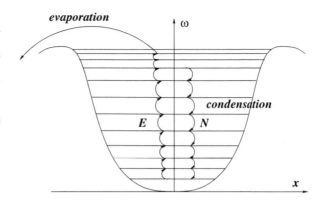

evaporation

condensation

The KZ spectra are power-law steady state solutions,

$$n_k = A\,k^v,$$

with constant dimensional pre-factors A and exponents $v = v_E$ and $v = v_N$ for the direct energy and the inverse particle cascades given by formulae (9.36) and (9.37) respectively. For NLS, the interaction coefficient is constant, i.e. its homogeneity degree is $\beta = 0$, whereas the frequency index is $\alpha = 2$. Thus in this case

$$v_E = -d, \qquad (15.19)$$

$$v_N = -d + 2/3, \qquad (15.20)$$

Again, the KZ solutions are only meaningful if they are local, i.e. when the collision integral in the original kinetic equation converges.

In 3D ($d = 3$) the inverse $\mathcal{N}$-cascade spectrum is local, whereas the the direct E-cascade spectrum is log-divergent at the infrared (IR) limit (i.e. at $k \to 0$) [17]. As usual, the log-divergence can be remedied by a log-correction to the spectrum, in the spirit of the well-known Kraichnan's log-correction to the enstrophy cascade spectrum of 2D turbulence [18]. First, for the energy flux in the four-wave systems we have $\varepsilon \sim n_k^3$, where n_k is in the integrand of the integral defining ε. Thus, the log-dependence of ε, arising from the log-divergence of the integral defining ε, cancels out if we make the prefactor A logarithmically slowly dependent on k [17]:

$$n_k \sim \frac{1}{[\ln(k/k_f)]^{1/3}}\, k^{v_E}, \qquad (15.21)$$

where k_f is an IR cutoff provided by the forcing scale. Note that this expression is only valid for $k \gg k_f$ (it is singular for $k = k_f$).

The 2D case ($d = 2$) appears to be even more tricky. It turns out that formally the $\mathcal{N}$-cascade spectrum is local, but the $\mathcal{N}$-flux appears to be positive [17], in contradiction with the Fjørtoft's argument. What spectrum can we expect in this case at the scales larger than the forcing scale? Further, for the

E-cascade spectrum, the exponent ν_E coincides with the one of the thermodynamic E-equipartition spectrum. The energy flux of the pure power-law spectrum with index ν_E is zero due to such a root degeneracy. What spectrum should we expect then in the forced/dissipated 2D NLS turbulence for the scales smaller than the forcing scale? Both of these questions are easier to resolve using so-called differential approximation model, which we will now consider.

15.5   Differential Approximation Model

Differential approximation model (DAM) for wave turbulence is a nonlinear differential equation for the waveaction spectrum n_k, which has the same scalings properties and solutions as the kinetic equation (see Sect. 16.1). In this section, we will follow an approach the details of which can be found in [19, 20].

15.5.1   DAM for NLS Wave Turbulence

DAM for NLS turbulence mimics the kinetic equation (15.2); it is given by the following equation [17]:

$$\frac{\partial n}{\partial t} = \frac{\partial^2 R}{\partial \omega^2},$$
(15.22)

$$\text{with} \qquad R = S\,\omega^{2+3d/2}\,n^4\,\frac{\partial^2}{\partial \omega^2}\frac{1}{n},$$
(15.23)

where S is a positive constant. This equations can be obtained by multiplying the integrand in (15.2) by an interactions coefficient which is strongly peaked at "super-local" interactions with $\omega \approx \omega_1 \approx \omega_2 \approx \omega_3$, and Taylor expanding with respect to the differences of these frequencies, as it was done in [17]. This gives an expression for constant S. However, there is no rigorous justification for leaving exclusively such super-local interactions, and DAM should be treated as a simplified model only and not as an asymptotically rigorous equation.

Equation (15.22) can be written as a conservation law of $\mathcal{N}$,

$$\frac{\partial n}{\partial t} + \frac{\partial \zeta}{\partial \omega} = 0,$$
(15.24)

with the waveaction flux

$$\zeta = -\frac{\partial R}{\partial \omega},$$
(15.25)

or as a conservation law of E,

$$\frac{\partial(\omega n)}{\partial t} + \frac{\partial \varepsilon}{\partial \omega} = 0,$$
(15.26)

Fig. 15.2 The energy and
the waveaction fluxes as a
function of the spectral index
for 3D NLS turbulence

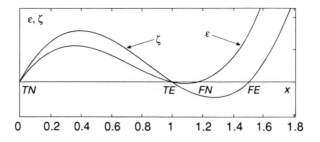

Fig. 15.3 The energy and
the waveaction fluxes as a
function of the spectral index
for 2D NLS turbulence

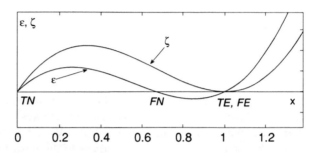

with the energy flux

$$\varepsilon = R - \omega \frac{\partial R}{\partial \omega}. \tag{15.27}$$

For power-law spectra $n = c \; \omega^{-x}$ we have:

$$\zeta = -c^3 S \, (y-1) \, x \, (x-1) \omega^{-y}, \tag{15.28}$$

$$\varepsilon = c^3 S \, y \, x \, (x-1) \omega^{-y+1}, \tag{15.29}$$

$$\text{where} \quad y = 3x - 3d/2 + 1. \tag{15.30}$$

From these expressions we recover the four power-law solutions. For this, it is
instructive to plot the fluxes ζ and ε as functions of x for fixed ω; see Figs. 15.2 and
15.3 for the 3D and the 2D systems respectively.

Two thermodynamic RJ spectra are given by $x = 0$ (equipartition of $\mathcal{N}$) and
$x = 1$ (equipartition of E). In Figs. 15.2 and 15.3, these spectra are denoted as *TN*
and *TE* respectively. Both of the fluxes, ζ and ε are zero on these spectra. KZ spectra
with constant $\mathcal{N}$-flux and E-flux are given by $y = 0$ (i.e. $x = d/2 - 1/3$) and $y = 1$
(i.e. $x = d/2$) respectively; they are denoted as *FN* and *FE* in Figs. 15.2 and 15.3.

The 3D case is quite straightforward: on the KZ spectra the $\mathcal{N}$-flux and the
E-flux are in accordance with the Fjørtoft's argument (inverse cascade of $\mathcal{N}$ and
forward cascade of E), see Fig. 15.2.

The 2D case is less obvious. First, the indices of the E-flux KZ and the
E-equipartition RJ spectra coincide. Using DAM one can easily remedy this
degeneracy. The double root for the exponent x in the 2D case points at existence
of a log-corrected solution, as formulated in the following exercise.

Exercise 15.2 *Show that the differential model for 2D NLS has a log-corrected energy cascade solution*

$$n = \frac{C}{\omega} \ln^{1/3}\left(\frac{\omega_*}{\omega}\right) \left[1 + O\left(1/\ln\frac{\omega_*}{\omega}\right)\right], \qquad (15.31)$$

where ω_* is a constant frequency representing an UV cutoff and having a physical meaning of an energy dissipation scale.

Show that in the leading order in small ω/ω_* the energy flux ε is constant and positive, as it should be for the energy cascade solutions in four-wave systems.[3]

Secondly, the 2D NLS system is degenerate in that the waveaction cascade KZ solution is characterized by a positive $\mathcal{N}$-flux. Indeed, in Fig. 15.3 we see that $\zeta > 0$ at point *FN*.

Remark 15.5.1 The results about the flux directions are general and they do not depend on the DAM we have used in this section to obtain them. Moreover, to obtain these results one does not need to evaluate the fluxes based on the kinetic equation. Indeed, as we explained in Sect. 9.2.2, the $\mathcal{N}$—an E-flux directions are predetermined by the ordering of the four indices of the power-law solutions, i.e. the zero crossings in the Figures of type 15.2 and 15.3. In particular, the "bad" direction of the $\mathcal{N}$-flux arises because the index of the respective KZ spectrum got in between of the two thermodynamic spectra, thereby violating the "good" ordering (9.38).

15.5.2 What Happens When a Pure KZ Spectrum Corresponds to "Wrong" Flux Direction?

Formally, KZ's with "wrong" flux directions are valid mathematical solutions of DAM, if the boundary conditions (e.g. for n and for $\partial_\omega n$ on two ends of a finite inertial interval) are chosen to agree precisely with this KZ solution. However, these boundary conditions are artificial, and one would not be able to match such a KZ spectrum to real forcing and dissipation regions. Indeed, a slight decrease of the spectrum's index near the dissipation scale, according to (15.29) and (15.28), would immediately lead to an infinite flux (dissipation rate) of the companion invariant. Thus, pure flux solution is impossible in this case and there must be a finite thermal component present in this case,—so called "warm cascade"; see Sect. 9.2.2. Roughly, the farther one has to "push" the relevant invariant to the dissipation scale ω_d, the larger the thermal part must be. Within DAM, this effect is expressed by the fact that for any fixed flux, temperature and chemical potential,

---

[3] The opposite statement about the negative energy flux was made in [17] based on the spectrum $n = (c/\omega)\ln^{1/3}(\omega)$ thus effectively assuming an IR (rather than an UV) cutoff, which is unnatural because one should not assume a zero spectrum at the forcing scale.

the solution must terminate at a finite frequency ω_d: $n(\omega_d) = 0$. Generally, for a fixed flux of the relevant cascading invariant, and for a fixed dissipation scale ω_d, one can find relations

$$T = f(\varepsilon, \omega_d) \qquad \text{for direct cascades,} \tag{15.32}$$

$$T/\mu = g(\zeta, \omega_d) \qquad \text{for inverse cascades.} \tag{15.33}$$

In particular, for the inverse particle cascade in 2D NLS, we have a finite-front solution describing the cutoff at the dissipation scale

$$n = A \left(\omega - \omega_d\right)^{2/3}, \tag{15.34}$$

where $A = \text{const}$.

Exercise 15.3 *Substitute expression (15.34) into equation $-\zeta \, \omega = R$ and prove that it is a valid solution for $\omega - \omega_d \ll \omega_d$ with*

$$A = \left(-\frac{9\zeta}{10S\omega_d^4}\right)^{1/3}. \tag{15.35}$$

Far enough from this front, the spectrum can be found in the form of a RJ spectrum with a small correction,

$$n = \frac{T}{\omega + \mu + \theta(\omega)}. \tag{15.36}$$

Substituting this expression into equation $-\zeta\omega = R$ with R given by (15.23) and $d = 2$, and linearizing with respect to the disturbance θ we have:

$$\theta''(\omega) = -\frac{\zeta}{T^3} \frac{(\omega + \mu)^4}{\omega^4}. \tag{15.37}$$

Integrating this equation twice, we get:

$$\theta(\omega) = -\frac{\zeta}{T^3}\left(\frac{\omega^2}{2} + 4\mu(\omega \ln \omega - \omega) - 6\mu^2 \ln \omega + \frac{2\mu^3}{\omega} + \frac{\mu^4}{6\omega^2}\right) + C_1\omega + C_2,$$

where C_1 and C_2 are integration constants. Constant and linear in ω terms can be ignored as they simply renormalize μ and T. Therefore,

$$\theta(\omega) = -\frac{\zeta}{T^3}\left(\frac{\omega^2}{2} + 4\mu\omega \ln \omega - 6\mu^2 \ln \omega + \frac{2\mu^3}{\omega} + \frac{\mu^4}{6\omega^2}\right).$$

Close to the dissipation/cutoff scale the correction must fail to remain small. Assuming $\omega_d \ll \mu$, we can retain the largest negative powers of ω, which gives a relation of the type (15.33),

$$\mu/T = \left(\frac{6\omega_d^2}{\zeta}\right)^{1/3}. \tag{15.38}$$

Numerical solution of the DAM for 2D NLS (15.22), (15.23) corresponding to the stationary inverse cascade is shown in Figs. 15.4, 15.5. One can clearly see the structure of the solution described above with a wide thermal range and a sharp cutoff which is in an excellent agreement with the asymptotical solution (15.34), (15.35).

15.5.3 Extending BEC Description to Include Thermal Clouds

Even more interesting picture arises in the full quantum kinetics described by the Nordheim equation (15.6), which includes both the classical wave kinetics (arising from NLS (15.2)) and the classical particle kinetics described by the Boltzmann equation (15.7). The Boltzmann particles in this setup can be interpreted as a *"thermal clould"* over the NLS condensate. Here too, most interesting effects can bee studied within DAM, which can be easily extended to mimic the Nordheim equation (15.6). This amounts to re-defining the function R in (15.22) as

$$R = S\,\omega^{2+3d/2}\,n^2(n+1)^2\,\frac{\partial^2}{\partial\omega^2}\ln\left(1+\frac{1}{n}\right). \tag{15.39}$$

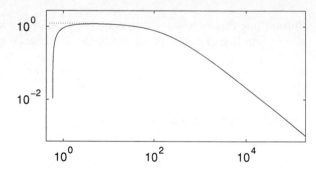

Fig. 15.4 Steady state solution of the DAM for the warm inverse cascade in 2D NLS in log-log coordinates. The *dashed line* corresponds to the RJ spectrum

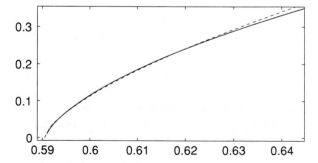

Fig. 15.5 Zoom on the front of the warm inverse cascade in 2D NLS. The *dashed line* corresponds to the asymptotic solution (15.34) with $A = 2.53$, which agrees within 5% with the value $A = 2.66$ calculated (15.35)

This equation has a general thermodynamic solution: Bose-Einstein distribution (15.4). In limit $n_k \to \infty$, we get the classical Boltzmann limit:

$$R = -S\,\omega^{2+3d/2}\,n^2\,\frac{\partial^2}{\partial\omega^2}\ln n. \qquad (15.40)$$

Exercise 15.4 *Find the constant flux solutions of the Boltzmann equation using the DAM (15.22), (15.40). Find the signs of the fluxes ε and ζ and show that they disagree with the Fjørtoft argument.*

Thus in this case too, we do not expect the pure KZ solutions to form because they cannot be matched to the physical forcing and dissipation. Instead, the spectrum will tend to a thermodynamic solution with a flux correction. Consider, for example a general setup with forcing at wavenumber ω_f and a strong dissipation at ω_{min} and ω_{max}, such that $\omega_{min} \ll \omega_f \ll \omega_{max}$ (dissipation is strong in a sense that the spectrum turns into zero at ω_{min} and ω_{max},—within DAM description). A detailed study of this system can be found in [20], whereas here we will only present the main results. Spectrum obtained by numerical solution of the steady-state DAM for Boltzmann is shown in Fig. 15.6. In both the direct and the inverse cascade ranges the spectrum is close to the thermodynamic MB distribution $n = \exp[-(\omega + \mu)/T]$, (with the same T and μ in both ranges), and sharp cutoffs at ω_{min} and ω_{max} of form

$$n = (-\zeta/S)^{1/2}\,\omega_{min}^{-\frac{2+3d}{4}}\,(\omega - \omega_{min}), \qquad (15.41)$$

$$n = (\varepsilon/S)^{1/2}\,\omega_{max}^{-1-3d/4}\,(\omega_{max} - \omega), \qquad (15.42)$$

where the energy and the particle fluxes are related via $\varepsilon = \omega_f\eta$.

Exercise 15.5 *Obtain asymptotical solutions (15.41) and (15.42) for $\omega \to \omega_{min}$ and $\omega \to \omega_{max}$ respectively.*

Fig. 15.6 Dual cascade in the Boltzmann gas; figure courtesy of Davide Proment. The *arrow* shows the forcing scale. The MB distribution is shown by the *dashed curve*. The *right* and *left insets* show zooms on the IR an the UV cutoff fronts corresponding to the solutions (15.41) and (15.42) respectively

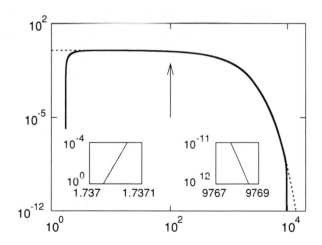

The values of T and μ are determined by ω_{max} and ω_{min} [20]:

$$T = \frac{2\omega_{max}}{(\frac{3d}{2} - 1) \ln \frac{\omega_{max}}{\omega_{min}} + \ln \frac{\omega_{max}}{\omega_f}}, \tag{15.43}$$

$$\mu = \frac{T}{2} \ln \frac{\omega_{min}^{\frac{3d}{2}-1} \omega_f}{\varepsilon}. \tag{15.44}$$

Remark 15.5.2 Surprisingly, according to formula (15.43), temperature is independent of the forcing rate ε, and is determined by the position of the forcing and the dissipation scales only (mostly on the later because $\ln \frac{\omega_{max}}{\omega_{min}} \gg \ln \frac{\omega_{max}}{\omega_f}$). On the other hand, according to formula (15.44), μ/T does depend on the forcing scale (as well as on ω_{min} and ω_f but not on ω_{max}). Thus, no matter how hard or weakly we pump energy into the system, the mean energy per particle stays constant, whereas the total number of particles varies.

15.5.4 Wave-Particle Crossover in Turbulent BEC Cascades

Let us now consider the direct energy cascade for the general quantum Bose gas which originates in the (NLS) condensate at low ω's (where $n \gg 1$) and crosses over into the (Boltzmann) thermal range at high ω's (where $n \ll 1$). Since the NLS is a model for classical waves, and the Boltzmann is a model for classical particles, we can view the crossover between the condensate and the thermal-cloud as a *wave-particle crossover*.

How to reconcile the facts that on the pure KZ solution the E-flux is positive in NLS and negative in the Boltzmann range? From the examples we have considered above, we can guess that the E-flux will start at the NLS range with a scaling close to pure KZ and will develop a thermal component when approaching to the the crossover scale where $n \sim 1$.

This system can be studied within DAM (15.22) with R for the general quantum Bose gas defined in (15.39). In the steady state E-cascade we have $\varepsilon = R = $ const. Numerical solution of this equation is shown in Fig. 15.7. Indeed, one can see a gradual transition from the E-cascade KZ scaling in the NLS range to the BE distribution (15.4) with $\mu = 0$ (i.e. Plank's distribution), and further to a cutoff front spectrum described by (15.42).

Remark 15.5.3 Because the cutoff spectrum is deeply in the Boltzmann range, one can use the formulae obtained for the Boltzmann case for the cutoff front spectrum, (15.42).

For the flux-temperature-cutoff relation, in this ($\mu = 0$) case we have [19]:

$$T = \frac{2\omega_{max}}{\ln \frac{\omega_{max}^{3d/2}}{\varepsilon}}. \tag{15.45}$$

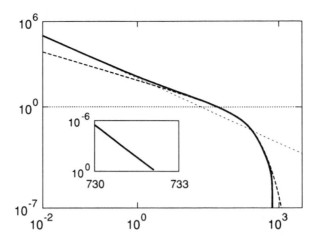

Fig. 15.7 Steady solution of the DAM for the quantum kinetics (15.39) for $d = 3$ in log-log coordinates. The wave-particle crossover is marked by the *dotted line*. Thin *dashed line* corresponds to the E-flux KZ solution; see (15.19). *Thick dashed curve* is the BE distribution (15.4) with $\mu = 0$. Inset is a zoom on the front solution in the lin-lin coordinates; see (15.42)

Remark 15.5.4 Note a bottleneck bump in the wave-particle crossover region in Fig. 15.7. This bottleneck is associated with presence of the thermal component, and from formula (15.45) it is clear that the bottleneck is bigger for larger ω_{max}. Such bottlenecks are typical for turbulent cascades crossing between the ranges characterized by a continuous and a discrete dynamics, e.g. crossover from the hydrodynamic to Kelvin wave cascade in superfluid turbulence [21], as well as in hydrodynamic turbulence simulated in Fourier space. In the latter case, the Fourier-space cutoff k_{max} introduces an effective physical space lattice with spacing $\delta x \sim 2\pi/k_{max}$.

Remark 15.5.5 In a similar way one can consider the particle-wave crossover in the inverse cascade setting [19]. In this case, forcing is at high ω's corresponding to the Boltzmann particle range, and dissipation is at low wave-dominated ω's. In 3D, the distribution will gradually change from MB to KZ, whereas in 2D it will be BE in both particle and the wave ranges with a ω_{min} cutoff (because of the "bad" flux direction of the pure 2D KZ).

15.6 Transient Evolution, Self-Similar Spectra

Obviously, if there is no dissipation at the lowest levels in the trap, the inverse cascade can only be a transient state. In fact, this transient state is characterized not by KZ but by an evolving self-similar solution [22–24] describing a spectrum front propagating toward low k's and with a power-law solution behind the front with a (numerically found) exponent ≈ 2.48 (for $d = 3$) which is different from the one of the KZ inverse cascade, 7/6 (for $d = 3$) [23–25]. This situation is typical for the finite-capacity spectra, as was explained in Sect. 9.2.3 (see also the MHD turbulence example, Sect. 14.4.6).

15.7 Breakdown of the Weak Four-Wave Turbulence and Transition to a Three-Wave Regime

As we already discussed in Sect. 15.3, the WT description based on the four-wave kinetic equation (15.2) inevitably breaks down for small μ, see the estimate (15.18). This explain the failure of the non-condensation argument for 2D BEC based on the four-wave kinetic equation (15.2).

Breakdown of the four-wave WT occurs also in the non-equilibrium inverse-cascade setting. In 2D, the argument remains the same because, as we saw above, the inverse cascade in this case is "warm", i.e. takes a form of a thermodynamic state weakly modified by the particle flux. In 3D, one can find the breakdown condition by substituting the inverse-cascade KZ spectrum $n \sim \zeta^{1/3} \omega^{-3/2}$ (see (15.20)) into the estimate (15.17). This gives for the nonlinear-to-linear frequency ratio:

$$\Gamma_K / \omega_k \sim n_k^2 k^6 / \omega_k^2 \sim \zeta^{2/3} \omega^{-2}. \tag{15.46}$$

We see that nonlinearity becomes strong at low frequencies (provided that there is no dissipation at low ω's which could absorb the WT cascade before it becomes strong).

15.7.1 WT on Background of Strong Condensate

What happens when WT becomes strong at large scales? It was suggested in [17] that this will eventually lead to a different type of WT with a strong coherent condensate component $\psi = \Psi_0 e^{-i|\Psi_0|^2 t}$, $\Psi_0 = \text{const}$ (uniform in the physical space), and weak random disturbances $\phi(\mathbf{x}, t)$ on the background of this condensate,

$$\psi(\mathbf{x}, t) = \Psi_0 (1 + \phi(\mathbf{x}, t)), \quad \phi \ll 1. \tag{15.47}$$

To develop the weak WT closure in this case, one has to diagonalize the linear part of the dynamical equation with respect to condensate perturbations ϕ. Such a diagonalization procedure is called Bogoliubov transformation, which in the NLS case is as follows [16, 17, 26],

$$a_k = \frac{\sqrt{\rho_0}}{2} \left[\left(\frac{\omega_k^{1/2}}{k} - \frac{ik}{\omega_k^{1/2}} \right) \hat{\phi}_k + \left(\frac{\omega_k^{1/2}}{k} + \frac{ik}{\omega_k^{1/2}} \right) \hat{\phi}_{-k}^* \right], \tag{15.48}$$

or conversely

$$\hat{\phi}_k = \frac{1}{2\sqrt{\rho_0}} \left[\left(\frac{i\omega_k^{1/2}}{k} + \frac{k}{\omega_k^{1/2}} \right) a_k - \left(\frac{i\omega_k^{1/2}}{k} - \frac{k}{\omega_k^{1/2}} \right) a_{-k}^* \right], \tag{15.49}$$

where $\rho_0 = |\Psi_0|^2$ is the condensate density, a_k are the new normal amplitudes, and ω_k is the new frequency of the linear waves,

$$\omega_k = k\sqrt{k^2 + 2\rho_0}, \qquad (15.50)$$

which is called the Bogoliubov dispersion relation [27].

For strong condensate, $\rho_0 \gg k^2$, this dispersion relation corresponds to sound,

$$\omega_k = c_s k, \quad c_s = \sqrt{2\rho_0}. \qquad (15.51)$$

Because of the non-zero background, the dominant nonlinearity is quadratic with respect to the condensate perturbations. Further, the Bogoliubov frequency (15.50) allows three-wave resonances. Thus, the lowest order of the resonant interaction for WT on background of the condensate is three-wave. It is described by the kinetic equations (6.69) and (6.70) with the following interaction coefficient [17, 26],

$$V_{12}^3 = \frac{\sqrt{\rho_0\omega_1\omega_2\omega_3}}{(2\pi)^{d/2}} \left[\frac{6}{\sqrt{\alpha_1\alpha_2\alpha_3}} + \frac{1}{2}\left(\frac{\mathbf{k_1}\cdot\mathbf{k_2}}{k_1k_2\alpha_3} + \frac{\mathbf{k_2}\cdot\mathbf{k_3}}{k_2k_3\alpha_1} + \frac{\mathbf{k_3}\cdot\mathbf{k_1}}{k_3k_1\alpha_2} \right) \right], \qquad (15.52)$$

where $\alpha_k = 2\rho_0 + k^2$.

Remark 15.7.1 Another approach of deriving a three-wave kinetic equation for WT on background of condensate was used in [28] and [29] by substituting $n_k = \mathcal{N}_0 \delta(\mathbf{k}) + \tilde{n}_k$ into the four-wave kinetic equation (15.2). The resulting three-wave equation for $\tilde{n}_k$ is obviously different: it has the dispersion relation $\omega_k = k^2$ rather than Bogoliubov (15.50) and an interaction coefficient which is different from (15.52). However, such an approach could only work if the original four-wave equation (15.2) remained valid, i.e. if the condensate was incoherent and weakly nonlinear. As we saw above, these conditions are violated when condensation occurs.

For strong condensate, $\rho_0 \gg k^2$, we have:

$$V_{12}^3 \sim \frac{\sqrt{k_1k_2k_3}}{\rho_0^{1/4}}, \qquad (15.53)$$

i.e. again like in the thee-wave acoustic WT. Thus, the energy cascade spectrum is of the Zakharov-Sagdeev type [30], $E_k^{(1D)} \sim \sqrt{\varepsilon c_s}\, k^{-3/2}$, see (3.13).

Paper [26] considered a coupled system of the coherent condensate and the acoustic-type Bogoliubov WT, which led to a power-law prediction for the condensate growth $\rho_0 \sim t^2$. In this work, the WT forcing was assumed to be of an instability type, and the condensate growth rate will be different for forcings of other types (e.g. for an external white-noise forcing). Numerically, the condensate growth was studied in 2D NLS turbulence in [31].

15.7.2 Strongly Nonlinear Transition Between the Two Weakly Nonlinear Regimes

The described above picture of acoustic WT relies on two major assumptions.

1. Condensate is coherent enough so that its spatial variations are slow and it can be treated as uniform when evolution of the perturbations about the condensate is considered. In the other words, a scale separation between the condensate and the perturbations occurs.
2. Coherent condensate is much stronger than the chaotic acoustic disturbances. This allows to treat nonlinearity of the perturbations around the condensate as small.

Both of these assumptions require validation, and this task was undertaken with the aid of numerical simulations of the 2D NLS equation in [13] (for forced WT) and in [14] (for freely-decaying WT). The transition stage that lies in between of the four-wave and the three-wave turbulence regimes is characterized by strong nonlinearity and numerical simulations become an indispensable tool to study this process.

On the other hand, we believe that the main stages of the condensation, i.e. transition from a four-wave process, through vortex annihilations, to three-wave acoustic turbulence, are robust under a wide range of forcing types. Below, we will summarize the findings of the papers [13] and [14] about such a transition. We start with introducing NLS vortices.

15.7.2.1 Vortices

Vortices appear to be at the heart of the condensation process in the NLS model and the four-wave to three-wave transition, and therefore now we will introduce them into our description.

First of all, the vortices are usually associated with fluid flows. Why do they appear in the NLS model? The link between the NLS and the fluid dynamics is provided by Madelung transformation [32, 33],

$$\psi = \sqrt{\rho}\, e^{i\varphi}. \tag{15.54}$$

Then one can interpret ρ and $\mathbf{u} = 2\nabla\varphi$ as a fluid density and a fluid velocity respectively. These quantities satisfy the following equations,

$$\frac{\partial \rho}{\partial t} + \nabla \cdot (\rho \mathbf{u}) = 0, \tag{15.55}$$

$$\frac{\partial \varphi}{\partial t} + (\nabla \varphi)^2 + \rho - \frac{\nabla^2 \sqrt{\rho}}{\sqrt{\rho}} + U = 0. \tag{15.56}$$

Fig. 15.8 Pitaevskii vortex

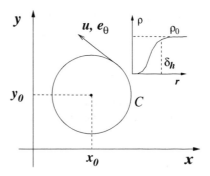

Equation (15.55) corresponds to the fluid mass conservation, and (15.56) is a Bernoulli-type relation.

Exercise 15.6 *Differentiate $\rho = |\psi|^2$ and $\varphi = \Im \ln \psi$ with respect to time and use the defocusing version of (15.1) (i.e. with the minus sign) to obtain (15.55) and (15.56).*

Differentiating equation (15.56), we obtain the fluid momentum equation,

$$\frac{\partial \mathbf{u}}{\partial t} + (\mathbf{u} \cdot \nabla)\mathbf{u} = -\frac{\nabla p}{\rho} - 2\nabla U + 2\nabla \frac{\nabla^2 \sqrt{\rho}}{\sqrt{\rho}}, \tag{15.57}$$

where $p = \rho^2$ is pressure and the last term on the RHS is so-called "quantum pressure term". Apart for the quantum pressure, (15.55) and (15.57) formally coincide with the isentropic compressible Euler equations with adiabatic index $\gamma = 2$.

Pitaevskii vortex [4] is a solution of the defocusing 2D NLS (without U) of the form (15.54) with $\rho \equiv \rho(r)$ and

$$\varphi = \pm\theta - \rho_0 t, \tag{15.58}$$

where

$$r = \sqrt{(x - x_0)^2 + (y - y_0)^2} \tag{15.59}$$

and

$$\theta = -i \ln \frac{(x - x_0) + i(y - y_0)}{r} \tag{15.60}$$

are the distance and the polar angle with respect to the vortex center (x_0, y_0), see Fig. 15.8. The vortex center is defined as a point where $\rho = 0$. At $r \to \infty$, the value of ρ tends that the density of surrounding condensate, $\rho \to \rho_0$.

With θ as in (15.60), the vortex velocity becomes

$$\mathbf{u} = \pm \frac{2\mathbf{e}_\theta}{r}, \tag{15.61}$$

where $\mathbf{e}_\theta$ is the unit vector in azimuthal direction, see Fig. 15.8. Respectively, for the velocity circulation of the vortex we have

$$\Gamma = \oint_C \mathbf{u} \cdot d\mathbf{l} = \pm 4\pi, \qquad (15.62)$$

where the integration was taken along a closed contour C around the vortex in the counterclockwise direction, see Fig. 15.8. Thus we see that the circulation Γ is quantized.

The vortex density profile is described by (15.56) with the phase substituted from (15.58),

$$-\rho_0 + \frac{1}{r^2} + \rho - \frac{1}{r\sqrt{\rho}} \partial_r (r \partial_r \sqrt{\rho}) = 0, \qquad (15.63)$$

with boundary conditions $\rho(0) = 0$ and $\rho(\infty) = \rho_0$. Solution for the (15.63) can be obtained either numerically, or analytically as a series in r the terms in which are obtained recursively [34] (in the latter case the solution is only available up to a finite r corresponding the the series convergence radius). The vortex density profile has a characteristic width $\delta_h \sim 1/\sqrt{\rho_0}$ called the *healing length*, see Fig. 15.8.

When several Pitaevskii vortices are present in the system simultaneously, and they are separated from each other by distances much greater than the healing length δ_h, they move like point vortices in the 2D Euler fluids. This is because the quantum pressure term is negligible at large scales (it is only important at the short distances of order of the healing length). However, the fact that at the short scales the quantum pressure is important results in substantial differences between the NLS and the Euler vortex dynamics. In particular, in NLS a pair of oppositely signed vortices can annihilate or, conversely, get created "from nothing". This is because the quantum pressure term breaks the Kelvin circulation theorem which forbids such creation and annihilation events in Euler.

In 3D NLS, the vortex centers are located on 1D curves in the 3D space, and the creation-annihilation processes take the form of creation or shrinking (to nothing) of vortex loops. In addition, in 3D NLS the vortex lines can reconnect [34, 35].

Finally, let us mention interesting solutions called Jones-Roberts (JR) solitons [36]. These are direct generalizations of the Euler vortex pairs (in 2D) and vortex rings (in 3D). The JR solitons represent a one-parametric family of solutions in which the vortex separation (or the ring diameter) can become of the order or less than the healing length, and the weaker solitons have no vortices at all. In the latter case, the JR solitons become acoustic, and isomorphic to Kadomtsev-Petviashvilli solitons in the weak amplitude limit.

15.7.2.2 Transition Stages

We will now outline the main findings about the route to condensation found by numerical simulations of the defocusing 2D NLS equation in [13, 14]. Several setups were studied in these works, including the ones corresponding to the direct cascade and to dissipated inverse cascade. Here, we will only describe the results for the non-dissipated inverse cascade, which is most appropriate to the condensation study. In this setup, the NLS equation was computed in Fourier space (pseudo-spectral method) with forcing at large k's and dissipation at even larger k's (to prevent UV bottleneck) and no dissipation at low k's to allow condensation [13]. Alternatively, in the decaying turbulence setup, the large-k forcing was replaced with an initial excitation at large k's [13]. In both cases, it was ensured that WT is weak at the initial stage.

In both setups (forced and un-forced), the initial stage is characterized by a spectrum with a front propagating toward low k's, and having scaling $n_k \sim k^{-2}$ behind the front. This is the thermodynamic energy-equipartition scaling, and its observation is natural considering that the pure KZ particle cascade spectrum has "wrong" flux direction (see discussion in Sect. 15.5.2). In the physical space, the corresponding field $\psi(\mathbf{x})$ has a lot of zeroes which can be formally identified as vortices. However, these are not fully nonlinear Pitaevskii vortices because the corresponding field $\psi(\mathbf{x})$ remains uniformly small so that the nonlinearity is small. For this reason, they are called *ghost vortices*.

When the spectral front propagates to sufficiently low k's, the weak nonlinearity assumption breaks down. Numerically, this is detected by measuring the quadratic and the quadric Hamiltonians, and finding the moment when their ratio becomes one. After this moment, the spectrum at lowest k's steepen further with respect to $n_k \sim k^{-2}$. A mode with $k = 0$ emerges and grows in the system, which corresponds to emergence and growth of a condensate component with $\rho = \rho_0$. In the physical space, the corresponding field $\psi(\mathbf{x})$ becomes more large-scale and the number of vortices (zero crossings) is decreasing. This is seen as a pairwise *annihilation* of vortices with opposite circulations, see Fig. 15.9.

Importantly, the moment of time when the weak nonlinearity assumption breaks down corresponds to the time when the vortices become strongly nonlinear. After this time, the vortices continue to annihilate, and the mean inter-vortex distance becomes larger that the healing length δ_h, and most of them can be identified as Pitaevskii vortices (i.e. locally they have the same density and the velocity profiles and in the prototype Pitaevskii vortex). The vortices move in a way similar to the point vortices in the 2D Euler. In particular, when two like-signed vortices are closer to each other than to the rest of the vortices, they rotate around each other, and when two countersigned vortices are close (vortex dipole), they move along a straight line perpendicular to the line connecting the vortices, see Fig. 15.9. One can also observe other motions typical for Euler, e.g. when a vortex dipole scatters off a third vortex (with occasional swapping of the vortex partners).

The vortex annihilation can be viewed as a result of JR soliton loosing its strength. Initially, such a JR soliton comprises a vortex dipole, in which the inter-vortex distance $d(t)$ is decreasing as the soliton weakens. Numerical observations reveal two ways in which this can happen. Firstly, the vortex dipole is *gradually* loosing its momentum by scattering the background acoustic disturbances (phonons). In liquid Helium, this would be similar to the mechanism of the mutual

Fig. 15.9 Vortices in 2D
Gross-Pitaevskii turbulence.
Initially, in small-scale weak
turbulence, there are many
vortices of both signs (*top
panel*). Evolution leads to
annihilation of positive and
negative vortices, so that their
total number is getting less,
and their nonlinearity is
increasing. At later times
close to complete annihilation
(*bottom panel*) there are fast
vortex pairs and relatively
slow isolated vortices

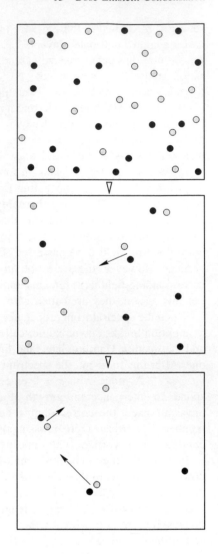

friction between the superfluid and the normal components.[4] For this process, the
following law was found in [14] for the dipole shrinking: $d(t) \propto \sqrt{t_0 - t}$ (with
$t_0 = $ const). Secondly, one can also see *sudden* losses when vortex dipoles collide
with a third vortex producing strong acoustic waves taking the momentum/energy
from the vortices. Even after two vortices in a dipole annihilate, the JR soliton
often continues to exist as an acoustic pulse, with occasional transient vortex re-
appearances within as it moves along. At some moment, however, the JR acoustic

---

[4] The normal component in Helium is representing, on a course-grained level, motion of the gas
of phonons.

soliton looses stability, blows up emitting a shock, and disintegrates into the ambient acoustic background.

The decay of the number of vortices was found numerically in [13] to follow the law

$$N_{\text{vortices}} = A - B \log t \tag{15.64}$$

(with $A = 3.36$ and $B = 0.9223$) which sets in at a certain time and proceeds until time t^* when all the vortices have annihilated, $t^* = \exp A/B$. To date, this dependence does not have a theoretical explanation.

Note that the vortices are phase defects (i.e. the phase is not defined at the vortex centers). Thus, the inter-vortex distance can serve as a measure of the phase coherence. The law of the vortex number decay (15.64) indicates that the condensate phase becomes coherent over the length of the entire computational box in finite time t^*. It is likely that time t^* remains finite in the infinite box limit, although this issue remains to be clarified.

As it was shown in [13], sound plays a crucial role in the vortex annihilations. Indeed, numerical experiments with dissipated sound led to an incomplete condensation with a finite vortex number left in the system at large time. Moreover, strong dissipation of sound at some intermediate time was shown to reverse the annihilation process (i.e. leading to an increase of the number of vortices in time). This means that presence of sound is essential for maintaining stability of the coherent condensate component.

The described above WT regime change, accompanied by vortex annihilations, is very similar to the Kibble-Zurek mechanism of the second-order phase transition [37, 38]. This mechanism, originally suggested in cosmology, implies that at an early inflation stage, Higgs fields experience a symmetry breaking transition from "false" to "true" vacuum, and this transition is accompanied by a reconnection-annihilation of "cosmic strings" which are 3D analogs of the 2D point vortices considered here.

At the final stage, when all the vortices have annihilated, the system comprises a coherent uniform component Ψ_0 and sound-like perturbations of this state. For weak perturbation of the condensate, the evolution takes the form of a three-wave WT process described by the kinetic equations (6.69), (6.70) with the interaction coefficient (15.52).

Checking whether the condensate perturbations at the latest evolution stage are weakly nonlinear and behave like sound with the Bogoiubov dispersion relation (15.50) was done in [13] via finding the space-time Fourier transform of ψ, see Fig. 15.10.

The normal variable for the Bogoiubov sound is given in terms of ψ by expressions (15.47) and (15.48) , and, therefore, when plotting the Bogoiubov dispersion (15.50), we should add a constant frequency of the condensate oscillations, $\omega_0 = \langle |\psi|^2 \rangle$. One can see that the main branch of the space-time spectrum is quite narrowly concentrated around the Bogoiubov curve, thus indicating presence of weakly nonlinear Bogoiubov sound. Note that the lower branch in Fig. 15.10 is related to the a_k^* contribution to expression (15.48) which vanishes at

Fig. 15.10 Squared modulus
of the space-time Fourier
transform of ψ at late evolution stages. Figure courtesy of
Miguel Onorato. Bogoiubov
dispersion relation (15.50) is
shown by the *solid line*

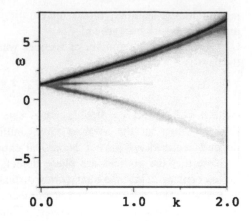

larger k. Importantly, we can also see the middle (horizontal) branch with frequency ω_0 which quickly fades away at finite k's and which corresponds to the
coherent large-scale condensate component.

15.8 Direct Cascade in 3D NLS

Let us now consider turbulence described by the 3D NLS model. Unfortunately, the most relevant to BEC studies of the inverse cascade within the 3D
NLS have not been done done numerically yet,—they are underway. Thus, we
will restrict ourselves to describing the numerical simulations of the 3D NLS
within the direct cascade settings. Starting with pioneering simulations of Nore
et al. [39, 40] and in more recent simulations [41–44] an evidence was found
for a $-5/3$-spectrum which was interpreted as K41. An argument was given in
[39, 40] that the K41 spectrum is natural for 3D NLS turbulence because of the
similarity of the NLS system to the classical Euler fluid observed via Madelung
transformation, see Sect. 15.7.2.1. On the other hand, the situation is far from
obvious because strictly speaking one can only expect similarity with hydrodynamic eddies a the scales greater than the mean inter-vortex separation ℓ
where the NLS vortex discreteness is unimportant. However, the $-5/3$-scaling
in simulations [41–44] was observed at much smaller scales, almost down to
the healing length δ_h, where one expects the energy cascade be carried by
Kelvin waves rather than hydrodynamic-type eddy interactions. Accidentally,
the $-5/3$-spectrum was recently derived for Kelvin WT, see expression (3.24)
and discussion around it. Thus possibly it was this Kelvin wave spectrum that
was observed numerically.

The simulations [39–44] were done in the strong turbulence regime. To test
predictions of the weak WT, particularly the direct cascade spectrum (15.19),
numerical simulations were performed with weak forcing in [45]. To observe

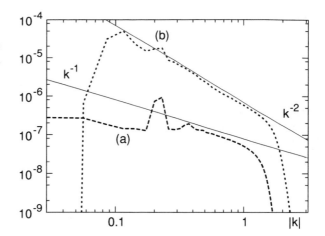

Fig. 15.11 1D waveaction spectra $(4\pi k^2 \langle |\psi_k|^2 \rangle = n_k^{(1D)} = E_k^{(1D)}/\omega_k)$ of 3D NLS turbulence for two types of IR dissipation. Figure courtesy of Davide Proment. **a** IR friction, **b** hypo-viscosity

the direct cascade, in addition to the high-k dissipation there was also dissipation at k's lower than the forcing scales for absorbing the inverse cascade. It turned out that the direct cascade spectrum is sensitive to the type of such low-k dissipation. KZ spectrum (15.19) was indeed observed for the low-k dissipation of friction type (i.e. k-independent), see Fig. 15.11. On the other hand, for steeper dissipation of hypo-viscosity type the direct cascade spectrum obeyed the critical balance (CB) scaling, see Fig. 15.11. We described the CB states and possible reasons for its formation (e.g. wave breaking) in Sect. 3.2. In the NLS simulations, CB is likely to be related with IR bottleneck. Indeed, IR bottleneck arises due to inefficiency of dissipation of the particle cascade by hypo-viscosity. This leads to accumulation of WT near the forcing scale at low k's until its nonlinear timescale will become comparable to the linear one. At this point the dual cascade behavior will be arrested because the energy invariant will not be strictly quadratic (as required by Fjørtoft) since the interaction Hamiltonian will be as big as the free Hamiltonian. However, at larger k's the nonlinearity would still be small at this point and the IR accumulation will continue higher k's until the nonlinear time would reduce to the linear time there too. This process will continue with the IR bottleneck scale moving to higher and higher k's until the CB state forms throughout the entire direct cascade inertial range.

In [45], simulations were also done without IR dissipation, but with some inverse cascade range present. The wave spectrum in these simulations is shown in Fig. 15.12. One can see that KZ forms at early stages while the propagating low-k front has not reached $k = 0$. After this condensation at $k = 0$ occurs, and transition to Zakharov-Sagdeev acoustic spectrum (3.13) is observed when the condensate becomes strong, as it could be expected from Sect. 15.7.1. Note the set of peaks corresponding to the harmonics of the forcing on this spectrum in Fig. 15.12: they are typical for the acoustic-type waves due to deficiency of dispersion.

Fig. 15.12 1D spectrum
$4\pi k^2 \langle |\psi_k|^2 \rangle$ of 3D NLS
turbulence in absence of IR
dissipation **a** early time,
b late time. Note that for
(**a**) this spectrum is still the
1D waveaction spectrum,
whereas for (**b**) it is proportional to the 1D energy spec-
trum, $E_k^{(1D)} = 4\pi\rho k^2 \langle |\psi_k|^2 \rangle$.
Figure courtesy of Davide
Proment

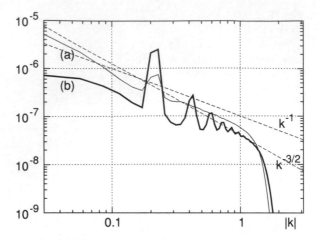

15.9 Inhomogeneous WT in a Trapping Potential

Now let us take into account the trapping potential $U(\mathbf{x})$. Obviously, trapped WT
cannot be homogeneous. Inhomogeneity will result from both the $\mathbf{x}$-dependence of
U and the $\mathbf{x}$-dependence of the condensate Ψ_0. Firstly, such an inhomogeneity will
make the nontrivial dynamics and transfers in the $\mathbf{k}$-space possible even in the
linear approximation. This can be viewed as stretching or squashing of wave
packets due to velocity differences in different wave-packet parts. Secondly, like
in any bounded system, the set of linear eigenmodes will become discrete and the
set of resonantly interacting modes will be depleted with respect to the unbounded
homogeneous WT. Thirdly, the wavenumber change due to the linear dynamics
will make the the wave resonances last only over *finite* time or/and space. The
system of wave packets undergoing a series of finite-time resonant interaction
events invites an analogy with a gas of colliding molecules described by a suitable
Boltzmann-type equation for the collisions of the respective order (binary, three-
particle etc). However, such a description has not been yet developed. Thus, here
we will restrict ourselves to describing the linear dynamics only, following the
approach of papers [16, 46].

Consider weak perturbations $\tilde{\psi}(\mathbf{x}, t)$,

$$\psi = \Psi_0 + \tilde{\psi}, \tag{15.65}$$

where $\tilde{\psi} \ll \Psi_0$. For the perturbation $\tilde{\psi}$ we have a linearized defocusing NLS
equation,

$$i\,\partial_t \tilde{\psi} + \nabla^2 \tilde{\psi} - |\Psi_0|^2 \tilde{\psi} - (\tilde{\psi}\Psi_0^* + \tilde{\psi}^*\Psi_0)\Psi_0 = 0. \tag{15.66}$$

WKB theory deals with linear superposition of small-scale wave packets, whose
distribution depends on the wavenumber and on the physical coordinate. To
describe such wave-packet distributions one can use Gabor transform,

$$\hat{\psi}(\mathbf{x},\mathbf{k},t) = \int f(\sigma|\mathbf{x} - \mathbf{x}_0|)\, e^{i\mathbf{k}\cdot(\mathbf{x}-\mathbf{x}_0)}\, \tilde{\psi}(\mathbf{x}_0,t)\, d\mathbf{x}_0, \qquad (15.67)$$

where f is an arbitrary function rapidly decaying at infinity, e.g. $f(\mathbf{x}) = (2\pi)^{-d}\, e^{-x^2}$. In WKB, perturbations $\tilde{\psi}(\mathbf{x},t)$ are small-scale, in the sense that their characteristic scale l is much less than the characteristic trap size L. The parameter σ must be chosen such that $1/L \ll \sigma \ll 1/l$. Hence, the kernel f varies at the intermediate-scale, and the Gabor transform can be viewed a local (window) Fourier transform, in which the fast scale is represented by the wavenumber argument $\mathbf{k}$ whereas the slow scale is described by the wave-packet positions in the physical space $\mathbf{x}$.

The linear WKB theory, which is valid with or without the presence of a condensate, was derived in papers [16, 46], and interested readers are referred these papers for details. In particular, in paper [46] one can find a re-formulation of the Hamiltonian description for the WKB setup, and respective generalization of the Bogoliubov transformation (diagonalizing the quadratic Hamiltonian) to the inhomogeneous WT systems. Here, we will only present the results for the generalized wave action,

$$n(\mathbf{k},x,t) = \frac{1}{2}\frac{\omega\rho}{k^2}\left|\widehat{\Re\phi} - \frac{ik^2}{\omega}\widehat{\Im\phi}\right|^2, \qquad (15.68)$$

which corresponds to the wave-packet density in the $(\mathbf{k}, \mathbf{x})$ phase space. The evolution equation for $n(\mathbf{k}, x, t)$, called the WKB transport equation, takes the form of conservation along trajectories (wave rays) in the $(\mathbf{k}, \mathbf{x})$ space:

$$D_t n(\mathbf{x},\mathbf{k},t) = 0, \qquad (15.69)$$

where

$$D_t \equiv \partial_t + \dot{\mathbf{x}}\cdot\nabla + \dot{\mathbf{k}}\cdot\partial_k \qquad (15.70)$$

is the full time derivative along the trajectories and

$$\dot{\mathbf{x}} = \partial_k\omega, \quad \dot{\mathbf{k}} = -\nabla\omega \qquad (15.71)$$

are the ray equations with

$$\omega = k\sqrt{k^2 + 2\rho} + U + \rho. \qquad (15.72)$$

Equations (15.69), (15.70), (15.71) and (15.72) are valid with or without condensate present. In the case of strong condensate, such that $\rho_0 \gg 1/L^2$, the linear term is much less than the nonlinear term in the NLS equation (15.1): this is so-called *Thomas-Fermi regime*. In this case the condensate follows the shape of the well, $U + \rho = $ const, see Fig. 15.13.

Fig. 15.13 High-frequency
excitations on background of
a strong condensate
component

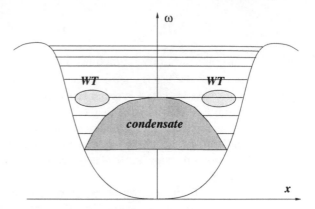

According to the ray equations, for time independent U, we have $\omega = \text{const}$ along the rays. Thus we can solve (15.72) for the wavenumber,

$$k^2 = \sqrt{\rho^2 + (\omega - U - \rho)^2} - \rho,$$

from which if follows that k^2 remains positive for any value of ρ, i.e. that presence of the condensate does not lead to any new wave-packet reflection points (where $k = 0$). Therefore, wave packets can penetrate into the center of the potential well. However, the group velocity increases when the condensate becomes stronger, $\partial_k \omega \sim \sqrt{\rho}$ (for $\rho \gg k^2$). This means that the density of wave packets decreases towards the center of well. Therefore, the condensate tends to push the WT away from the center, towards the edges of the potential trap, see Fig. 15.13.

Remark 15.9.1 Effect of the condensate on the high-wavenumber excitations is sometimes considered in literature as a modification of the shape of the trapping potential [47, 48]. Within the NLS model, this is partially true, and is reflected by the second and the third terms in the expression (15.72). However, there is also Bogoliubov-type contribution in the first (square-root) term of (15.72), which cannot be described as an effective potential.

Now let us discuss the nonlinear effects. One has to distinguish two cases where the nonlinear resonance broadening Γ is respectively greater or less than the frequency spacings between the adjacent eigenmodes in the trap. One can think of these cases as an opaque and a transparent wave-packet gas correspondingly, i.e. when the wave-packet mean-free-paths are smaller than the trap or of the same size as the trap. The first case is simpler. In this case, the linear and nonlinear effects can be treated independently and the resulting kinetic equation will be the one as in the homogeneous WT (four-wave or three-wave, depending on the condensate strength) with the left hand side (LHS) replaced by the waveaction derivative along the wave rays, as in LHS of (15.69). The second case is more complicated and mostly unstudied. As was mentioned in the beginning of this section, the wave packets in this case experience short-lasted resonant "collisions", and developing a kinetic theory to this process would be an interesting subject for future research.

15.10 Condensation in 1D Systems: Optical Turbulence

Finally, let us discuss the 1D systems. Because the 1D NLS is integrable, there will be no energy transfers unless extra effects are introduced that break the integrability. One way to do this is to take into account the corrections to the NLS nonlinearity, as it was done in (6.87) arising in the nonlinear optics context (see discussion below (6.87)). A detailed study of WT arising in such 1D optical turbulence, including experimental and numerical investigations, was done in [49, 50], and interested reader should read these papers. Here, we will only present highlights of these studies.

First of all, the pure KZ solutions appear to be un-realizable in this system, as we learn from the following exercise.

Exercise 15.7 *Find the E-cascade and the $\mathcal{N}$-cascade spectra for the 1D optical model* (6.87) *using the dimensional argument applied to the kinetic equation* (6.89) *with interaction coefficient* (6.88). *Find the flux directions for the energy and the particles and show that they disagree with Fjørtoft.*

Such a situation have already been seen in this chapter for the Boltzmann gas. Thus, we expect behavior similar to the Boltzmann system, namely the spectra in the forced-dissipated systems will be close to the thermodynamic RJ spectra in most of the inertial ranges except near the dissipation scales where they'd experience sharp fall-off. Also like in Boltzmann, the temperature and the chemical potentials of the resulting RJ states will be determined by the forcing rate and the positions of the forcing and the dissipation scales.

On the other hand, the KZ solutions are still useful because they allow to construct a DAM model for the 1D optical turbulence, as formulated in the following exercise.

Exercise 15.8 *Use the KZ solutions found in the exercise* 15.7 *to construct DAM for this system.*

We saw before that DAM's are useful for predicting the steady state spectra. They are also useful for predicting unsteady spectra and for studying the ways the steady spectra form. In the Boltzmann example we saw that within DAM the temperature T in the emerging steady state is independent of the forcing rate and is determined by the position of the forcing and the dissipation scales whereas the chemical potential μ depends on both the forcing/dissipation scales and on the forcing rate, see expressions (15.43) and (15.44). Further, according to these expressions the particle equipartition part of the RJ spectrum is situated to the left of the forcing frequency, whereas the energy equipartition part is on the right of this frequency, see Fig. 15.6. Similar features are expected for the 1D optical

Fig. 15.14 Squared modulus
of the space-time Fourier
transform of ψ at late evolution stages. Bogoiubov dispersion relation (15.50) is
shown by the *solid line*

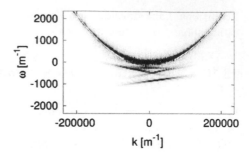

turbulence. In particular, since here we are mostly concerned with the condensation process, we should concentrate on the inverse cascade range, for which we
anticipate a "warm cascade" spectrum close to the particle equipartition,
$n_k = $ const. Such a spectrum was indeed observed in numerical simulations performed in the inverse cascade setting in [49, 50].

However, since values of T and μ become large for wide inertial ranges, we
expect breakdown of the weak WT description for the 1D optical turbulence if the
inertial ranges are wide, even if forcing is weak. When this happens, strong
coherent structures—solitons—arise in the system, which is clearly observed in
numerical simulations and in experiments [49, 50]. In fact, the spectrum appears to
be a rather poor tool for detecting solitons and distinguishing them from the
random waves because they appear to yield the same spectrum as the particle-
equipartition RJ spectrum, $n_k = $ const. Separation of the weak random wave
component and strong coherent solitons can be effectively done using simultaneous space and time Fourier transforms, leading to (k, ω)-plots which we already
used before for separating the waves and the condensate, see Fig. 15.10.
Figure 15.14 shows a similar plot for the 1D optical turbulence. Here we can see a
large upper branch following the linear-wave dispersion relation (possibly with
small Bogoliubov corrections) and a set of straight lines with different slopes on
the bottom of the graph. These straight lines are signatures of the solitons with the
slopes corresponding to the soliton speeds.

The solitons are also seen directly in the physical space, see Fig. 15.15. We can
see here that the solitons can pass through each other, but not freely as in the
integrable system[5]: they can exchange energy, loose energy by emitting random
waves and gain energy by absorbing waves. Typically, the weaker solitons loose
energy whereas the strong ones absorb it. Occasionally, the solitons merge upon
collision, and their total number in the system on average gets smaller. If there is
no dissipation at the largest scales and if the system is allowed to evolve long
enough, a single strong soliton immersed in a sea of random waves will remain.

The described above processes present yet another example of the WT life
cycle in which random waves and coherent structures coexist, interact and get

---

[5] In 1D NLS the solitons pass each other almost freely,—without change of their amplitudes or
shapes but with small shifts of the original trajectories.

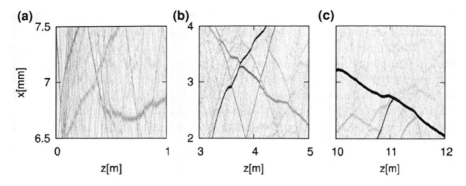

Fig. 15.15 Squared modulus of ψ as a function of z (*horizontal*) and x (*vertical*)

Fig. 15.16 The forcing is at k_f and the main cascade follows the dotted line. WT life cycle on the wavenumber-amplitude plane in the inverse cascade system. The *dashed line* corresponds to solitons

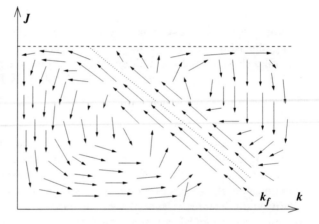

transformed into each other. From the example of the water gravity waves we remember that such processes of the WT cycle can be effectively described using the wavenumber-amplitude space, see Fig. 11.6 and discussion around it. Similarly, one can sketch (k, J)-fluxes for the optical WT, by adopting the approach of Fig. 11.6 to the inverse cascade system, see Fig. 15.16. On this sketch we show that weak WT is generated at some high wavenumber k_f and an inverse cascade starts with a flux directed toward low k's. At the same time, the same flux has a component in along the intensity axis J in such a way that WT seises to be weak at a sufficiently small k marked in Fig. 15.16 as an intersection of the dotted line (representing the mean direction of the inverse cascade) and the horizontal dashed line (representing the spectrum of solitons). After that the flux will spread along the soliton (dotted) line, possibly in a nonlocal way, because all k-modes in a soliton are correlated. Occasionally, the solitons will return their energy to the weak incoherent waves (due to soliton collisions etc) which is indicated on Fig. 15.16 as reversal of the flux in the (k, J)-plane toward low J's.

Note that the general features of the WT cycle described above should be expected for many inverse cascade systems, although the details may vary. In fact, the first description of a WT cycle of this kind was put forward in the context of the 3D focusing NLS turbulence in [17]. In this case, the coherent structures arising in the WT cycle from the inverse cascade are wave collapses. The collapses are coherent non-stationary structures which become singular in a finite time [1, 2]. Physically, the time of singularity corresponds to a time when such a coherent structure breaks down due to a physical process unaccounted by the NLS model. At this time, the waveaction partially "burns out" (gets converted into an internal energy) and partially gets transferred back to the high-frequency weak incoherent waves.

15.11 Summary

In this chapter, we have considered Bose-Einstein condensation (BEC) process using the NLS model. We have also extended our description to include the quantum statistics and thermal clouds by invoking the Nordheim kinetic equation and a corresponding simplified model—DAM. This formulation allowed us to study the wave-particle crossover in the energy cascade, when energy is pumped into low-momentum wave-like condensate and it cascades to high-momentum thermal particle motions. Conversely, the system can be pumped into Boltzmann particles and the inverse cascade will excite the wave-like condensate.

We have studied the role of pure KZ solutions in this dual cascade system. We showed that the pure KZ states are not always realizable, in which case a thermal component of the spectrum should be present, with characteristic "bottleneck" and sharp cutoff features. Using DAM, we derived relationships between the temperature, the chemical potential, the forcing rate and the forcing and dissipation scales. In the Boltzmann limit, we saw an interesting feature that the temperature is independent of the forcing rate and is determined by the positions of the forcing and dissipation scales only.

We saw that the inverse cascade evolution leads to breakdown of the weak WT theory based on the four-wave kinetic equation. This leads to BEC and non-equilibrium transition to another weak WT regime: three-wave interaction of Bogoliubov sound on background of a strong uniform condensate. The transition between the two weak WT regimes, four-wave and three-wave, is strongly turbulent. In 2D, it is realized as a motion of a "gas" of strongly nonlinear Pitaevskii vortices which undergo pairwise annihilations. Respectively, in 3D one observes shrinking and disappearing vortex rings.

Finally, we considered 1D optical turbulence based on the focusing 1D NLS with a correction in the nonlinear term breaking the integrability. Here again, we saw an important role played by the warm cascades. We saw that in this system the condensation leads to formation of a gas of strongly nonlinear solitons which become stronger as their number decreases and so that the only strong soliton remains at large time. Using the 1D optical turbulence example, we discussed realization of a WT life cycle corresponding to the inverse cascade settings.

Inevitably, many important details had to be skipped in our description of this exciting subject, and interested readers are advised to read the original papers cited in the bibliography.

15.12 Solutions to Exercises

15.12.1 Direct Cascade in 2D NLS: Exercise 15.2

Consider DAM (15.22) with R given (15.23), and seek solution in form $n = \frac{c}{\omega} \ln^z\left(\frac{\omega}{\omega_*}\right)$. We have:

$$\partial_\omega \frac{1}{n} = \frac{1}{c} \partial_\omega \left[\omega \ln^{-z}\left(\frac{\omega}{\omega_*}\right)\right] = \frac{1}{c} \ln^{-z}\left(\frac{\omega}{\omega_*}\right) - \frac{z}{c} \ln^{-z-1}\left(\frac{\omega}{\omega_*}\right). \tag{15.73}$$

Here, the second term is subdominant for large $\ln[\omega/\omega_*]$ (i.e. for both $\omega \gg \omega_*$ and $\omega \ll \omega_*$) and, therefore, will be neglected. Thus, we have:

$$R = S \omega^5 n^4 \partial_{\omega\omega} \frac{1}{n} = c^3 S \omega \ln^{4z}\left(\frac{\omega}{\omega_*}\right) \frac{(-z) \ln^{-z-1}\left(\frac{\omega}{\omega_*}\right)}{\omega} = c^3 S \ln^{3z-1}\left(\frac{\omega}{\omega_*}\right).$$

In the pure direct cascade, $\varepsilon = \text{const}$, $\zeta = 0$, we have $\varepsilon = R$. Thus, for ε to be constant, we must have $\boxed{z = 1/3}$. Also we see that $\varepsilon > 0$ for $c < 0$ and $\varepsilon < 0$ for $c > 0$.

What is the meaning of ω_*? Since our spectrum is zero at $\omega = \omega_*$, it makes sense to think that ω_* represents the dissipation rather than the forcing scale. Thus, in the direct cascade range $\omega < \omega_*$, so it is best to rewrite our spectrum as

$$n = -\frac{c}{\omega} \ln^{1/3}\left(\frac{\omega_*}{\omega}\right).$$

But the spectrum must always be positive, so we conclude that $c < 0$ and, therefore, the $\varepsilon > 0$, as it should be. Changing $-c \to C$, we get the answer (15.31).

15.12.2 Front Solution for Inverse Cascade in 2D NLS: Exercise 15.3

Let us consider DAM (15.22) with R given by (15.23), and let us suppose WT is forced at ω_f and dissipated at $\omega_d \ll \omega_f$, so that $n(\omega_d) = 0$. From Fjørtoft theorem,

Fig. 15.17 The energy and
the particle fluxes as a function of the spectral index for
the Boltzmann kinetics

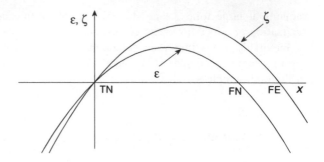

we know that in case $\omega_d \ll \omega_f$ the particle cascade dominates between ω_d and ω_f, so we can take $\zeta = $ const and $\varepsilon = 0$. In this case $R = -\zeta\omega$, so that in the 2D case from (15.23) we have:

$$\zeta = -R/\omega = S\,\omega^4\,n^4\,\frac{\partial^2}{\partial\omega^2}\frac{1}{n}. \tag{15.74}$$

Now substitute $n = A\,(\omega - \omega_d)^\alpha$ with $A = $ const, and consider the limit $\omega \to \omega_d$, thereby writing $\omega = \omega_d$ everywhere except where ω comes via the combination $(\omega - \omega_d)$. Balancing powers of $(\omega - \omega_d)$ on both sides of the equation, we immediately get $\alpha = 2/3$ and for A we get (15.35).

15.12.3 KZ Solutions and Flux Directions for Boltzmann Gas: Exercise 15.4

For Boltzmann gas, we will use DAM (15.22) with R given by (15.40). As usual, the pure energy cascade will be given by $\varepsilon = R = $ const ($\zeta = 0$) and the pure particle cascade—by $\zeta = R/\omega = $ const ($\varepsilon = 0$). Substituting $n = c\,\omega^{-x}$ we get

$$\zeta = c^2 S x (3d/2 - 2x)\,\omega^{3d/2 - 2x - 1}, \tag{15.75}$$

$$\varepsilon = c^2 S x (1 + 3d/2 - 2x)\,\omega^{3d/2 - 2x}. \tag{15.76}$$

From these expressions we recover the three power-law solutions, one thermodynamic[6] with $x = x_{TN} = 0$, the ζ-flux KZ with $x = x_{FN} = 3d/4 - 1/2$ and the ε-flux KZ with $x = x_{FE} = 3d/4$. Plot the fluxes ζ and ε as functions of x for fixed ω is shown in Fig. 15.17 (cf. Figs. 15.2 and 15.3). We can see on this figure that,

---

[6] Index $x = 0$ corresponds to a formal solution with particle equipartition in the momentum space. However, this solution is not normalizable.

because $x_{TN} < x_{FN} < x_{FE}$, we have $\varepsilon < 0$ and $\zeta > 0$ on the respective KZ solutions, in disagreement with the Fjørtoft argument.

15.12.4 Front Solutions for Boltzmann: Exercise 15.5

Substituting $n = A\,(\omega_{max} - \omega)$ into $\varepsilon = R = $ const, R given by (15.40), we have

$$\varepsilon = SA^2\omega^{2+3d/2} \to SA^2\omega_{max}^{2+3d/2} \quad \text{for} \quad \omega \to \omega_{max}. \tag{15.77}$$

Solving for A, we get (15.42).

Similarly, one can obtain (15.41) by substituting $n = B\,(\omega - \omega_{min})$ into $\zeta = R/\omega = $ const and finding the constant B.

15.12.5 Madelung Transformation: Exercise 15.6

Multiply Eq. (15.1) (with the minus sign) by ψ^* and subtract from it its complex conjugate. This gives

$$i\rho_t + (\psi^*\nabla^2\psi - c.c.) = 0,$$

or

$$\rho_t + 2\sqrt{\rho}\left[\nabla\varphi \cdot \sqrt{\rho} + \nabla(\sqrt{\rho}\varphi)\right] = 0,$$

or, taking into account $\mathbf{u} = 2\nabla\varphi$, we obtain (15.55).

Now divide (15.1) (with the minus sign) by ψ and take the real part of the result. This gives

$$-\varphi_t + \frac{1}{\sqrt{\rho}}\left[\nabla^2\sqrt{\rho} - \sqrt{\rho}(\nabla\varphi)^2\right] - \rho - U = 0,$$

which immediately gives (15.56).

15.12.6 KZ Spectra for 1D Optical Turbulence: Exercise 15.7

From equations defining the particle and the energy fluxes,

$$\partial_t n_k + \partial_k \zeta_k = 0$$

and

$$\partial_t(\omega_k n_k) + \partial_k \varepsilon_k = 0,$$

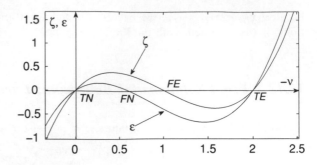

Fig. 15.18 The energy and the waveaction fluxes as a function of the spectral index in 1D optical WT

we have

$$\varepsilon \sim k\omega_k\, St \quad \text{and} \quad \zeta \sim k\, St, \tag{15.78}$$

where St is the six-wave collision integral on the RHS of the kinetic equation (6.89). Taking into account that the interaction coefficient (6.88) is of zero-degree homogeneous in k, we have for St:

$$St \sim k^4 n_k^5 / \omega_k. \tag{15.79}$$

Thus,

$$\varepsilon \sim k^5 n_k^5 \quad \text{and} \quad \zeta \sim k^5 n_k^5 / \omega_k = k^3 n_k^5. \tag{15.80}$$

From these expressions we see that the constant $\mathcal{N}$-flux and the constant E-flux KZ solutions look as follows,

$$n_k^{(FN)} \sim \zeta^{1/5} k^{-3/5} \quad \text{and} \quad n_k^{(FE)} \sim \varepsilon^{1/5} k^{-1}. \tag{15.81}$$

The flux directions can be found as in Sect. 9.2.2.1 based on the relative positions of the indices of the RJ and KZ spectra,

$$-v_{TN} = 0 < -v_{FN} = 3/5 < -v_{FE} = 1 < -v_{TE} = 2. \tag{15.82}$$

Namely, using the argument of Sect. 9.2.2.1 we can sketch the fluxes ζ and ε on power-law spectra $n_k \sim k^v$ as functions of $-v$, see Fig. 15.18 (cf. Fig. 9.3). As we see in this figure, the flux directions for both the energy and the particles on the pure KZ solutions (*FE* and *FN* in Fig. 15.18 respectively) disagree with the Fjørtoft prescription. Thus, they cannot be matched to any physical forcing and dissipation at the ends of the inertial ranges and, therefore, are not realizable. Instead, a "warm cascade" solution with mixed flux and temperature components should form.

15.12.7 DAM for 1D Optical Turbulence: Exercise 15.8

For constructing DAM we note that the general form for such a model for six-wave
dual-cascade systems is

$$\dot{n}_k = \partial_{\omega\omega}[\omega^a n_k^6 \partial_{\omega\omega}(1/n_k)]. \tag{15.83}$$

To find constant a, it suffices to postulate that one of the pure KZ spectra is a
solution, e.g. the energy cascade with $n_k \sim k^{-1} = \omega^{-1/2}$. This gives $a = 9/2$. Easy
to check that the other KZ, $n_k \sim k^{-3/5} = \omega^{-3/10}$, is automatically a solution for
this choice of a.

References

1. Bergé, L.: Wave collapse in physics: principles and applications to light and plasma waves. Phys. Rep. **303**, 259 (1998)
2. Sulem, C., Sulem, P.: The Nonlinear Schrödinger Equation. Springer, Berlin (1999)
3. Gross, E.P.: Nuovo Cimento **20**, 454 (1961)
4. Pitaevskii, L.P.: Zh. Eksp. Teor. Fiz. **40**, 646 (1961) [Sov. Phys. JETP **13**, 451 (1961)]
5. Zakharov, V.E., Musher, S.L., Rubenchik, A.M.: Hamiltonian approach to the description of nonlinear plasma phenomena. Phys. Rep. **129**, 285 (1985)
6. Connaughton, C., Josserand, C., Picozzi, A., Pomeau, Y., Rica, S.: Condensation of classical nonlinear waves. Phys. Rev. Lett. **95**, 263901 (2005)
7. Berloff, N.G., Svistunov, B.V.: Scenario of strongly non-equilibrated Bose-Einstein condensation. Phys. Rev. A **66**, 013603 (2002)
8. Nordheim, L.W.: On the kinetic method in the new statistics and its application in the electron theory of conductivity. Proc. R. Soc. London Ser. A **119**, 689 (1928)
9. Bose-Einstein condensate, Wikipedia
10. Pitaevskii, L.P., Stringari, S.: Bose-Einstein Condensation, Oxford University Press, International Series of Monographs on Physics No. 116 (2004)
11. Hohenberg, P.C.: Phys. Rev. **158**, 383 (1967)
12. Mullin, W.J.: J. Low Temp. Phys. **106**, 615 (1997)
13. Nazarenko, S., Onorato, M.: Wave turbulence and vortices in Bose-Einstein condensation. Physica D **219**, 1–12 (2006)
14. Nazarenko, S., Onorato, M.: Freely decaying turbulence and Bose-Einstein condensation in Gross-Pitaevski model. J. Low Temp. Phys. **146**(1/2), 31–46 (2007)
15. Proment, D., Nazarenko, S., Onorato, M.: Bose-Einstein condensation in 2D NLS model (in preparation)
16. Lvov, Y., Nazarenko, S.V., West, R.J.: Wave turbulence in Bose-Einstein condensates. Physica D **184**(1–4), 333–351 (2003)
17. Dyachenko, A., Newell, A.C., Pushkarev, A., Zakharov, V.E.: Optical turbulence: weak turbulence, condensates and collapsing fragments in the nonlinear Schrodinger equation. Physica D **57**(1–2), 96 (1992)
18. Kraichnan, R.: Phys. Fluids **10**, 1417 (1967)
19. Nazarenko, S., Onorato M., Proment, D.: Wave turbulence and thermal effects in non equilibrium Bose-Einstein condensation, (in preparation)
20. Onorato, M., Proment D., Nazarenko, S.: Nonequilibrium statistics in a forced-dissipated Boltzmann equation (in preparation)
21. L'vov, V.S., Nazarenko, S.V., Rudenko, O.: Phys. Rev. B **76**, 024520 (2007)

22. Kagan, Y., Svistunov, B.V., Shlyapnikov, G.V.: JETP **75**, 387 (1992)
23. Semikoz, D.V., Tkachev, I.I.: Phys. Rev. Lett. **74**, 3093 (1995)
24. Semikoz, D.V., Tkachev, I.I.: Phys. Rev. D **55**, 489 (1997)
25. Lacaze, R., Lallemand, P., Pomeau, Y., Rica, S.: Physica D **152–153**, 779 (2001)
26. Zakharov, V.E., Nazarenko, S.V.: Dynamics of the Bose-Einstein condensation. Physica D **201**, 203–211 (2005)
27. Bogoliubov, N.: J. Phys. USSR **11**, 23 (1947)
28. Lacaze, R., et al.: Physica D **152**, 779 (2001)
29. Connaughton, C., Pomeau, Y.: C. R. Physique **5**, 91 (2004)
30. Zakharov, V.E., Sagdeev, R.Z.: Spectrum of acoustic turbulence. Soviet Physics: Doklady **15**, 439 (1970)
31. Dyachenko, A., Falkovich, G.: Condensate turbulence in two dimensions. Phys. Rev. E **54**, 5095 (1996)
32. Madelung, E.: Die Mathematischen Hilfsmittel des Physikers. Springer, Berlin (1957)
33. Spiegel, A.: Fluid dynamical form of the linear and nonlinear Schrödinger equations. Physica D **1**, 236 (1980)
34. Nazarenko, S.V., West, R.J.: Analytical solution for NLS vortex reconnection. J. Low Temp. Phys. **132**(1–2), 1–10 (2003)
35. Koplik, J., Levine, H.: Phys. Rev. Lett. **71**(9), 1375 (1993)
36. Jones, C.A., Roberts, P.H.: Motions in a Bose condensate. IV. Axisymmetric solitary waves. J. Phys. A **15**, 2599 (1982)
37. Kibble, T.W.B.: Topology of cosmic domains and strings. J. Phys. A: Math. Gen. **9**, 1387 (1976)
38. Zurek, W.H.: Cosmological experiments in superfluid helium? Nature **317** 505, (1985); Acta Physica Polonica **B24**, 1301 (1993)
39. Nore, C., Abid, M., Brachet, M.E.: Kolmogorov turbulence in low-temperature superflows. Phys. Rev. Lett. **78**, 3896 (1997)
40. Nore, C., Abid, M., Brachet, M.E.: Decaying Kolmogorov turbulence in a model of superflow. Phys. Fluids **9**, 2644 (1997)
41. Kobayashi, M., Tsubota, M.: Kolmogorov spectrum of superfluid turbulence: numerical analysis of the Gross-Pitaevskii Equation with a small-scale dissipation. Phys. Rev. Lett. **94**, 065302 (2005)
42. Kobayashi, M., Tsubota, M.: Kolmogorov spectrum of quantum turbulence. J. Phys. Soc. Jpn. **74**, 3248–3258 (2005)
43. Kobayashi, M., Tsubota, M.: Quantum turbulence in a trapped Bose-Einstein condensate. Phys. Rev. A **76**, 045603 (2007)
44. Yepez, J., Vahala, G., Vahala, L., Soe, M.: Superfluid turbulence from quantum kelvin wave to classical kolmogorov cascades. Phys. Rev. Lett. **103**, 084501 (2009)
45. Proment, D., Nazarenko, S., Onorato, M.: Quantum turbulence cascades in the Gross-Pitaevskii model. Phys. Rev. A **80**, 051603 (R) (2009)
46. Gershgorin, B., Lvov, Y.V., Nazarenko, S.: Canonical Hamiltonians for waves in inhomogeneous media. J. Math. Phys. **50**(1), 013527 (2009)
47. Gardiner, C.W., et al.: Phys. Rev. Lett. **81**, 5266 (1998)
48. Gardiner, C.W., et al.: Phys. Rev. Lett. **79**, 1793, (1998)
49. Bortolozzo, U., Laurie, J., Nazarenko, S., Residori, S.: Optical wave turbulence. J. Opt. Soc. Am. B **26**(12), 2280–2284 (2009)
50. Bortolozzo, U., Laurie, J., Nazarenko, S., Residori, S.: Optical wave turbulence, to appear as a book chapter In: Localized States in Physics: Solitons and Patterns. Springer, Berlin (2010)

Chapter 16
List of Projects

In this chapter we will mention several further themes which could be be used for research projects or essays.

16.1 Differential Approximation Models for WT and for Strong Turbulence

Differential Approximation model (DAM) refer to a nonlinear differential equation for the turbulence spectrum evolution in time and wavenumber (or frequency) which preserve the symmetries, the conservation laws and the scaling solutions of a higher-level turbulence closure (e.g. kinetic equations in case of WT). DAM's are close relatives to the *shell models* of turbulence [1, 2]. Indeed, discretizing a DAM would lead to a valid shell model, and taking the continuous limit of a shell model could lead to a DAM. (Obtaining a shell model by discretizing a DAM and simulating it numerically could make a good research project).

We have already discussed DAM's for the NLS, Nordheim and Boltzmann systems in Chap. 15. Thus, the approaches and methods used for analyzing DAM in these examples should be adopted for the examples discussed in this section.

The very first "mother of DAM's" appeared in Lewis F. Richarson's 1926 paper [3]. In this paper he used experimental data to deduce a diffusion equation for the PDF of the particle separations ℓ in turbulence with his famous eddy diffusivity $\sim \ell^{4/3}$. In fact, this was the first closure satisfying the Kolmogorov scaling 15 years before Kolmogorov presented his theory. Solving the Richardson's PDF equation for a δ-distribution initial condition and taking the second moment, one can derive for the r.m.s. of ℓ dependence $\sim t^{3/2}$ which is known as Richardson's law (check this!). You are also recommended to consider a forced/dissipated system where small clusters of particles (with small ℓ's) are injected into turbulence and taken out when they reach large ℓ's (e.g. due to burning or another chemical reaction). By

S. Nazarenko, *Wave Turbulence*, Lecture Notes in Physics, 825,
DOI: 10.1007/978-3-642-15942-8_16, © Springer-Verlag Berlin Heidelberg 2011

considering steady state solutions of the Richardson's PDF equation with a constant ℓ-flux, you can get PDF's which are power-law in ℓ.

Second time DAM was spotted in Iroshnikov 1963 paper [4], where it was used to obtain famous Iroshnikov-Kraichnan (IK) spectrum of MHD turbulence, see Eq. (3.14). Iroshnikov's DAM is actually of WT type,—its scaling corresponds to three-wave interactions. It has a form of a nonlinear diffusion equation which has two independent power-law solutions: the energy equipartition RJ spectrum and the energy cascade IK spectrum, which is of the KZ-type.

DAM for 3D Navier–Stokes turbulence is known at Leith 1967 model [9, 10]. This model was used to formulate the concept of "warm cascades", i.e. turbulent states which involve both cascading and thermalized components with characteristic "bottlenecks" near the end of the inertial range [11]. Such warm cascades were found in [11] as exact analytical solutions of the Leith model. This paper also analyzed non-stationary solutions corresponding to formation of the steady K41 spectrum. Such solutions are interesting because they exhibit an anomalous scaling on the front of the propagating solution which is steeper than K41. One of the themes of a research project could be study of non-stationary evolution within DAM for a particular application using the analytical and numerical approaches of paper [11].

Another major example of DAM can be found in the theory of the water surface gravity waves. It was introduced in 1985 independently by Hasselmann and Hasselmann [5] and by Iroshnikov [6] and further studied, e.g., in [7, 8]. In its full form, this DAM is a fourth-order equation with respect to the frequency, because this four-wave system has two conserved quantities, the energy and the waveaction. Thus, it should have four fundamental power-law solutions: two RJ's and two KZ's. However, one can also use reduced versions of DAM based on lower-order equations, like in [6], if one sacrifices some solutions (e.g. RJ), or one of the conservation laws (e.g. one can sacrifice the waveaction conservation while studying the direct energy cascade).

For 2D Navier–Stokes turbulence, DAM was developed in [12]. Since this is also a dual cascade system, this DAM is of the fourth order too. In fact, it is very similar to Hasselmanns-Iroshnikov DAM, and in particular it has a full RJ solution rather than only its two power-law limits. Remarkably, DAM can correctly predict the ratio of the constants of the forward and the inverse cascades [12]. This prediction is lost if one considers a reduced second-order DAM which ignores the RJ solution. However, such reduced DAM is still useful: it was used by Lily for describing 2D turbulence in a range between two forcing scales in order to explain the Nastrom-Gage spectrum of atmospheric turbulence [13].

For quantum turbulence, various DAM's were developed in [14–17]. DAM allowed us to find new scaling regimes for the systems including both the normal and superfluid components with mutual friction [14, 17], e.g. the k^{-3} energy spectrum in 3D superfluid experiencing friction with a still normal fluid (realizable in Helium-3). For smaller scales in superfluid turbulence, where the cascade is carried by interacting Kelvin waves on quantized vortex lines, DAM was developed in [15, 16] including forcing by vortex reconnections and dissipation by radiating phonons. A particularly interesting effect arises in the crossover range between the large eddy-dominated scales and the small Kelvin-wave scales. Using DAM allowed to predict a bottleneck in this crossover range and to describe it in terms of the warm cascade solutions [18, 19]. See further description of superfluid turbulence in Sect. 16.5.

In summary, a research project could involve studying the cited papers on DAM, developing and using them further or/and applying them to new turbulent systems.

16.2 Collapses and Their Role in WT Cycle

In Chap. 15, we saw that in the defocusing NLS model, the inverse cascade leads to generation of a stable coherent condensate which changes WT from a four-wave to a three-wave system. In the end of Chap. 15, we also briefly mentioned how the situation is changing when the NLS nonlinearity is focusing. Namely, there will be coherent collapsing events which effectively eliminate condensate from the large scales, partially "burn" it and partially return its waveaction to the smaller scales. This is a characteristic picture of the WT life cycle in this case, with random waves and coherent structures coexisting and getting transformed into each other [20]. As it was argued in [20], such elimination of the condensate by collapses leads to better attainability of the KZ spectra of the weak WT theory.

A research project theme here could consist of computing the focusing NLS turbulence (in 2D or 3D) in both the direct and the inverse cascade settings and quantifying the relative roles of the random waves and the coherent structures in the WT cycle. Some of such numerical modeling was done in [20] for 2D NLS, and now it is time to revise these studies using the numerical power which has significantly increased over the eighteen years since the publication of [20]. It is particularly timely to compute the 3D case too.

16.3 Modulational Instability and Its Role in WT

We mentioned in Sect. 16.2 that the inverse waveaction cascade in focusing NLS leads to strongly nonlinear coherent collapses. They appear due to breakdown of weak WT at large scales. It is believed that the first stage of such breakdown could be associated with *modulational instability* (MI). The classical setting for MI is to consider a high-frequency monochromatic *carrier wave* whose amplitude is shaped by a long-wave sinusoidal *modulation*. Taking into account that in the optics context the NLS model is already written for modulations of a monochromatic wave, one can simply consider instability of a state which is homogeneous in the physical space, i.e. of a uniform condensate. Easy to see that this will lead to a dispersion relation similar to (15.50) (obtained for the defocussing NLS) but now with a minus sign, i.e. $\omega_k = k\sqrt{k^2 - 2\rho_0}$, where ρ_0 is the condensate density and k is the wavenumber of the modulation. Thus, modulations with $k^2 < 2$ ρ_0 will be unstable.

MI is quite common in wave systems. For the gravity water waves it the famous Benjamin-Feir instability [21, 22]. It was argued that MI may cause rogue waves [23, 24]. For the gravity water waves too, MI may arise due to an inverse wave-action cascade. However, since MI is obtained for a coherent carrier wave, it is not *a priori* obvious how relevant this instability is in the cases when the wave spectrum is broad. It was argued in [24, 25] that MI is relevant when waves are sufficiently strong and have a sufficiently narrow spectrum. They introduced so-called Benja-min-Fair index (BFI) as a ratio of the wave nonlinearity to the spectrum width, to characterize probability of triggering MI and, therefore, probability of a rogue wave. The BFI turned out to be quite efficient for predicting extreme wave events, and it is presently used by ECMWF for the operational wave weather forecast.

Part of this project could be studying the cited literature about the classical MI instability in the focusing NLS and in the gravity water wave systems, and about its role and manifestations in WT in these systems.

Another important example of MI can be found in the theory of Rossby waves in GFD and drift waves in plasmas [26–30]. These two applications are often described by the same nonlinear model,—CHM Eq. (6.13). In these systems, MI results in generation of zonal jets whose practical importance is to block the turbulent transport, as we already discussed in Chap. 13. On the other hand, energy cascade to zonal scales also follows from the Fjørtoft's argument illustrated in Fig. 8.2. Which mechanism of zonation will dominate in reality, the MI or the anisotropic inverse cascade? By similarity with the gravity wave example of the previous paragraph, one could guess that the dominant zonation mechanism for narrow wave spectra should be the MI, whereas for wider spectra we expect the anisotropic cascade mechanism to me more efficient. However, these ideas have not been confirmed or quantified yet. A project here could be numerical simula-tions of the CHM model with initial spectra of various widths and strengths with the aim of finding the dominant mechanisms, MI or cascade, and possibly iden-tifying a defining quantity similar to BFI.

16.4 Interacting Particle Systems

A considerable part of WT consist of analyzing the wave kinetic equations, par-ticularly finding power-law KZ solutions. Since the wave kinetic equations are in many ways similar to the kinetic equations for particles, WT approaches were applied to the interacting particle systems (IPS) in a number of works [31–37]. We already discussed application of WT to the Boltzmann gas in Sect. 15.5.3 of Chap. 15. In particular we saw that the pure power-law KZ solutions are not realizable in this system, and we found warm cascade solutions, which are close to the Maxwell–Boltzmann distribution in most of the inertial range, and which have sharp cutoffs at the dissipation scales.

Applications of WT to another type of IPS, "sticky particles", were considered in [34–37]. In [34], a "gas" of merging galaxies was considered, and the other systems of this kind include merging water droplets in fog, suspension of sticky clay particles in dirty water, merging air bubbles in water. All of these systems are described by *Smoluchowsky* kinetic equation for the mass distribution function of $2 \rightarrow 1$ kind (i.e. 2 particles produce 1 upon merging). Different processes, e.g. gravitational sedimentation or merging in an external shear, lead to different kernels (interaction coefficients) in this equation. KZ spectra were found and analyzed in various settings. In particular, for some kernels the KZ spectra were found to be nonlocal [34, 37], and a nonlocal theory was developed and confirmed numerically [37]. It is clear that RJ solutions are impossible for the sticky particles because the inverse process, $1 \rightarrow 2$ particle splitting, is absent. Respectively, we do not expect any warm cascades as in the Boltzmann IPS, and all the stationary states for the forced/dissipated Smoluchowsky IPS are always pure cascades. Some IPS exhibit so-called *gelation*, i.e. formation of infinitely massive particles in finite time. In the WT language, gelation means that the system is of the finite-capacity type.

The project here could consist of comparative analysis of different IPS, conditions of realizability of KZ, RJ and mixed states, effect of the nonlocal iteractions, non-stationary evolution, etc.

One could also consider a system of "splitting" particles, e.g. small solid clusters subjected to shear and strain in a turbulent flow. One could expect such systems to be described by $1 \rightarrow n$ kinetic equations with $n \in \mathbb{N}$. Another interesting system of this kind is an *air shower*. In Wikipedia we read: "An air shower is an extensive (many kilometers wide) cascade of ionized particles and electromagnetic radiation produced in the atmosphere when a primary cosmic ray (i.e. one of extraterrestrial origin) enters the atmosphere. The term cascade means that the incident particle, which could be a proton, a nucleus, an electron, a photon, or (rarely) a positron, strikes a molecule in the air so as to produce many high energy ions (secondaries), which in turn create more, and so on." The particles in such a cascade typically have power law energy distributions [38], which invites analogy with WT systems. So far, no application of WT ideas has been made to the atmospheric showers or other cascades in energetic (GeV–TeV) particle systems. It would make a good research project to learn about the existing particle cascade models for these systems, and to explore if the WT approach could be used to predict the particle distributions.

16.5   Superfluid Turbulence

Superfluid turbulence at zero temperature is a fascinating system in which a turbulent cascade transfers energy from macroscopic classical to microscopic quantum scales, thereby crossing through intermediate scales characterized by a gradual transition from the classical to the quantum laws of physics [39–41]. The basic model for such turbulence comprises a tangle ("spaghetti") of quantized vortex lines. This tangle is not completely random, but is polarized (partially co-aligned) in such a way that at the scales greater that the mean intervortex separation ℓ the coarse-grained motions organize themselves into classical fluid eddies. It is believed (but not confirmed experimentally or numerically yet) that turbulence at these scales is classical K41. At scales $\sim\ell$ the vortex lines in the tangle reconnect thereby generating *Kelvin waves* propagating along the lines, nonlinearly interacting and transferring energy further down-scale, until the energy of these waves is lost into radiating phonons. Thus, the problem of superfluid turbulence at zero temperature can be divided into three sub-problems: (1) large-scale turbulence which is presumably classical, (2) small-scale turbulence of Kelvin waves, and (3) crossover range with coexisting and interacting eddies and waves.

Challenge for the large-scale range is to understand the mechanisms which could lead to the vortex line polarizations consistent with the K41 spectrum. Numerical challenge is to compute a sufficiently large tangle with a sufficiently large range of scales above ℓ while resolving the individual vortex line motions in a reasonable range below ℓ too. Such theoretical and numerical studies are yet to be done, and the K41 picture for the large-scale range remains so far unconfirmed. A good research project could explore various ideas and to suggest possible ways forward for establishing the nature of the large-scale superfluid turbulence.

At the small scales, we deal with typical WT. We have already discussed such 1D turbulence of Kelvin waves in Sect. 3.1.3.6 and in Sect. 3.2.6. In Sect. 3.1.3.6 we mentioned that the energy cascade KZ state is nonlocal within the six-wave WT theory, which leads to an effective four-wave theory with local spectrum (3.24). The current state of the Kelvin WT theory can be found in [42, 43] (see also references therein). One element which is still missing, and which could be a good research theme, is a numerical verification of the WT predictions for this system. In the simplest setting this could be done by exciting long waves on a straight vortex line and dissipating the system at the small scales (e.g. using the Biot Savart or the defocusing NLS models).

Most intriguing in superfluid turbulence is the crossover range between the large eddy-dominated scales and the small Kelvin-wave scales. When discussing DAM in Sect. 16.1, we mentioned the prediction of a bottleneck in this crossover range and its description in terms of the warm cascades [18, 19]. There is an indirect experimental evidence in terms of so-called effective viscosity measurements (described in [18, 19]) in favor of the bottleneck scenario. However, we still lack a more direct confirmation of this picture, e.g. numerical, and this could be a good theme for a research project.

Further issues in superfluid turbulence which could be explored in research projects were already mentioned in Sect. 16.1, e.g. extending the theory to finite temperatures by including both the normal and superfluid components with mutual friction [14, 17], including the vortex reconnections and phonon radiation [15, 16].

16.6 Gravity Water Wave Turbulence

Gravity waves on surface of deep water is clearly the most important system which has motivated and stimulated the development of the WT approach. This is the system for which (along with the capillary water waves) the concept of the KZ spectra was developed by Zakharov and Filonenko [44]. We have already discussed this example in several places in this book: in Sects. 3.1.3.4, 3.2.2, 6.9.1, 7.12, 6.9.2.1, 11.1, 11.2, various sections in Chap. 10. In particular, in these sections we have discussed the KZ cascades (ZF spectra), wavebreaking, critical balance and Philips spectrum, WT intermittency and power-law PDF tails, coexisting and interacting random and coherent components in the WT cycle, finite size effects, discrete and mesoscopic WT, sandpile behavior.

Here is an incomplete list of recent paper on the subject: [5–7, 45–65]. There is a great variety of possible research projects on this area, with a common element of studying the relevant parts of the present book and the cited literature.

An interesting theme for a research project could be learning about the role of coherent structures in forming WT spectra and in causing intermittent behavior, classifying the different types of wave breaking and finding ways of identifying them in the experimental and in the numerical data. This line of enquiry was recently pursued in works [52–54, 56–58, 60], and the project could continue developing these directions.

Another interesting theme could be exploring the finite-size effects, signatures of which were seen in both the experiment [52–54, 56, 57] and in the numerics [47, 51]. It would be interesting to analyze the detailed experimental and the numerical data to identify the signatures of discrete wave resonances and their clusters, e.g. the ones described theoretically in [59].

Finally, interaction of the gravity WT with wind could make an interesting theme for a research project or an essay [55, 64, 65]. We have not covered this subject in the book, and those interested in this project should read the original research papers. An interesting new effect and an evidence of fast WT evolution (at the dynamical rather than the kinetic equation time scale) arising due to sudden gusts of wind were recently described in [55]. Further, there remains uncertainty about relevance of several previously suggested wind wave forcing mechanisms for realistic sea conditions. In particular, it is not clear to what extent the nonlinear effects and wave breaking contribute to the wave forcing. Thus, finding the ways of identification of different forcing mechanisms in the experimental data and developing realistic wind forcing models for numerical simulations would be an important and interesting project.

16.7 Metal-plate Wave Turbulence

Studies of WT on metal plates were conducted relatively recently, theoretical in [66] and experimental in [67–69]. This system turned out to be a very good test

bed for the WT theory because it allows detailed and precise experimental diag-
nostics which is not yet available in other WT experiments, including the
($\mathbf{k}$, ω)-Fourier transforms with the full 2D $\mathbf{k}$-space. Yet, no confirmation for the
energy cascade KZ spectrum has been obtained so far, and the measured spectrum
appears to be significantly steeper than the theoretical prediction. The reason for
this discrepancy is not known yet, although several possibilities were suggested,
including influence of dissipation, a Bose-type condensation at the large scales
with subsequent influence of the condensate onto the small-scale spectrum,
influence of large-scale plate imperfections which could transform the four-wave
process into an effectively three-wave process, finite-size effects. The condensate
possibility could probably be eliminated because no signature of a condensate is
seen on the experimental (k, ω)-plots (it would be seen as an extra line or lines as
e.g. on the numerical plots in Fig. 15.10 or Fig. 15.14) There was also a special
care taken for not having large-scale imperfections on the experimental plate. The
dissipation remains a possibility, as no exact mechanism or rate of the dissipation
is currently found. The finite size effects seem quite an important factor, because
the latest results indicate that the dispersion curve broadening is of the same order
as the distance between the discrete Fourier modes [70]. Such a balance is a
signature of the critical spectrum arising in a "sandpile" scenario due to the finite
size, see Sect. 10.3 in Chap. 10.

This project could consist of studying the cited literature, and exploring the
subject further, in particular attempting to establish the reason for the deviation of
the experimental spectrum from the weak WT prediction and modifying the
existing theory for explaining the experiment.

References

1. Gledzer, E.B.: Dokl. Akad. Nauk. SSSR **200**, 1043 (1973)
2. Yamada, M., Ohkitani, K.: J. Phys. Soc. Jpn. **56**, 4210 (1987)
3. Richardson, L.F.: Atmospheric diffusion shown on a distance-neighbour graph atmospheric diffusion shown on a distance-neighbour graph. Proc. R. Soc. Lond. Ser. A **110**(756), 709–737 (1926)
4. Iroshnikov, R.S.: Turbulence of a conducting fluid in a strong magnetic field, Astronomicheskii Zhurnal **40**, 742 (1963); translation: Sov. Astron **7**, 566 (1964)
5. Hasselmann, S., Hasselmann, K.: Computations and parametrizations of the nonlinear energy transfer in gravity wave spectrum. Part 1 J. Phys. Oceanogr. **15**, 136977 (1985)
6. Iroshnikov, R.S.: Possibility of a non-isotropic spectrum of wind waves by their weak interaction. Sov. Phys. Dokl. **30**, 1268 (1985)
7. Zakharov, V.E., Pushkarev, A.N.: Diffusion model of interacting gravity waves on the surface of deep fluid. Nonlin. Proc. Geophys. **6**, 110 (1999)
8. Nazarenko, S.: Sandpile behaviour in discrete water-wave turbulence, J. Stat. Mech. L02002 (2006). doi:10.1088/1742-5468/2006/02/L02002
9. Leith, C.: Diffusion approximation to inertial energy transfer in isotropic turbulence. Phys. Fluids **10**, 1409 (1967)
10. Leith, C.: Phys. Fluids **11**, 1612 (1968)

11. Connaughton, C., Nazarenko, S.: Warm cascades and anomalous scaling in a diffusion model of turbulence. Phys. Rev. Lett. **92**, 044501 (2004)
12. Lvov, V., Nazarenko, S.: Differential models for 2D turbulence, JETP Lett. **83**(12), 635–639 (2006) (arXiv: nlin.CD/0605003)
13. Lilly, D.K.: Two-dimensional turbulence generated by energy sources at two scales. J. Atmos. Sci. **46**(13), 2026–2030 (1989)
14. Lvov, V.S., Nazarenko, S., Volovik, G.: Energy spectra of developed superfluid turbulence. JETP Lett. **80**(7), 535–539 (2004)
15. Nazarenko, S.: Differential approximation for Kelvin-wave turbulence, JETP Lett. **83**(5), 198–200 (2005) (arXiv: cond-mat/0511136)
16. Nazarenko, S.: Kelvin wave turbulence generated by vortex reconnections. JETP Lett. **84**(11), 585–587 (2006)
17. Lvov, V.S., Nazarenko, S.V., Skrbek, L.: Energy spectra of developed turbulence in Helium superfluids. JLTP **145**(1–4), 125–142 (2006)
18. L'vov, V.S., Nazarenko, S.V., Rudenko, O.: Bottleneck crossover between classical and quantum superfluid turbulence. Phys. Rev. B **76**(2), 024520 (2007)
19. L'vov, V.S., Nazarenko, S.V., Rudenko, O.: Gradual eddy-wave crossover in superfluid turbulence. J. Low Temp. Phys. **153**(5–6), 140–161 (2008). doi:10.1007/s10909-008-9844-0
20. Dyachenko, A., Newell, A.C., Pushkarev, A., Zakharov, V.E.: Optical turbulence: weak turbulence, condensates and collapsing fragments in the nonlinear Schrodinger equation. Phys. D **57**(1–2), 96 (1992)
21. Benjamin, T.B.: Instability of periodic wavetrains in nonlinear dispersive systems. Proc. R. Soc. Lond. **A299**, 59–75 (1967)
22. Benjamin, T.B., Feir, J.E.: The disintegration of wave trains on deep water, Part 1. J. Fluid Mech. **27**, 417–430 (1967)
23. Osborne, A.R., Onorato, M., Serio, M.: The nonlinear dynamics of rogue waves and holes in deep-water gravity wave trains. Phys. Lett. A **275**(5–6), 386–393 (2000)
24. Onorato, M., Osborne, A.R., Serio, M., Berton, S.: Freak waves in random oceanic sea states. **86**, 5831–5834 (2001)
25. Janssen, P.A.E.M.: Nonlinear four-wave interactions and freak waves. J. Phys. Oceanogr. **33**, 863–884 (2003)
26. Lorentz, E.N.: Barotropic instability of rossby wave motion. J. Atmos. Sci. **29**, 258–269 (1972)
27. Gill, A.E.: The stability on planetary waves on an infinite beta-plane. Geophys. Fluid Dyn. **6**, 2947 (1974)
28. Manin, D.Yu., Nazarenko, S.V.: Nonlinear interaction of smallscale Rossby waves with an intense largescale zonal flow. Phys. Fluids **6**(3), 1158–1167 (1994)
29. Diamond, P.H., Itoh, S.-I., Itoh, K., Hahm, T.S.: Zonal flows in plasma—a review. Plasma Phys. Control. Fusion **47**(5), R35–R161 (2005)
30. Connaughton, C.P., Nadiga, B.T., Nazarenko, S.V., Quinn, B.E.: Modulational instability of Rossby and drift waves and generation of zonal jets. J. Fluid Mech. **654**, 207–231 (2010)
31. Kats, A.V., Kontorovich, V.M., Moiseev, S.S., Novikov, V.E.: Power-law solutions of the Boltzmann kinetic equation, describing the spectral distribution of particles with fluxes. ZhETF Pisma **21**, 13 (1975)
32. Kats, A.V.: Direction of transfer of energy and quasi-particle number along the spectrum in stationary power-law solutions of the kinetic equations for waves and particles. Sov. Phys. JETP **44**, 1106 (1976)
33. Karas, V.I., Moiseev, S.S., Novikov, V.E.: Nonequilibrium stationary distribution of particles in a solid plasma. Zhurnal Eksperimental'noi i Teoreticheskoi Fiziki **71**, 1421–1433 (1976)
34. Kontorovich, V.M.: Zakharovs transformation in the problem of galaxy mass distribution function. Phys. D 152–153, 676–681 (2001)
35. Connaughton, C., Rajesh, R., Zaboronski, O.V.: Stationary Kolmogorov solutions of the Smoluchowski aggregation equation with a source term, Part 1. Phys. Rev. E **69**(6), 061114 (2004)

36. Connaughton, C., Rajesh, R., Zaboronski, O.: Cluster-cluster aggregation as an analogue of a turbulent cascade: Kolmogorov phenomenology, scaling laws and the breakdown of self-similarity. Phys. D **222**(1–2), 97–115 (2006)
37. Horvai, P., Nazarenko, S.V., Stein, T.H.M.: Coalescence of particles by differential sedimentation. J. Stat. Phys. **130**(6), 1177–1195 (2008)
38. Gaisser, T.K.: Cosmic Rays and Particle Physics, p. 296. Cambridge University Press, UK (1991)
39. Quantized vortex dynamics and superfluid turbulence. In: Barenghi, C.F. et al. (eds.) Lecture Notes in Physics, vol. **571**. Springer, Berlin (2001)
40. Vinen, W.F., Niemela, J.J.: J. Low Temp. Phys. **128**, 167 (2002)
41. Vinen, W.F.: An introduction to quantum turbulence. J. Low Temp. Phys. **145**, 7 (2006)
42. Laurie, J., L'vov, V.S., Nazarenko, S., Rudenko, O.: Interaction of Kelvin waves and non-locality of the energy transfer in superfluids. Phys. Rev. B **81**, 104526 (2010)
43. L'vov, V.S., Nazarenko, S.: Spectrum of Kelvin-wave turbulence in superfluids. JETP Lett. (Pis'ma v ZhETF) **91**, 464-470 (2010)
44. Zakharov, V.E., Filonenko, N.N. Energy spectrum for stochastic oscillations of a fluid surface, Doklady Acad. Nauk SSSR **170**, 1292–1295 (1966) (translation: Sov. Phys. Dokl. **11**, 881–884 (1967))
45. Dyachenko, A.I., Korotkevich, A.O., Zakharov, V.E.: Weak turbulent Kolmogorov spectrum for surface gravity waves. Phys. Rev. Lett. **92**, 13 (2004)
46. Dyachenko, A.I., Korotkevich, A.O., Zakharov, V.E.: Weak turbulence of gravity waves. JETP Lett. **77**(10), 546–550 (2003)
47. Zakharov, V.E., Korotkevich, A.O., Pushkarev, A.N., Dyachenko, A.I.: Mesoscopic wave turbulence. JETP Lett. **82**(8), 487 (2005)
48. Annenkov, S.Y., Shrira, V.I.: Numerical modeling of water wave evolution based on the Zakharov equation. J. Fluid. Mech. **449**, 341–371 (2001)
49. Annenkov, S.Y., Shrira, V.I.: Direct numerical simulation of downshift and inverse cascade for water wave turbulence. Phys. Rev. Lett. **96**, 204501 (2006)
50. Annenkov, S., Shrira, V.: Role of non-resonant interactions in evolution of nonlinear random water wave fields. J. Fluid Mech. **561**, 181–207 (2006)
51. Lvov, Y., Nazarenko, S., Pokorni, B.: Discreteness and its effect on water-wave turbulence. Phys. D **218**(1), 24–35 (2006)
52. Denissenko, P., Lukaschuk, S., Nazarenko, S.: Gravity surface wave turbulence in a laboratory flume. Phys. Rev. Lett. **99**, 014501 (2007)
53. Lukaschuk, S., Nazarenko, S., McLelland, S., Denissenko, P.: Gravity wave turbulence in wave tanks: space and time statistics, PRL **103**(4), 044501 (2009)
54. Nazarenko, S., Lukaschuk, S., McLelland, S., Denissenko, P.: Statistics of surface gravity wave turbulence in the space and time domains. J. Fluid Mech. **642**, 395–420 (2010)
55. Annenkov, S.Y., Shrira, V.I.: Fast nonlinear evolution in wave turbulence. Phys. Rev. Lett. **102**, 024502 (2009)
56. Falcon, E., Laroche, C., Fauve, S.: Observation of gravitycapillary wave turbulence. Phys. Rev. Lett. **98**, 094503 (2007)
57. Falcón, C., Falcon, E., Bortolozzo, U., Fauve, S.: Capillary wave turbulence on a spherical fluid surface in zero gravity. EPL **86**, 14002 (2009)
58. Falcon, E., Fauve, S., Laroche, C.: Observation of intermittency in wave turbulence. Phys. Rev. Lett. **98**, 154501 (2007)
59. Kartashova, E., Nazarenko, S., Rudenko, O.: Resonant interactions of nonlinear water waves in a finite basin. Phys. Rev. E **78**, 016304 (2008)
60. Zakharov, V.E., Korotkevich, A.O., Pushkarev, A.N., Resio, D.: Coexistence of weak and strong wave turbulence in a swell propagation. Phys. Rev. Lett. **99**(16), 164501 (2007)
61. Korotkevich, A.O.: On the Doppler distortion of the sea-wave spectra. Phys. D **237**(21), 2767–2776 (2008). doi:10.1016/j.physd.2008.04.005

62. Korotkevich, A.O.: Simultaneous numerical simulation of direct and inverse cascades in wave turbulence. Phys. Rev. Lett. **101**(7), 074504 (2008). doi:10.1103/PhysRevLett.101.074504

63. Korotkevich, A.O., Pushkarev, A.N., Resio, D., Zakharov, V.E.: Numerical verification of the weak turbulent model for swell evolution. Eur. J. Mech. B/Fluids **27**(4), 361–387 (2008). doi:10.1016/j.euromechflu.2007.08.004

64. Janssen, P.A.E.M.: The Interaction of Ocean Waves and Wind. Cambridge University Press, Cambridge (2004)

65. Badulin, S.I., Babanin, A.V., Resio, D., Zakharov, V.E.: Weakly turbulent laws of wind-wave growth. J. Fluid Mech. **591**, 339–378 (2007)

66. Düring, G., Josserand, C., Rica, S.S.: Phys. Rev. Lett. **97**, 025503 (2006)

67. Boudaoud, A., Cadot, O., Odille, B., Touze, C.: Phys. Rev. Lett. **100**, 234504 (2008)

68. Mordant, N.: Phys. Rev. Lett. 100, 234505 (2008)

69. Cobelli, P., Petitjeans, P., Maurel, A., Pagneux, V., Mordant, N.: Phys. Rev. Lett. **103**(20), 204301 (2009)

70. Mordant, N.: Private communication (2010)

Lightning Source UK Ltd.
Milton Keynes UK
UKOW031338081111

181705UK00003B/58/P